PREFACE

In this little volume the publishers are offering to all who are interested a compact collection of principles, methods, formulas, and tables pertaining to the different branches of civil engineering. It is intended as a ready reference manual for the student as well as for the technical man engaged in practical work. For this reason, whenever there was a choice of rules or methods, only the simplest and those best suited to practical use were selected. For the same reason, wherever possible, examples such as would occur in practice have been given, together with their solutions, thus illustrating the different steps and processes to be performed in order to obtain practical results.

Attention is called to the tables, which are very numerous. Many of these can be found elsewhere only in special works, and many are original, being found only in this book. Of the latter kind, are the Hydraulic Tables, giving the discharge, velocity, and head per unit of length for cast-iron pipes from 4 to 72 inches in diameter; and the Reinforced-Concrete Tables, by means of which rapid computation of unit stresses in reinforced-concrete beams can be made for any combination of steel and concrete.

This handbook was prepared by Mr. C. K. Smoley, Principal of the School of Civil Engineering of the International Correspondence Schools.

INTERNATIONAL CORRESPONDENCE SCHOOLS

April, 1913

The (

A CONV

Chainmen,

Surveyors, as well as Draftsmen, Computers, and Railroad, Municipal, and Hydraulic Engineers

BY

International Correspondence Schools
SCRANTON, PA.

1st Edition, 22d Thousand, 3d Impression

SCRANTON, PA.
INTERNATIONAL TEXTBOOK COMPANY

PRESS OF
INTERNATIONAL TEXTBOOK COMPANY
SCRANTON, PA.

27002

INDEX

G

H

T

The Civil Engineer's Handbook

TABLES OF WEIGHTS AND MEASURES

THE ENGLISH SYSTEM

LINEAR MEASURE

12	inches (in.)	= 1 foot	ft.
3	feet	= 1 yard	yd.
5½	yards	= 1 rod	rd.
40	rods	= 1 furlong	fur.
8	furlongs	= 1 mile	mi.

in.		*ft.*		*yd.*		*rd.*		*fur.*		*mi.*
36	=	3	=	1						
198	=	16½	=	5½	=	1				
7,920	=	660	=	220	=	40	=	1		
63,360	=	5,280	=	1,760	=	320	=	8	=	1

SURVEYOR'S MEASURE

7.92	inches	= 1 link	li.
25	links	= 1 rod	rd.
4 100 66	rods links feet	= 1 chain	ch.
80	chains	= 1 mile	mi.

mi.		*ch.*		*rd.*		*li.*		*in.*
1	=	80	=	320	=	8,000	=	63,360

SQUARE MEASURE

144	square inches (sq. in.)	= 1 square foot	sq. ft.
9	square feet	= 1 square yard	sq. yd.
30¼	square yards	= 1 square rod	sq. rd.
160	square rods	= 1 acre	A.
640	acres	= 1 square mile	sq. mi.

sq. mi.	*A.*	*sq. rd.*	*sq. yd.*	*sq. ft.*	*sq. in.*
1 =	640 =	102,400 =	3,097,600 =	27,878,400 =	4,014,489,600

SURVEYOR'S SQUARE MEASURE

625	square links (sq. li.)	= 1 square rod	sq. rd.
16	square rods	= 1 square chain	sq. ch.
10	square chains	= 1 acre	A.
640	acres	= 1 square mile	sq. mi.
36	square miles (6 mi. square)	= 1 township	Tp.

sq. mi.	*A.*	*sq. ch.*	*sq. rd.*	*sq. li.*
1 =	640 =	6,400 =	102,400 =	64,000,000

The *acre* contains 4,840 sq. yd., or 43,560 sq. ft., and is equal to the area of a square measuring 208.71 ft. on a side.

Other terms are the *pole* or *perch* (P.), which is equal to 1 sq. rd. and the *rood* (R.), which is equal to 40 sq. rd.

CUBIC MEASURE

1,728	cubic inches (cu. in.)	= 1 cubic foot	cu. ft.
27	cubic feet	= 1 cubic yard	cu. yd.
128	cubic feet	= 1 cord	cd.
24¾	cubic feet	= 1 perch	P.

cu. yd.	*cu. ft.*	*cu. in.*
1 =	27 =	46,656

MEASURE OF ANGLES OR ARCS

60	seconds(″)	= 1 minute	′
60	minutes	= 1 degree	°
90	degrees	= 1 rt. angle or quadrant	∟
360	degrees	= 1 circle	cir.

1 cir. = 360° = 21,600′ = 1,296,000″

AVOIRDUPOIS WEIGHT

437½	grains (gr.)	= 1 ounce	oz.
16	ounces	= 1 pound	lb.
100	pounds	= 1 hundredweight	cwt.
20	cwt., or 2,000 lb.	= 1 ton	T.

T. cwt. lb. oz. gr.
1 = 20 = 2,000 = 32,000 = 14,000,000

The avoirdupois pound contains 7,000 gr.

LONG-TON TABLE

16 ounces	= 1 pound	lb.
112 pounds	= 1 hundredweight	cwt.
20 cwt., or 2,240 lb.	= 1 ton	T.

TROY WEIGHT

24 grains (gr.)	= 1 pennyweight	pwt.
20 pennyweights	= 1 ounce	oz.
12 ounces	= 1 pound	lb.

lb. oz. pwt. gr.
1 = 12 = 240 = 5,760

DRY MEASURE

2 pints (pt.)	= 1 quart	qt.
8 quarts	= 1 peck	pk.
4 pecks	= 1 bushel	bu.

bu. pk. qt. pt.
1 = 4 = 32 = 64

The *U. S. struck bushel* contains 2,150.42 cu. in. = 1.2444 cu. ft. By law, its dimensions are those of a cylinder 18½ in. in diameter and 8 in. deep. The *heaped bushel* is equal to 1¼ struck bu., the cone being 6 in. high. For approximations, the bushel may be taken at 1¼ cu. ft.; or 1 cu. ft. may be considered ⅘ bu.

The *British bushel* contains 2,218.19 cu. in. = 1.2837 cu. ft. = 1.032 U. S. bu.

The *dry gallon* contains 268.8 cu, in., being ⅛ struck bu.

LIQUID MEASURE

4	gills (gi.)	= 1 pint	pt.
2	pints	= 1 quart	qt.
4	quarts	= 1 gallon	gal.
31½	gallons	= 1 barrel	bbl.
2	barrels, or 63 gallons	= 1 hogshead	hhd.

hhd. bbl. gal. qt. pt. gi.
1 = 2 = 63 = 252 = 504 = 2,016

The *U. S. gallon* contains 231 cu. in. = .134 cu. ft., nearly; or 1 cu. ft. contains 7.481 gal. The following cylinders contain the given measures very closely:

	Diam. Inches	*Height Inches*		*Diam. Inches*	*Height Inches*
Gill	1¾	3	Gallon	7	6
Pint	3½	3	8 gal.	14	12
Quart	3½	6	10 gal.	14	15

When water is at its maximum density, 1 cu. ft. weighs 62.425 lb. and 1 gal. weighs 8.345 lb.

For approximations, 1 cu. ft. of water is considered equal to 7½ gal., and 1 gal. as weighing 8⅓ lb.

The *British imperial gallon*, both liquid and dry, contains 277.463 cu. in. = .16046 cu. ft., and is equivalent to the volume of 10 lb. of pure water at 62° F.

To reduce British to U. S. liquid gallons, multiply by 1.2. Conversely, to convert U. S. into British liquid gallons, divide by 1.2; or, decrease the number of gallons one-sixth.

MISCELLANEOUS TABLE

12 articles	= 1 dozen	20 quires	= 1 ream
12 dozen	= 1 gross	1 league	= 3 miles
12 gross	= 1 great gross	1 fathom	= 6 feet
2 articles	= 1 pair	1 hand	= 4 inches
20 articles	= 1 score	1 palm	= 3 inches
24 sheets	= 1 quire	1 span	= 9 inches
1 meter	= 3 feet 3⅜ inches (nearly)		

THE METRIC SYSTEM

The *metric system* is based on the *meter*, which, according to the U. S. Coast and Geodetic Survey Report of 1884, is equal to 39.370432 in. The value commonly used is 39.37 in., and is authorized by the U. S. government. The meter is defined as one ten-millionth of the distance from the pole to the equator, measured on a meridian passing near Paris.

There are three principal units—the meter, the liter (pronounced lee-ter), and the gram, the units of length, capacity, and weight, respectively. Multiples of these units are obtained by prefixing to the names of the principal units the Greek words *deca* (10), *hecto* (100), and *kilo* (1,000); the submultiples, or divisions, are obtained by prefixing the Latin words *deci* ($\frac{1}{10}$), *centi* ($\frac{1}{100}$), and *milli* ($\frac{1}{1000}$). These prefixes form the key to the entire system. In the following tables, the abbreviations of the principal units of these submultiples begin with a small letter, and those of the multiples begin with a capital letter; they should always be written as here printed.

MEASURES OF LENGTH

10 millimeters (mm.)	= 1 centimeter	cm.
10 centimeters	= 1 decimeter	dm.
10 decimeters	= 1 meter	m.
10 meters	= 1 decameter	Dm.
10 decameters	= 1 hectometer	Hm.
10 hectometers	= 1 kilometer	Km.

MEASURES OF SURFACE (NOT LAND)

100 square millimeters (sq. mm.)	= 1 square centimeter	sq. cm.
100 square centimeters	= 1 square decimeter	sq. dm.
100 square decimeters	= 1 square meter	sq. m.

MEASURES OF VOLUME

1,000 cubic millimeters (cu. mm.)	= 1 cubic centimeter	c. c. or cu. cm.
1,000 cubic centimeters	= 1 cubic decimeter	cu. dm.
1,000 cubic decimeters	= 1 cubic meter	cu. m.

MEASURES OF CAPACITY

10 milliliters (ml.)	= 1 centiliter	cl.
10 centiliters	= 1 deciliter	dl.
10 deciliters	= 1 liter	l.
10 liters	= 1 decaliter	Dl.
10 decaliters	= 1 hectoliter	Hl.
10 hectoliters	= 1 kiloliter	Kl.

The *liter* is equal to the volume occupied by 1 cu. dm.

MEASURES OF WEIGHT

10 milligrams (mg.)	= 1 centigram	cg.
10 centigrams	= 1 decigram	dg.
10 decigrams	= 1 gram	g.
10 grams	= 1 decagram	Dg.
10 decagrams	= 1 hectogram	Hg.
10 hectograms	= 1 kilogram	Kg.
1,000 kilograms	= 1 ton	T.

The *gram* is the weight of 1 c. c. of pure distilled water at a temperature of 39.2° F.; the *kilogram* is the weight of 1 l. of water; the *ton* is the weight of 1 cu. m. of water.

CONVERSION TABLES

By means of the accompanying tables metric measures can be converted into English, and vice versa, by simple addition. All the figures of the values given are not required, except in very exact calculations; as a rule, 4 or 5 digits only are used. To change 6,471.8 ft. into meters, consider 6,471.8 as 6,000+400+70+1+.8; also, 6,000=1,000×6; 400=100×4; etc. Hence, looking in the first column of the table entitled English Measures Into Metric, for 6, opposite it in the column headed Feet to Meters is found the number 1.8287838. Using but five digits and increasing the fifth digit by 1 (as the next is greater than 5), gives 1.8288. In other words, 6 ft. =1.8288 m.; hence, 6,000 ft.=1,000×1.8288=1,828.8, simply moving the decimal point three places to the right. Likewise, it is found that 400 ft.=121.92 m.; 70 ft.=21.336 m.; 1 ft.=.3048 m.; and .8 ft.=.2438 m. Adding as shown gives 1,972.6046 m. as the value of 6,471.8 ft.

1,828.8
121.92
21.336
.3048
.2438
1,972.6046

CONVERSION TABLE

English Measures Into Metric

English	Metric Inches to Meters	Metric Feet to Meters	Metric Pounds to Kilos	Metric Gallons to Liters
1	.0253998	.3047973	.4535925	3.7853122
2	.0507996	.6095946	.9071850	7.5706244
3	.0761993	.9143919	1.3607775	11.3559366
4	.1015991	1.2191892	1.8143700	15.1412488
5	.1269989	1.5239865	2.2679625	18.9265610
6	.1523987	1.8287838	2.7215550	22.7118732
7	.1777984	2.1335811	3.1751475	26.4971854
8	.2031982	2.4383784	3.6287400	30.2824976
9	.2285980	2.7431757	4.0823325	34.0678098
10	.2539978	3.0479730	4.5359250	37.8531220

English	Metric Square Inches to Square Meters	Metric Square Feet to Square Meters	Metric Cubic Feet to Cubic Meters	Metric Pounds per Square Inch to Kilos per Square Meter
1	.000645150	.092901394	.028316094	703.08241
2	.001290300	.185802788	.056632188	1,406.16482
3	.001935450	.278704182	.084948282	2,109.24723
4	.002580600	.371605576	.113264376	2,812.32964
5	.003225750	.464506970	.141580470	3,515.41205
6	.003870900	.557408364	.169896564	4,218.49446
7	.004516050	.650309758	.198212658	4,921.57687
8	.005161200	.743211152	.226528752	5,624.65928
9	.005806350	.836112546	.254844846	6,327.74169
10	.006451500	.929013940	.283160940	7,030.82410

CONVERSION TABLE

Metric Measures Into English

Metric	English Meters to Inches	English Meters to Feet	English Kilos to Pounds	English Liters to Gallons
1	39.370432	3.2808693	2.2046223	.2641790
2	78.740864	6.5617386	4.4092447	.5283580
3	118.111296	9.8426079	6.6138670	.7925371
4	157.481728	13.1234772	8.8184894	1.0567161
5	196.852160	16.4043465	11.0231117	1.3208951
6	236.222592	19.6852158	13.2277340	1.5850741
7	275.593024	22.9660851	15.4323564	1.8492531
8	314.963456	26.2469544	17.6369787	2.1134322
9	354.333888	29.5278237	19.8416011	2.3776112
10	393.704320	32.8086930	22.0462234	2.6417902

Metric	English Square Meters to Square Inches	English Square Meters to Square Feet	English Cubic Meters to Cubic Feet	English Kilos per Square Meter to Pounds per Square Inch
1	1,550.03092	10.7641034	35.3156163	.001422310
2	3,100.06184	21.5282068	70.6312326	.002844620
3	4,650.09276	32.2923102	105.9468489	.004266930
4	6,200.12368	43.0564136	141.2624652	.005689240
5	7,750.15460	53.8205170	176.5780815	.007111550
6	9,300.18552	64.5846204	211.8936978	.008533860
7	10,850.21644	75.3487238	247.2093141	.009956170
8	12,400.24736	86.1128272	282.5249304	.011378480
9	13,950.27828	96.8769306	317.8405467	.012800790
10	15,500.30920	107.6410340	353.1561630	.014223100

As another example, convert 19.635 Kg. into pounds. Working according to the explanation just given, it is found that 19.635 Kg. = 43.2879 lb.

22.046
19.842
1.3228
.0661
.0110
43.2879

The only difficulty in applying these tables lies in locating the decimal point; it may always be found thus: If the figure considered lies to the left of the decimal point, count each figure in order, beginning with units (but calling units' place zero), until the desired figure is reached, then move the decimal point to the *right* as many places as the figure being considered is to the left of the unit figure. Thus, in the first example, 6 lies three places to the left of 1, which is in units' place; hence, the decimal point is moved three places to the right. By exchanging the words right and left, the statement will also apply to decimals. Thus, in the second example, the 5 lies three places to the right of units' place; hence, the decimal point in the number taken from the table is moved three places to the left.

MATHEMATICS

MENSURATION

In the following formulas, unless otherwise stated, the letters have the meanings here given.

D = larger diameter;
d = smaller diameter;
R = radius corresponding to D;
r = radius corresponding to d;
p = perimeter or circumference;
C = area of convex surface = area of flat surface that can be rolled into the shape shown;
S = area of entire surface = C + area of the end or ends;
A = area of plane figure;
π = 3.1416, nearly = ratio of any circumference to its diameter;
V = volume of solid.

The other letters used will be found on the illustrations.

CIRCLE

$$p = \pi d = 3.1416d$$
$$p = 2\pi r = 6.2832r$$
$$p = 2\sqrt{\pi A} = 3.5449\sqrt{A}$$
$$p = \frac{2A}{r} = \frac{4A}{d}$$
$$d = \frac{p}{\pi} = \frac{p}{3.1416} = .3183p$$
$$d = 2\sqrt{\frac{A}{\pi}} = 1.1284\sqrt{A}$$
$$r = \frac{p}{2\pi} = \frac{p}{6.2832} = .1592p$$
$$r = \sqrt{\frac{A}{\pi}} = .5642\sqrt{A}$$
$$A = \frac{\pi d^2}{4} = .7854d^2$$
$$A = \pi r^2 = 3.1416r^2$$
$$A = \frac{pr}{2} = \frac{pd}{4}$$

TRIANGLE

Case I.—Given the base b and the altitude h,

$$A = \frac{bh}{2}$$

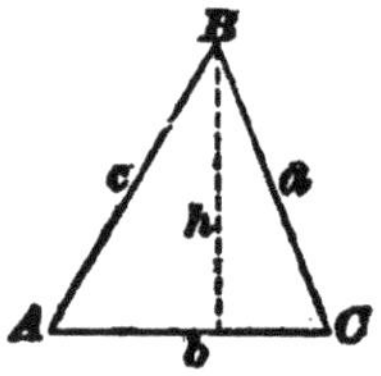

Case II.—Given the three sides a, b, and c,

$$A = \sqrt{s(s-a)(s-b)(s-c)}$$

in which $$s = \frac{a+b+c}{2}$$

Case III.—Given two sides a and c and the included angle B,

$$A = \tfrac{1}{2}a\,c\sin B$$

Case IV.—Given the side b and the angles A, B, and C,

$$A = \frac{b^2 \sin A \sin C}{2 \sin B}$$

also, $$A = \frac{b^2}{2(\cot A + \cot C)}$$

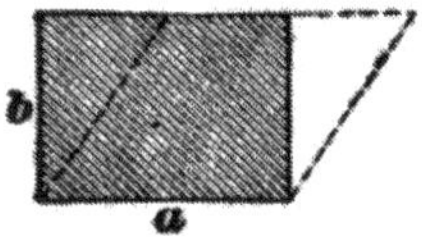

RECTANGLE AND PARALLELOGRAM

$$A = a\,b$$

TRAPEZOID

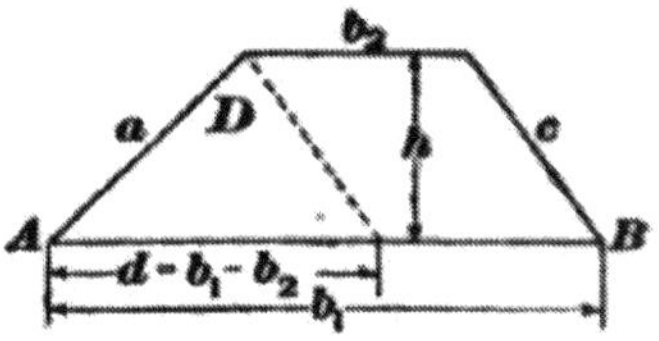

Case I.—Given the two bases b_1 and b_2 and the altitude h,

$$A = \frac{(b_1+b_2)h}{2}$$

Case II.—Given the bases and the angles adjacent to one of them,

$$A = \frac{b_1^2 - b_2^2}{2(\cot A + \cot B)};$$

or,

$$A = \frac{(b_1 - b_2)(b_1 + b_2)\sin A \sin B}{2 \sin (A+B)}$$

Case III.—Given the four sides,

$$A = \frac{b_1+b_2}{d}\sqrt{s(s-a)(s-c)(s-d)},$$

in which $s = \frac{1}{2}(a+c+d)$

TRAPEZIUM

Divide into two triangles and a trapezoid.

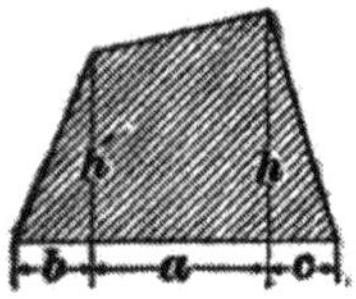

$$A = \tfrac{1}{2}bh' + \tfrac{1}{2}a(h'+h) + \tfrac{1}{2}ch;$$

$$\text{or, } A = \tfrac{1}{2}[bh' + ch + a(h'+h)]$$

Or, divide into two triangles by drawing a diagonal. Consider the diagonal as the base of both triangles, call its length l; call the altitudes of the triangles h_1 and h_2; then

$$A = \tfrac{1}{2}l(h_1+h_2)$$

OTHER POLYGONS

The area of any polygon can be determined by dividing the polygon into triangles and measuring in each triangle whatever parts are necessary for the determination of its area. The parts to be measured depend on special conditions. If in surveying a closed field the chain alone is used, the three sides of each triangle will have to be measured and the formula for Case II, page 10, used. If a transit or a compass is used, angles can be measured and the formulas of Cases III or IV,

page 10, applied. For the method of figuring areas of polygons by double longitudes, see page 60.

AREA INCLUDED BETWEEN A STRAIGHT LINE AND A CURVE

Method by Selected Ordinates.—Draw perpendiculars on AB from the points of the curve at which its direction changes

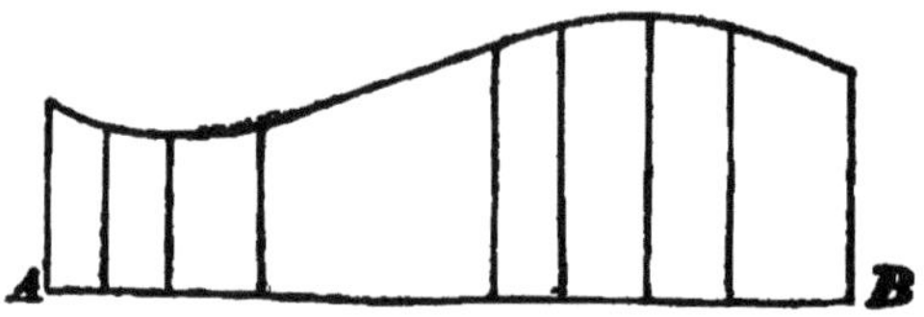

appreciably, and consider the portion of the curve between two consecutive perpendiculars to be a straight line. The figure is then treated as if divided into a number of trapezoids, whose areas can be computed by the rules already given.

Trapezoidal Rule.—The ordinates are measured at regular intervals d along the straight line as shown. The area is then equal to

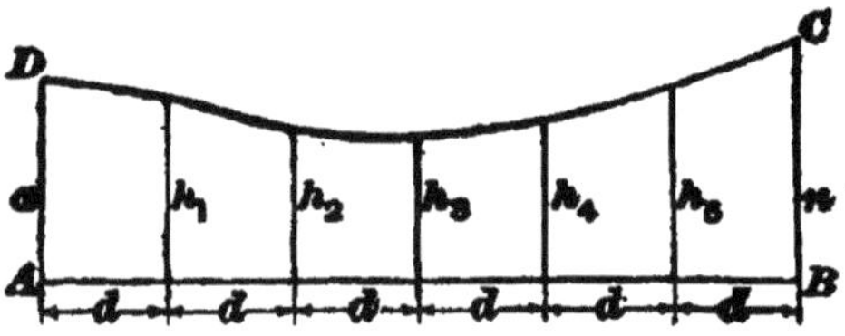

$$A=\left(\frac{a+n}{2}+\Sigma h\right)d$$

in which Σh is the sum of all the intermediate ordinates.

EXAMPLE.—If the ordinates from the straight line AB to the curved boundary DC, are 19, 18, 14, 12, 13, 17, and 23 li., respectively, and are at equal distances of 50 li., what is the area included between the curved boundary and the straight line?

SOLUTION.—Area $ABCD=\left(\frac{19+23}{2}+18+14+12+13+17\right)\times 50=4{,}750$ sq. li.

Simpson's Rule.—The base line must be divided into an *even* number of equal parts. The area is then equal to

$$A=(a+n+4\Sigma h_2+2\Sigma h_3)\frac{d}{3},$$

in which $a+n$ is the sum of the end ordinates; $4\Sigma h_2$ is four times the sum of all intermediate even-numbered ordinates; and $2\Sigma h_3$ is twice the sum of all intermediate odd-numbered ordinates. This rule is more accurate than the trapezoidal rule.

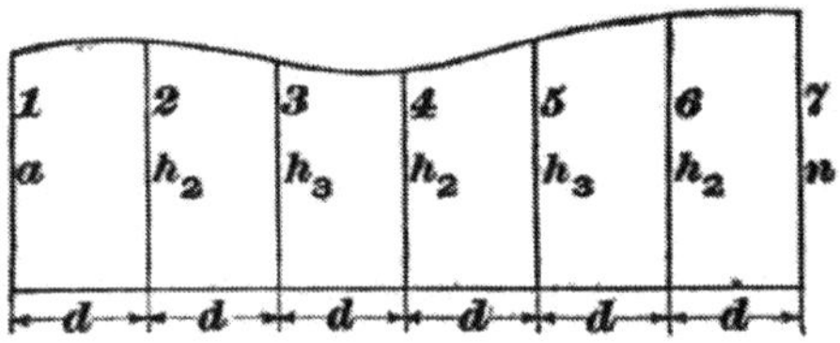

EXAMPLE.—Referring to the preceding example, what is the area *ABCD* according to Simpson's rule?

SOLUTION.— $A=[19+23+4(18+12+17)+2(14+13)]\times\frac{50}{3}$ $=4{,}733$ sq. li.

AREA BOUNDED BY AN IRREGULAR CURVE

Suppose that it is required to find the area enclosed by the heavy irregular curve shown in the accompanying illustration.

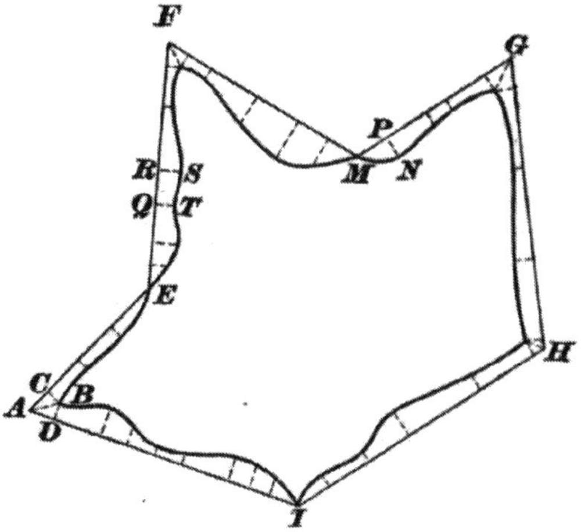

A broken line *AEFMGHIA* is drawn around the curved boundary line and as close to it as convenient. Ordinates to the straight lines thus drawn are measured from the points where the direction of the curved boundary changes materially as shown. The area of the polygon *AEFMGHIA* is calculated by one of the methods previously explained, and from it is subtracted the sum of the areas included between the curved boundary and the broken line, calculated in the manner just shown.

At such corners as *A*, the triangles *ABC* and *ABD* are computed from the measured bases *AC* and *AD* and the altitudes *BC* and *BD*. All the quadrilaterals, as *QRST*, are treated as trapezoids; and such three-sided figures as *MPN*, as triangles.

ELLIPSE

$$p^* = \pi\sqrt{\frac{D^2+d^2}{2} - \frac{(D-d)^2}{8.8}}$$

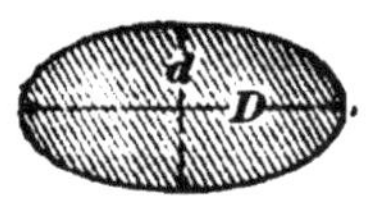

$$A = \frac{\pi}{4}Dd = .7854Dd$$

SECTOR

$$A = \tfrac{1}{2}lr$$

$$A = \frac{\pi r^2 E}{360} = .008727r^2E$$

l = length of arc

SEGMENT

$$A = \tfrac{1}{2}[lr - c(r-h)]$$

$$A = \frac{\pi r^2 E}{360} - \frac{c}{2}(r-h)$$

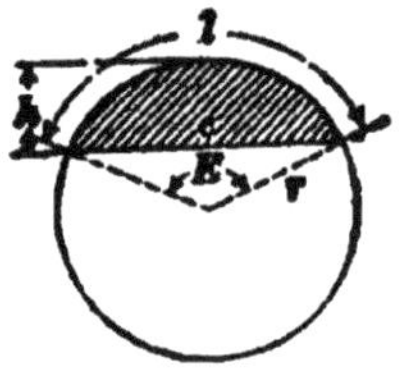

$$l = \frac{\pi r E}{180} = .0175rE$$

$$E = \frac{180l}{\pi r} = 57.2956\frac{l}{r}$$

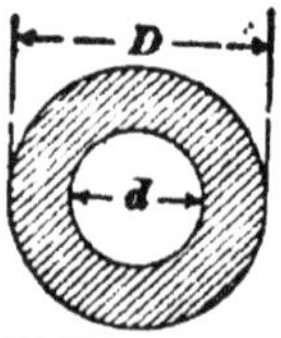

RING

$$A = \frac{\pi}{4}(D^2 - d^2)$$

* The perimeter of an ellipse cannot be exactly determined without a very elaborate calculation, and this formula is merely an approximation giving fairly close results.

CHORD

c = length of chord

$$r=\frac{c^2+4h^2}{8h}=\frac{e^2}{2\,h}$$

$$c=2\sqrt{2hr-h^2}$$

$$l=\frac{8e-c}{3}\text{, approximately}$$

HELIX

To construct a helix:

l = length of helix;
n = number of turns;
t = pitch.

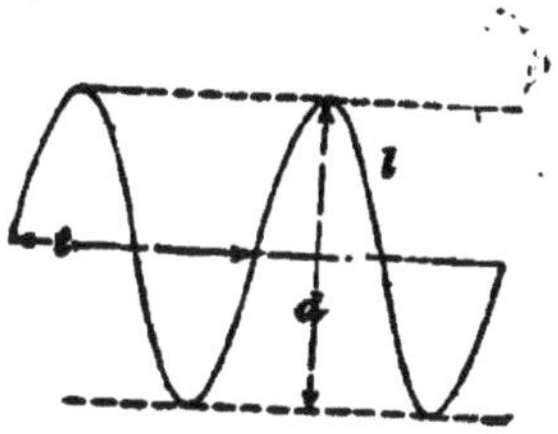

$$t=\sqrt{\frac{l^2}{n^2}-\pi^2d^2}$$

$$l=n\sqrt{\pi^2d^2+t^2}$$

$$n=\frac{l}{\sqrt{\pi^2a^2+t^2}}$$

CYLINDER

$$C=\pi dh$$

$$S=2\pi rh+2\pi r^2$$

$$=\pi dh+\frac{\pi}{2}d^2$$

$$V=\pi r^2h=\frac{\pi}{4}d^2h$$

$$V=\frac{p^2h}{4\pi}=.0796p^2h$$

FRUSTUM OF CYLINDER

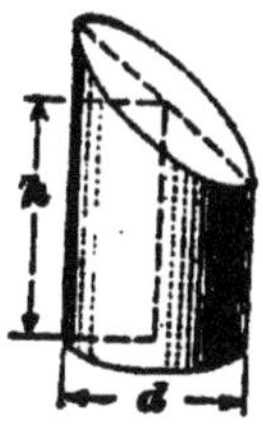

h = ½ sum of greatest and least heights

$$C=ph=\pi dh$$

$$S=\pi dh+\frac{\pi}{4}d^2+\text{area of elliptical top}$$

$$V=Ah=\frac{\pi}{4}d^2h$$

CONE

$$C = \tfrac{1}{2}\pi dl = \pi rl$$

$$S = \pi rl + \pi r^2 = \pi r\sqrt{r^2 + h^2} + \pi r^2$$

$$V = \frac{\pi d^2}{4} \times \frac{h}{3} = \frac{.7854d^2h}{3} = \frac{p^2h}{12\pi}$$

FRUSTUM OF CONE

$$C = \tfrac{1}{2}l(P + p) = \frac{\pi}{2}l(D + d)$$

$$S = \frac{\pi}{2}[l(D + d) + \tfrac{1}{2}(D^2 + d^2)]$$

$$V = \frac{\pi}{4}(D^2 + Dd + d^2) \times \tfrac{1}{3}h$$

$$= .2618h(D^2 + Dd + d^2)$$

SPHERE

$$S = \pi d^2 = 4\pi r^2 = 12.5664r^2$$

$$V = \tfrac{1}{6}\pi d^3 = \tfrac{4}{3}\pi r^3 = .5236d^3 = 4.1888r^3$$

CIRCULAR RING

D = mean diameter;
R = mean radius.

$$S = 4\pi^2Rr = 9.8696Dd$$

$$V = 2\pi^2Rr^2 = 2.4674Dd^2$$

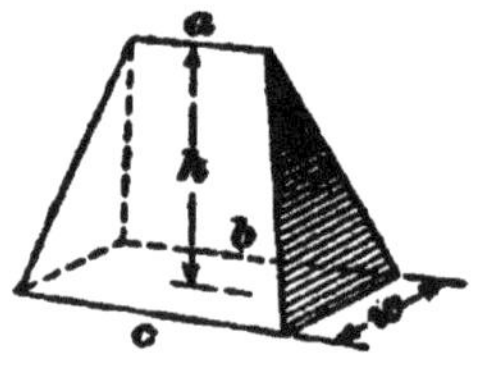

WEDGE

$$V = \tfrac{1}{6}wh(a + b + c)$$

PRISMOID

A *prismoid* is a solid having two parallel plane ends, the edges of which are connected by plane triangular or quadrilateral surfaces.

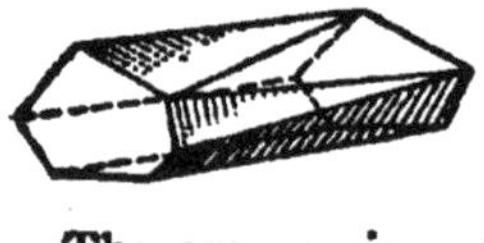

A = area of one end;
a = area of other end;
m = area of section midway between ends;
l = perpendicular distance between ends.

$$V = \tfrac{1}{6}l(A + a + 4m)$$

The area m is not in general a mean between the areas of the two ends, but its sides are means between the corresponding lengths of the ends.

Approximately, $V=\dfrac{A+a}{2}\times l$

REGULAR PYRAMID

P = perimeter of base;
A = area of base.

$$C=\tfrac{1}{2}Pl$$
$$S=\tfrac{1}{2}Pl+A$$
$$V=\frac{Ah}{3}$$

To obtain area of base, divide it into triangles, and find the sum of their areas.

The formula for V applies to any pyramid whose base is A and altitude h.

FRUSTUM OF REGULAR PYRAMID

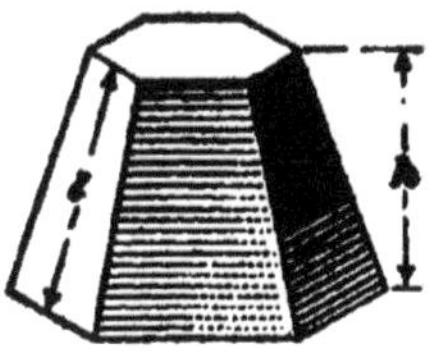

a = area of upper base;
A = area of lower base;
p = perimeter of upper base;
P = perimeter of lower base.

$$C=\tfrac{1}{2}l(P+p)$$
$$S=\tfrac{1}{2}l(P+p)+A+a$$
$$V=\tfrac{1}{3}h\left(A+a+\sqrt{Aa}\right)$$

The formula for V applies to the frustum of any pyramid.

LENGTH OF SPIRAL

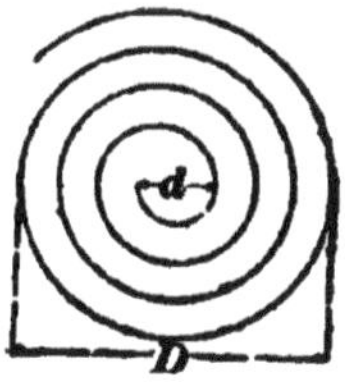

$$l=\pi n\left(\frac{D+d}{2}\right)$$
$$l=\frac{\pi}{t}(R^2-r^2)$$

n = number of coils;
l = length of spiral;
t = pitch.

PRISM OR PARALLELOPIPED

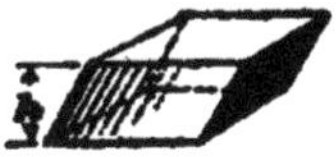

$$C=Ph$$
$$S=Ph+2A$$
$$V=Ah$$

For prisms with regular polygons as bases, P = length of one side $\times$ number of sides.

To obtain area of base, if it is a polygon, divide it into triangles, and find sum of partial areas.

FRUSTUM OF PRISM

If a section perpendicular to the edges is a triangle, square, parallelogram, or *regular* polygon,

$$V=\frac{\text{sum of lengths of edges}}{\text{number of edges}}\times\text{area of right section}$$

TRIGONOMETRY

Trigonometry is that branch of mathematics which treats of the properties of angular functions and their application to the solution of triangles. The *angular functions* are certain quantities, or ratios, depending on the magnitude of an angle, and serve for the determination of angles. There are six angular functions; namely, the *sine*, *cosine*, *tangent*, *cotangent*, *secant*, and *cosecant*. If, in any acute angle MAN, Fig. 1, a perpendicular BC be drawn to AN from any point on the side AM, forming the right triangle ABC, its three sides are named, with reference to the angle A, as follows: The side $AB=c$, the *hypotenuse*; the side $AC=b$, the *side adjacent*; and the side $BC=a$, the *side opposite*. The angular functions are then defined as follows:

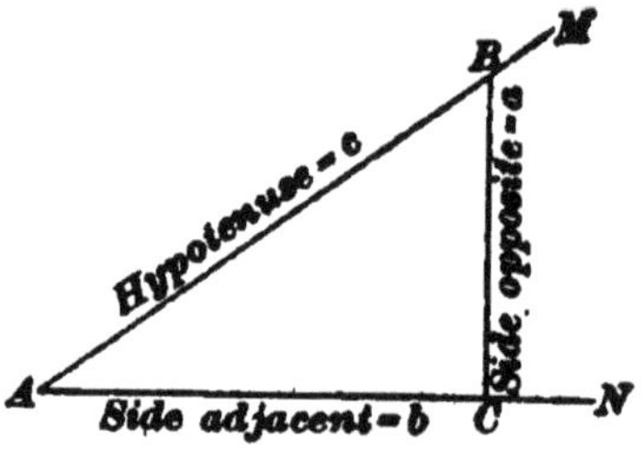

FIG. 1

$$\sin A=\frac{\text{side opposite}}{\text{hypotenuse}}=\frac{a}{c}$$

$$\cos A=\frac{\text{side adjacent}}{\text{hypotenuse}}=\frac{b}{c}$$

$$\tan A=\frac{\text{side opposite}}{\text{side adjacent}}=\frac{a}{b}$$

$$\cot A=\frac{\text{side adjacent}}{\text{side opposite}}=\frac{b}{a}=\frac{1}{\tan A}$$

$$\sec A=\frac{\text{hypotenuse}}{\text{side adjacent}}=\frac{c}{b}=\frac{1}{\cos A}$$

$$\csc A=\frac{\text{hypotenuse}}{\text{side opposite}}=\frac{c}{a}=\frac{1}{\sin A}$$

Besides these functions, use is sometimes made in railroad work of the *versed sine,* which is 1 minus the cosine of the angle, or $1-\frac{b}{c}$, and the *coversed sine,* which is 1 minus the sine of the angle, or $1-\frac{a}{c}$.

A good conception of the trigonometric functions may be formed from the diagram, shown in Fig. 2, in which the radius

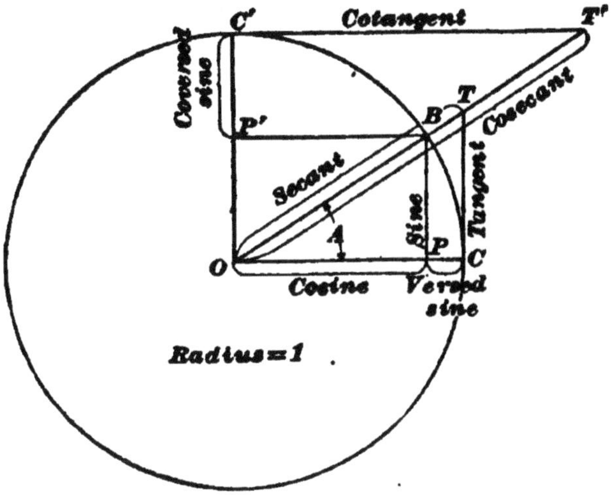

FIG. 2

of the circle is assumed as unity. Each ratio defining a trigonometric function is represented by a single line, as the denominator is in each case the radius, or unity.

The six angular functions are so related to each other as to enable the calculations of all when any one of them is known. These relations are given in the table on page 23.

For angles greater than 90°, the functions are expressed by those of acute angles. The rules and formulas relating thereto are given in the table on page 21. As an example, the cosine of 210° is found by formula 33; thus, cos $(180°+30°) = -\cos 30°$.

SOLUTION OF TRIANGLES

In every triangle there are six parts, three sides and three angles. The trigonometric functions of the angles are so related to the sides that when three parts of a triangle, one being a side, are known the other three parts, as well as the area of the triangle, may be computed. These relations are summed up in the tables on pages 24 and 25.

To facilitate calculations, tables of the trigonometric functions are used. Of these there are two kinds, namely, the table of *natural functions*, which gives the actual values of the functions, and the table of *logarithmic functions*, which gives the logarithms of their values.

TRIGONOMETRIC FORMULAS

Following are tabulated the principal formulas that are very useful in the solution of all kinds of problems requiring the application of trigonometry.

FUNCTIONS OF 0° AND 90°

1.	$\sin 0° = 0$	7.	$\sin 90° = 1$
2.	$\tan 0° = 0$	8.	$\tan 90° = \infty$
3.	$\cos 0° = 1$	9.	$\cos 90° = 0$
4.	$\cot 0° = \infty$	10.	$\cot 90° = 0$
5.	$\sec 0° = 1$	11.	$\sec 90° = \infty$
6.	$\csc 0° = \infty$	12.	$\csc 90° = 1$

FUNCTIONS OF NEGATIVE ANGLES

13.	$\sin(-A) = -\sin A$	16.	$\cot(-A) = -\cot A$
14.	$\tan(-A) = -\tan A$	17.	$\sec(-A) = \sec A$
15.	$\cos(-A) = \cos A$	18.	$\csc(-A) = -\csc A$

FUNCTIONS OF 90°+A

19.	$\sin(90°+A) = \cos A$	22.	$\cot(90°+A) = -\tan A$
20.	$\tan(90°+A) = -\cot A$	23.	$\sec(90°+A) = -\csc A$
21.	$\cos(90°+A) = -\sin A$	24.	$\csc(90°+A) = \sec A$

FUNCTIONS OF 180° − A AND OF 180° + A

25. $\sin (180° - A) = \sin A$
26. $\tan (180° - A) = -\tan A$
27. $\cos (180° - A) = -\cos A$
28. $\cot (180° - A) = -\cot A$
29. $\sec (180° - A) = -\sec A$
30. $\csc (180° - A) = \csc A$
31. $\sin (180° + A) = -\sin A$
32. $\tan (180° + A) = \tan A$
33. $\cos (180° + A) = -\cos A$
34. $\cot (180° + A) = \cot A$
35. $\sec (180° + A) = -\sec A$
36. $\csc (180° + A) = -\csc A$

FUNCTIONS OF 360° − A AND OF 360° + A

37. $\sin (360° - A) = -\sin A$
38. $\tan (360° - A) = -\tan A$
39. $\cos (360° - A) = \cos A$
40. $\cot (360° - A) = -\cot A$
41. $\sec (360° - A) = \sec A$
42. $\csc (360° - A) = -\csc A$
43. $\sin (360° + A) = \sin A$
44. $\tan (360° + A) = \tan A$
45. $\cos (360° + A) = \cos A$
46. $\cot (360° + A) = \cot A$
47. $\sec (360° + A) = \sec A$
48. $\csc (360° + A) = \csc A$

FUNCTIONS OF (A + B) AND OF (A − B)

49. $\sin (A + B) = \sin A \cos B + \cos A \sin B$
50. $\sin (A - B) = \sin A \cos B - \cos A \sin B$
51. $\cos (A + B) = \cos A \cos B - \sin A \sin B$
52. $\cos (A - B) = \cos A \cos B + \sin A \sin B$
53. $\tan (A + B) = \dfrac{\tan A + \tan B}{1 - \tan A \tan B}$
54. $\tan (A - B) = \dfrac{\tan A - \tan B}{1 + \tan A \tan B}$

FUNCTIONS OF 2A AND OF ½A

55. $\sin 2A = 2 \sin A \cos A$
56. $\cos 2A = \cos^2 A - \sin^2 A$
57. $\cos 2A = 2 \cos^2 A - 1$
58. $\cos 2A = 1 - 2 \sin^2 A$

59. $\tan 2A = \dfrac{2 \tan A}{1 - \tan^2 A}$

60. $\sin \frac{1}{2}A = \sqrt{\dfrac{1 - \cos A}{2}}$

61. $\cos \frac{1}{2}A = \sqrt{\dfrac{1 + \cos A}{2}}$

62. $\tan \frac{1}{2}A = \sqrt{\dfrac{1 - \cos A}{1 + \cos A}}$

63. $\tan \frac{1}{2}A = \dfrac{1 - \cos A}{\sin A}$

SUMS AND DIFFERENCES OF FUNCTIONS

64. $\sin A + \sin B = 2 \sin \frac{1}{2}(A+B) \cos \frac{1}{2}(A-B)$

65. $\sin A - \sin B = 2 \sin \frac{1}{2}(A-B) \cos \frac{1}{2}(A+B)$

66. $\cos A + \cos B = 2 \cos \frac{1}{2}(A+B) \cos \frac{1}{2}(A-B)$

67. $\cos A - \cos B = 2 \sin \frac{1}{2}(A+B) \sin \frac{1}{2}(B-A)$

68. $\tan A + \tan B = \dfrac{\sin (A+B)}{\cos A \cos B}$

69. $\tan A - \tan B = \dfrac{\sin (A-B)}{\cos A \cos B}$

70. $\sin^2 A - \sin^2 B = \sin (A+B) \sin (A-B)$

71. $\cos^2 A - \cos^2 B = \sin (A+B) \sin (B-A)$

72. $\cos^2 A - \sin^2 B = \cos (A+B) \cos (A-B)$

RELATIONS AMONG THE FUNCTIONS OF AN ANGLE

sin $A =$	tan $A =$	cos $A =$	cot $A =$	sec $A =$	csc $A =$
73. $\dfrac{\tan A}{\sqrt{1+\tan^2 A}}$	78. $\dfrac{\sin A}{\sqrt{1-\sin^2 A}}$	83. $\sqrt{1-\sin^2 A}$	88. $\dfrac{\sqrt{1-\sin^2 A}}{\sin A}$	93. $\dfrac{1}{\sqrt{1-\sin^2 A}}$	98. $\dfrac{1}{\sin A}$
74. $\sqrt{1-\cos^2 A}$	79. $\dfrac{\sqrt{1-\cos^2 A}}{\cos A}$	84. $\dfrac{1}{\sqrt{1+\tan^2 A}}$	89. $\dfrac{1}{\tan A}$	94. $\sqrt{1+\tan^2 A}$	99. $\dfrac{\sqrt{1+\tan^2 A}}{\tan A}$
75. $\dfrac{1}{\sqrt{1+\cot^2 A}}$	80. $\dfrac{1}{\cot A}$	85. $\dfrac{\cot A}{\sqrt{1+\cot^2 A}}$	90. $\dfrac{\cos A}{\sqrt{1-\cos^2 A}}$	95. $\dfrac{1}{\cos A}$	100. $\dfrac{1}{\sqrt{1-\cos^2 A}}$
76. $\dfrac{\sqrt{\sec^2 A-1}}{\sec A}$	81. $\sqrt{\sec^2 A-1}$	86. $\dfrac{1}{\sec A}$	91. $\dfrac{1}{\sqrt{\sec^2 A-1}}$	96. $\dfrac{\sqrt{1+\cot^2 A}}{\cot A}$	101. $\sqrt{1+\cot^2 A}$
77. $\dfrac{1}{\csc A}$	82. $\dfrac{1}{\sqrt{\csc^2 A-1}}$	87. $\dfrac{\sqrt{\csc^2 A-1}}{\csc A}$	92. $\sqrt{\csc^2 A-1}$	97. $\dfrac{\csc A}{\sqrt{\csc^2 A-1}}$	102. $\dfrac{\sec A}{\sqrt{\sec^2 A-1}}$

FORMULAS FOR THE SOLUTION OF RIGHT TRIANGLES

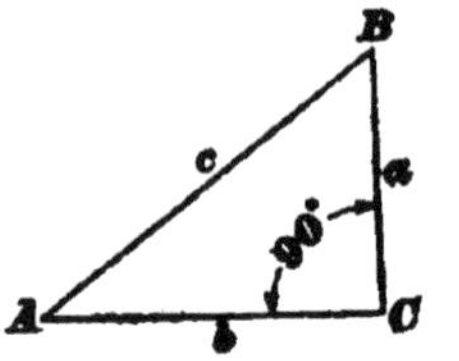

Given	Required	Formula
a, A	B, b, c	103. $B = 90° - A$
		104. $b = a \cot A$
		105. $c = \frac{a}{\sin A} = a \csc A$
a, B	A, b, c	106. $A = 90° - B$
		107. $b = a \tan B$
		108. $c = \frac{a}{\cos B} = a \sec B$
c, A	B, a, b	109. $B = 90° - A$
		110. $a = c \sin A$
		111. $b = c \cos A$
a, b	A, B, c	112. $\tan A = \frac{a}{b}$
		113. $\tan B = \frac{b}{a}$, or $B = 90° - A$
		114. $c = \sqrt{a^2 + b^2}$
		115. $c = \frac{a}{\sin A} = a \csc A$
a, c	A, B, b	116. $\sin A = \frac{a}{c}$
		117. $\cos B = \frac{a}{c}$ or $B = 90° - A$
		118. $b = \sqrt{c^2 - a^2} = \sqrt{(c+a)(c-a)}$
		119. $b = a \cot A$

FORMULAS FOR THE SOLUTION OF OBLIQUE TRIANGLES

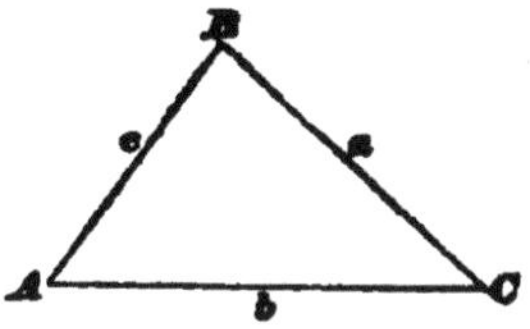

Given	Required	Formulas
a, b, C	A, B, c	120. $\tan \frac{1}{2}(A-B) = \frac{a-b}{a+b} \cot \frac{1}{2}C$ $A = (90° - \frac{1}{2}C) + \frac{1}{2}(A-B)$ $B = (90° - \frac{1}{2}C) - \frac{1}{2}(A-B)$ 121. $c = \frac{(a-b)\cos \frac{1}{2}C}{\sin \frac{1}{2}(A-B)}$ 122. $c = \sqrt{a^2+b^2-2ab \cos C}$
c, A, B	C, a, b	123. $C = 180° - (A+B)$ 124. $a = \frac{c}{\sin C} \sin A$ 125. $b = \frac{c}{\sin C} \sin B$
a, b, A	B, C, c	126. $\sin B = \frac{b}{a} \sin A$ 127. $C = 180° - A - B$ 128. $c = \frac{a}{\sin A} \sin C$
a, b, c $\frac{1}{2}(a+b+c) = s$	A	129. $\tan \frac{1}{2}A = \sqrt{\frac{(s-b)(s-c)}{s(s-a)}}$ 130. $\cos \frac{1}{2}A = \sqrt{\frac{s(s-a)}{bc}}$ 131. $\cos A = \frac{b^2+c^2-a^2}{2bc}$

CHAIN SURVEYING

INSTRUMENTS AND METHODS

Surveying is that branch of civil engineering which treats of the principles and methods employed for determining the relative positions of points on the earth's surface. Surveying is divided into three general branches, namely, *chain surveying*, in which no other measuring instrument is employed than a chain or tape for measuring distances; *angular surveying*, in which angle-measuring instruments are employed in connection with distance-measuring instruments; and *leveling*, which treats of the determination of elevations, or vertical distances.

The instruments used most commonly for measuring distances are the engineers' chain, the surveyors' chain, and the steel tape. Marking pins and range poles are used in connection with the chain, especially in measuring long lines.

The *engineers' chain* is 100 ft. long and is composed of 100 links of steel or iron wire, each two adjacent links being connected by small rings. The length of a link, including a ring at each end, is 1 ft. The engineers' chain is used chiefly in railroad surveying, but it is also used to some extent in other kinds of surveying where the foot is the unit of measurement.

The *surveyors' chain*, often called *Gunter's chain*, from the name of its inventor, is the same as the engineers' chain in every respect, except that its length is 66 ft., or 4 rd., instead of 100 ft. Like the engineers' chain, it is divided into 100 links, and consequently the length of each link is .66 ft., or 7.92 in. This chain is mainly used in land surveying, where the acre is the unit of area. It is very convenient for this purpose, as areas expressed in square chains can be expressed in acres by simply moving the decimal point one place to the left, there being 10 sq. ch. in 1 A. It is also well to remember that there are 80 ch. in 1 stat. mi.

The surveyors' chain is used in all United States land surveys, and whenever the word *chain* occurs in a legal document, it is understood to mean a surveyors' chain, or 66 ft.

Steel tapes are now used extensively in surveying and are largely superseding both the engineers' and the surveyors' chain. They can be obtained in any length from 1 yd. to 1,000 ft. and graduated to order. For city surveying, and for many other purposes, a tape 50 ft. long is generally preferred. For some purposes, tapes 300 or 500 ft. long and even of greater length are used. In some tapes, the handle forms part of the end division or graduation, the length of the tape counting from the outside of the handle. In others, the graduations begin on the inside of the handle, where the tape is attached, and in others the graduations begin on the tape itself, a short distance from the handle. When using a tape, the surveyor should ascertain where the graduations begin, as otherwise he may make serious errors. The tape has sometimes attached to it a handle that contains a spring balance for measuring the pull on the tape, a level bubble to guide in holding the tape so that it will be level, and a thermometer to show the temperature of the tape.

Correction for Erroneous Length of Chain.—The length of a chain or tape is altered by changes in temperature, and by wear and distortion. The variations due to temperature are very small, and need to be considered only in very accurate work. The alterations due to wear and distortion are sometimes considerable.

The length of the chain should be tested often. This is done either by comparing the chain with a chain or tape of standard length, or by stretching it between two points whose exact distance apart is known. It is advisable to have two such points marked permanently on an office floor, smooth pavement, curb, or some other convenient place.

If, after a line has been measured, the length of the chain (or tape) is found to be in error, the true length of the line can be easily determined by means of the following formula:

$$L_0 = L \pm eL,$$

in which L_0 = true length of line;

L = length of line as actually measured;

e = error in length of one unit of measure.

If, for instance, the length of a line is measured in feet, and the measurements are made with a 50-ft. tape that is found to be

.1 ft. too long, the error is $\frac{.1}{50}$, or .002 ft. in 1 ft. In this case, $e = .002$. If the measurement is recorded in chains, and the chain is found to be .1 li. too long, the error is .1 li., or .001 ch. per chain, and $e = .001$.

It should be understood that the correction eL expresses the same kind of units as e. If, for instance, e is 1.5 in. per ch., and the length of the line is expressed in chains, its true length is L ch. $\pm$ 1.5 L in.

If the chain is too long, the distance measured with it will be recorded as too short, and the correction eL should be added; and if the chain is too short, the distance measured will be recorded as too long, and the correction eL should be subtracted.

EXAMPLE.—The length of a line, measured with a 100-ft. chain, was found to be 1,048 ft. It was afterwards found that the chain was .19 ft. too long. What was the true length of the line?

SOLUTION.—If the error is .19 ft. in 100 ft., it is $\frac{1}{100}$ of .19 $= .0019$ ft., or, say, .002 ft. per ft. Then, $e = .002$, $L = 1,048$, and, therefore, $L_0 = 1,048 + .002 \times 1,048 = 1,050$ ft., nearly. The error is added, because, the chain being too long, the recorded length of the line was too small.

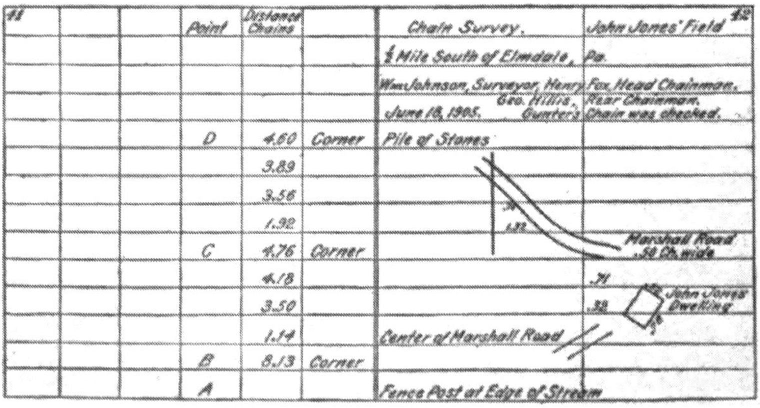

Keeping Notes.—The notes of a chain survey are usually kept in a transit book. The accompanying illustration shows a sample of notes of a chain survey. The right-hand page is used for sketches and remarks. The line that is being run is commonly represented by the red center line. In case more room is needed for sketching, the line that is being run may be represented by a line drawn on one side of the center line of the page and parallel to it. In sketching, it is better to face in the direction in which the line is being measured and to represent the line as running from the bottom to the top of the page in the notebook.

FIELD PROBLEMS

To Run a Line Over a Hill When the Ends of the Line Are Invisible From Each Other.—The points *A* and *B*, Fig. 1, are supposed to be on opposite sides of a hill, and to be invisible from each other. It is desired to run a line between them, or to locate some intermediate points.

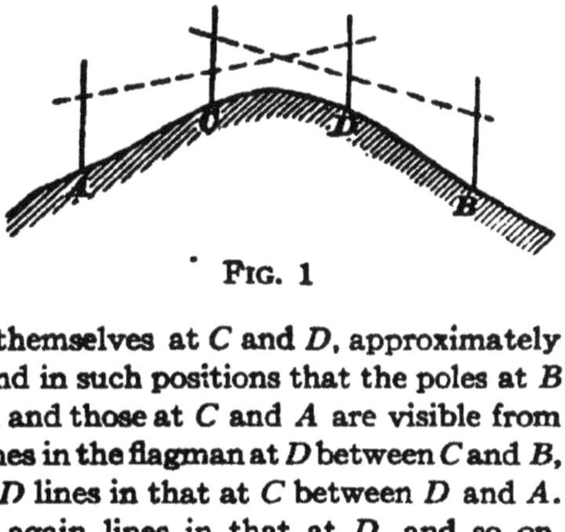

FIG. 1

Having set two poles at *A* and *B*, two flagmen with poles station themselves at *C* and *D*, approximately in line with *A* and *B*, and in such positions that the poles at *B* and *D* are visible from *C*, and those at *C* and *A* are visible from *D*. The flagman at *C* lines in the flagman at *D* between *C* and *B*, and then the flagman at *D* lines in that at *C* between *D* and *A*. Then the flagman at *C* again lines in that at *D*, and so on, until *C* is in line between *D* and *A* at the same time that *D* is in line between *C* and *B*. The points *C* and *D* will then be in line with *A* and *B*.

To Erect a Perpendicular to a Line at a Given Point.—Let it be required to erect a perpendicular to the line *AB* at the point *B*, Fig. 2. A triangle whose sides are in the proportion of 3, 4, and 5 is a right triangle, the longest side being the hypotenuse; for $5^2=4^2+3^2$. The following method is based on this principle: Lay off on *BA* a distance *BC* of 30 ft. (or li.). Fix one end of the chain at one of the extremities as C, and the end of the ninetieth link at the other extremity B. Hold the end of the fiftieth link and draw the chain until both parts are taut. The point D where the end of the fiftieth link is held will then be a point in the perpendicular, and the direction of the latter will therefore be BD.

The distance BC may be any other convenient multiple of 3. In general, if BC is denoted by $3a$, BD must be $4a$, and CD must be $5a$. Thus, BC may be made equal to 21 ($=3\times7$) li.; in which case BD must be $4\times7=28$, and CD must be $5\times7=35$, li. As $35+28=63$, one end of the chain must be fixed at one of the extremities of BC, the end of the sixty-third link at the other extremity, and the chain pulled from the end of the thirty-fifth link until both parts are taut.

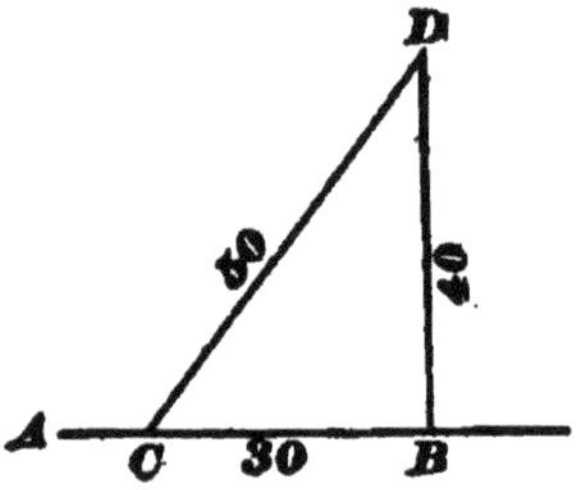

FIG. 2

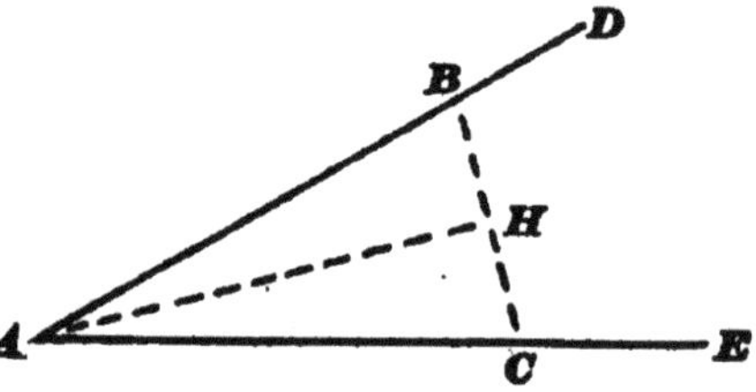

FIG. 3

To Determine the Angle Between Two Lines.—Let AD and AE, Fig. 3, be two lines on the ground. To determine the angle DAE, measure off from A on AD and AE equal distances AB and AC. Measure the distance BC. Then the angle DAE is calculated from the relation

$$\sin \tfrac{1}{2} DAE = \frac{\tfrac{1}{2} BC}{AB}$$

EXAMPLE.—If AB and AC are each 100 ft. and BC is 57.6 ft., what is the value of the angle DAE?

SOLUTION.—Substituting the values of BC and AB in the preceding equation,

$$\sin \tfrac{1}{2} DAE = \frac{\tfrac{1}{2}\times 57.6}{100} = .28800;$$

whence, $\tfrac{1}{2}DAE = 16° 44'$, nearly; and, therefore, $DAE = 16° 44' \times 2 = 33° 28'$.

To Determine the Distance to an Inaccessible Point.—Let it be required to determine the distance from the point B to an inaccessible point P, Fig. 4. Measure BC in any convenient direction and run a line $A'D'$ parallel to BC. Measure AD, the distance between the points where the lines PB and PC intersect $A'D'$. Measure also AB. Then,

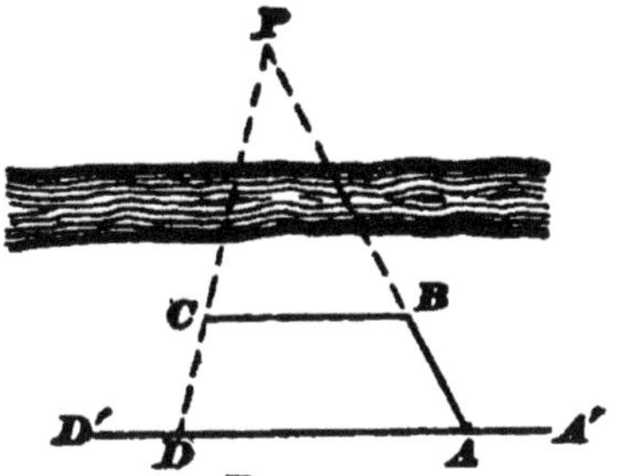

FIG. 4

$$BP = \frac{AB \times BC}{AD - BC}$$

EXAMPLE.—If, in Fig. 4, $BC = 100$ ft., $AB = 52.4$ ft., and $AD = 124.2$ ft., what is the distance BP?

SOLUTION.—Substituting these values in the preceding equation,

$$BP = \frac{52.4 \times 100}{124.2 - 100} = 216.5 \text{ ft.}$$

To Determine the Distance Between Two Points Invisible From Each Other.—Let it be required to find the distance between two points A and B, Fig. 5, that are invisible from each

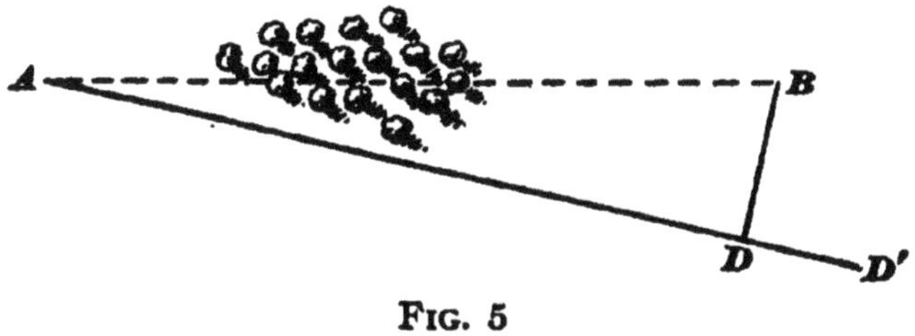

FIG. 5

other. First run a random line AD' in such a manner that it will pass as near B as can be estimated. From B drop a perpendicular BD on AD' and compute the required distance AB by the formula

$$AB = \sqrt{AD^2 + BD^2}$$

EXAMPLE.—If, in Fig. 5, the distance AD is 206.1 ft. and the distance BD is 35.1 ft., what is the distance from A to B?

SOLUTION.—Here $AD=206.1$ and $BD=35.1$; therefore, substituting in the preceding formula, $AB=\sqrt{206.1^2+35.1^2}=209.1$ ft.

Survey of a Closed Field.—If a closed field is to be surveyed without the aid of an angle-measuring instrument, the area is divided into triangles by means of diagonals, which are measured on the ground. The area of each triangle may then be determined by the formula

$$A=\sqrt{s(s-a)(s-b)(s-c)},$$

in which a, b, and c represent the three sides and s represents half of their sum, or $\frac{a+b+c}{2}$.

When obstacles make it impossible to measure directly the diagonals of a field, as, for instance, the diagonal BE, Fig. 6, a *tie-line* FG parallel to BE is run and measured. Then,

$$BE=\frac{GF\times AB}{AF}$$

To run the line FG, produce BA and select any convenient point F and measure AF. Then produce EA and locate G from the relation

$$AG=\frac{AF\times AE}{AB}$$

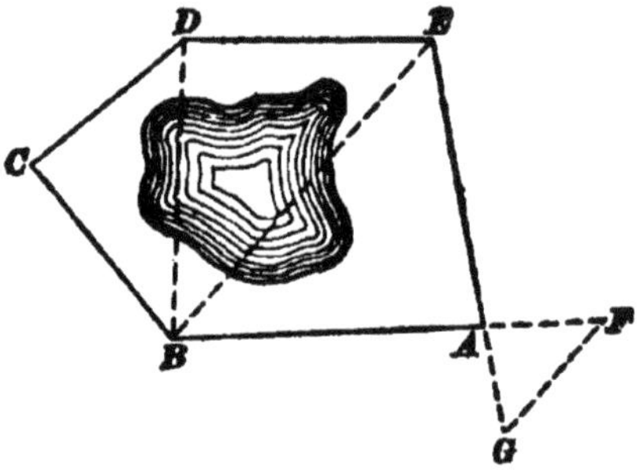

FIG. 6

EXAMPLE.—In Fig. 6, let the lengths of the sides be as follows: $AB=320$ ft., $BC=217$ ft., $CD=196$ ft., $DE=285$ ft., and $EA=304$ ft. It is required to calculate the length of the diagonal BE by means of a tie-line.

SOLUTION.—Let the line BA be prolonged 100 ft. beyond A; that is, make $AF=100$ ft. Then, AG must be equal to

$$\frac{AF\times AE}{AB}=\frac{100\times 304}{320}=95 \text{ ft.}$$

Let the length of GF, as found by measurement, be 125 ft.

Then, $$BE=\frac{GF\times AB}{AF}=\frac{125\times 320}{100}=400 \text{ ft.}$$

Precision.—In chain surveying, an error of 1 in 500 is generally permissible, and should not be exceeded; that is, two measurements of the same line should not give results differing by more than 1 ft. for every 500 ft. measured. If, however, the chaining is done carefully, and the ground is not rough, the error need not exceed 1 in 800 or 1,000.

ANGULAR SURVEYING

COMPASS SURVEYING

The *compass* used in surveying consists essentially of a magnetic needle supported freely on a pivot at the center of a horizontal graduated circle. To this circle is attached a pair of sights. The needle and graduated circle are enclosed in a brass case having a glass cover, and the whole is attached to a tripod, or Jacob's staff, by a ball-and-socket joint and is leveled by

FIG. 1

means of the plate levels. Fig. 1 illustrates the type of compass in general use.

Adjustments of the Compass.—Besides several conditions that are attended to by the instrument maker, the following are indispensable for accurate work:

1. The plane tangent to the level bubbles when at the centers of their respective tubes must be perpendicular to the vertical axis of the socket.

2. The two ends of the needle and the pivot must be in the same vertical plane.

3. The needle pivot must be in the center of the graduated circle.

A new compass made by a good manufacturer always satisfies these conditions, as the instrument is sold by the maker in perfect adjustment. Rough usage, however, a fall, or a hard blow may throw the compass out of order, and it is necessary that the surveyor should know how to test and readjust it.

To Adjust the Plate Levels.—Bring the bubbles to the centers of the level tubes by moving or rotating the plate carefully by means of the ball-and-socket joint; then revolve the compass horizontally through 180°; that is, turn it end for end. If the bubbles remain in the centers of the tubes, the levels are in adjustment. But if in turning the compass end for end, either bubble runs toward one end of the tube, lower that end and raise the opposite end sufficiently to bring the bubble half way back, by means of small screws that attach the ends of the tube to the plate; then bring the bubble to the center by moving the plate as before. Repeat the operation until both bubbles remain in the centers of the tubes in every position of the compass.

To Straighten the Needle.—Level the compass and turn it so that the north end of the needle points exactly toward or cuts some prominent graduation mark of the needle circle, and note the exact reading of the south end of the needle. In order to read either end of the needle accurately, the eye should be directly above a line in the prolongation of the opposite end of the needle. Then reverse the compass end for end and turn it so that the south end of the needle cuts the same graduation mark, and observe whether the north end reads the same as the south end did before reversing. If it does not read the same, correct one-half the error by bending the needle carefully, and repeat the operation, using different graduation marks, until exact reversals are obtained.

To Center the Needle Pivot.—Having, if necessary, straightened the needle, turn the compass so that the north end of the needle will exactly cut some prominent graduation mark, and observe whether the south end exactly cuts the opposite graduation mark. If it does not, find the position of the needle that shows the greatest difference in the readings of its opposite ends; then remove the needle from the pivot and bend the pivot carefully at right angles to this position an amount equal to one-half the error. Repeat the operation until the needle cuts accurately all opposite graduation marks.

Use of the Compass.—By means of the compass the angle between any line and the direction of the needle, or the *magnetic meridian*, can be measured directly. This angle is called the *magnetic bearing* of the line. The angle between two lines can be determined by either subtracting or adding their bearings, as the case may require. A rough sketch, showing the relative positions of the two lines with reference to the meridian, will enable one to determine by inspection the required arithmetical operation.

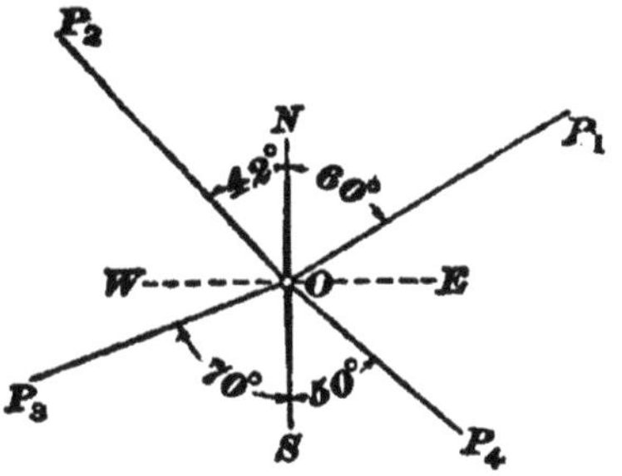

FIG. 2

Bearings are reckoned from 0° to 90° and indicate the amount by which a line is east or west of north or south. In Fig. 2, in which *NS* represents the magnetic needle, *N* and *S* being, respectively, the north and south ends, the line OP_1 makes, with the north half of the needle, an angle of 60°. As the line lies between the north point *N* and the east point *E*, its bearing is 60° northeast or 60° to the east of north. This is indicated by the notation N 60° E. Similarly, the bearings of OP_2, OP_3, and OP_4 are, respectively, N 42° W, S 70° W, and S 50° E.

To determine the magnetic bearing of a line, turn the compass, after it has been set and leveled, until the line *SN*, Fig. 3, which is the line of the sights, coincides with the line *OP* whose bearing is to be determined, the observer's eye being at the slit near *S*. The north end of the needle *FG* is then pointing to the bearing of the line. For example, in Fig. 3, the bearing is N 65° E. The north end of the needle may be recognized by the absence of the coil *s*. This coil is wound around the south half in order to balance the inclination of the needle in a vertical plane, called the ***dip of the needle***.

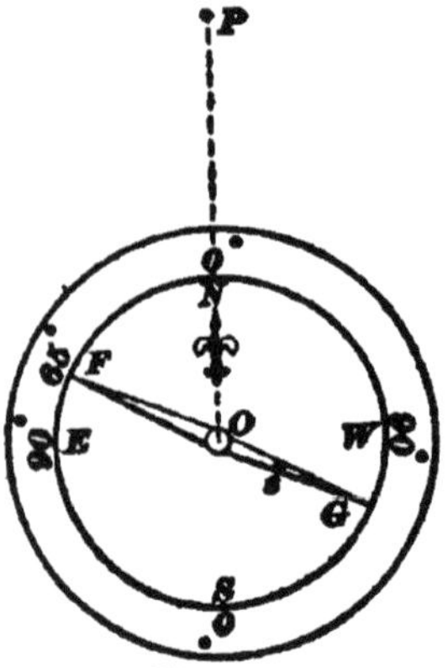

FIG. 3

Local Attraction.—The compass needle may be deflected from its natural direction by the attraction of any magnetic substance near it, such as iron ore, the rails of a railway, etc. This disturbing influence, called ***local attraction***, is very frequently met with, and the surveyor should take special care to avoid the errors to which it may give rise. When the bearing determined by a backsight does not equal that obtained by a foresight, with the letters N, S and E, W interchanged, the usual cause of the difference is local attraction. To determine whether the disturbing influence is at the end or the beginning of the line, set the compass at an intermediate point and take a sight on both points, when it will usually be found that the bearing thus obtained agrees with one of the bearings

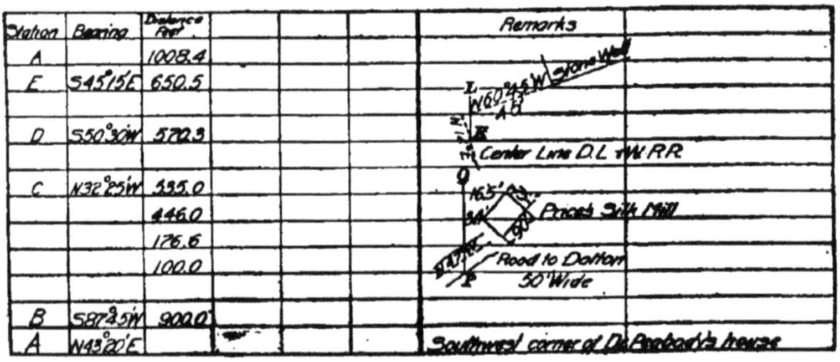

Station	Bearing	Distance Feet	Remarks
A		1008.4	
E	S45°15'E	650.5	
D	S50°30'W	570.3	
C	N32°25'W	335.0	
		446.0	
		176.6	
		100.0	
B	S87°45'W	900.0	
A	N43°20'E		Southwest corner of Dr. Peabody's house

FIG. 4

previously found. Should this not be the case, it would tend to show that local attraction exists at both the beginning and the end of the line, or also at the intermediate point, in which event the bearing of the line must be corrected by determining the angle by which the needle is deflected by the disturbing influences. This can be done by taking the foresight and backsight of a line formed by joining an outside point having no local attraction with the beginning or end of the line whose bearing is required.

Form for Compass Field Notes.—In Fig. 4 is shown a convenient form for keeping the notes of a compass survey. The left-hand half of the diagram represents the left-hand page of

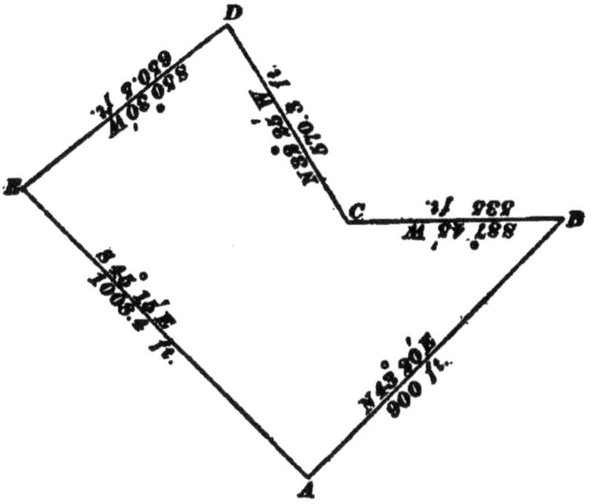

FIG. 5

the notebook; the right-hand half, the right-hand page. The notes are supposed to apply to the field *ABCDE*, Fig. 5. The corner, or station, *A* is the starting point of the survey, the courses being run from *A* to *B*, from *B* to *C*, etc. The notes read from bottom to top. Opposite the letter denoting a corner is given the bearing of the course running from that corner to the following one, in the order in which the survey was made. For instance, the bearing N 43° 20′ E horizontally opposite *A* denotes the bearing of the course *AB*. The number opposite a corner in the column of distances is the distance of this corner from the preceding one.

The right-hand page is used for remarks and sketches. When no objects are to be located along the line, as in the case from *A* to *B*, no sketch is necessary. Between *B* and *C*, a sketch is drawn showing the location of a road and mill with respect to the line *BC*. The line being run is usually represented by the center line on the right-hand page, unless objects are to be located at great distances on one side of the former line, in which case it is represented by a vertical line drawn near the right or the left edge of the page, as may be necessary. This is illustrated by the lines *PQ* and *KL*, which represent parts of *BC* and *DE*, respectively. A number written in the column of distances between two letters denoting corners, indicates the distance at which the point horizontally opposite it in the sketch is from the immediately preceding station or corner. Thus, the number 100, horizontally opposite *P*, indicates that the distance from *B* to *P* is 100 ft.

Declination of the Needle.—The angle that the magnetic meridian or the direction of the needle is making with the true meridian is called the *declination of the needle*. When this declination is known, the *true bearing* of a line, that is, the angle that it makes with the true meridian, can be determined from its magnetic bearing by adding or subtracting the declination, as the case may require.

The declination of the needle has different values in different localities, and also varies from year to year in a given locality. The approximate declination of the needle in a given locality at a given time can be determined from charts published by the United States Coast and Geodetic Survey. They show lines passing through all points where the declination of the needle is the same (*isogonic lines*) and also lines passing through all points where the declination is zero (*agonic lines*). These charts give also the yearly variation of the isogonic lines, and may be used for obtaining approximate values of declination for dates other than those for which the chart is prepared.

TRANSIT SURVEYING

The *engineers'* or *surveyors' transit* is now used almost exclusively in surveying. This instrument is primarily intended for measuring angles in a horizontal plane, but some transits

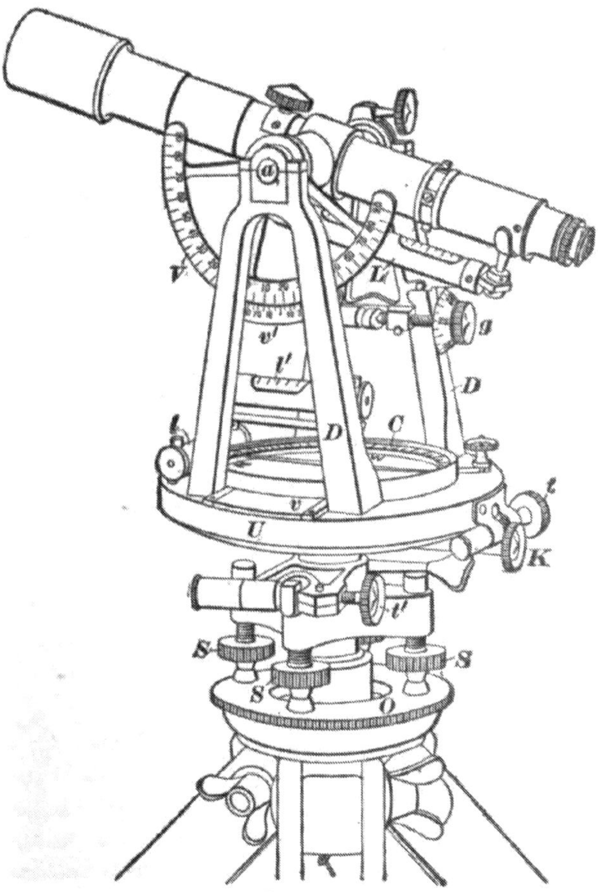

FIG. 1

have also a vertical circle, or arc, for measuring angles in a vertical plane. Fig. 1 shows a transit of this kind with a vertical arc *V* and a level *L* on the telescope.

The transit generally has a magnetic needle and a graduated needle circle *C*, and can therefore be used as an ordinary compass. The line of sight, however, instead of being given by a pair of sights is defined by the axis of the telescope. The telescope revolves in a vertical plane on the transverse axis *a*, and is supported by the standards *D*. These are attached to the upper, or vernier, plate *U*. The lower plate carries a graduated circle called the horizontal limb. These plates rotate independently around the vertical axis of the instrument.

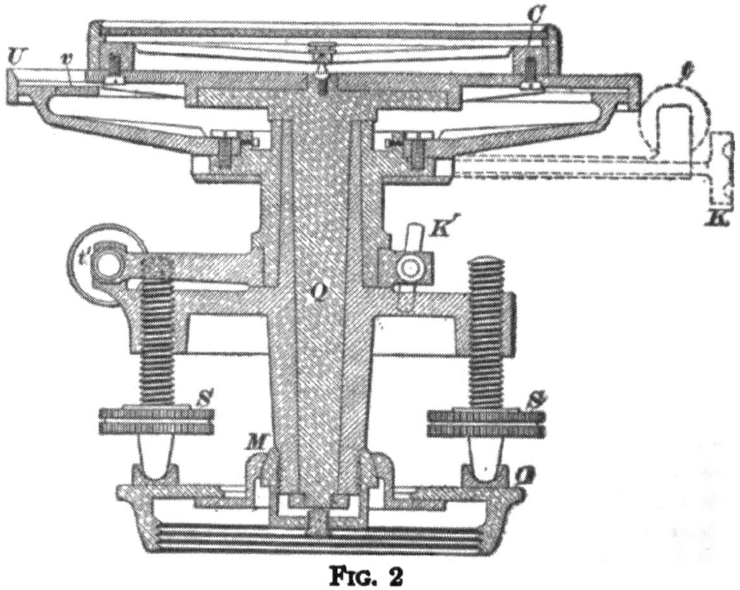

FIG. 2

The vernier plate rotates within and above the other, and to the former are attached two verniers *v* that travel along the graduated circle of the lower plate. The vernier plate can be clamped firmly to the lower plate by means of the clamp screw *K*, called the upper, or vernier, clamp; and by means of the upper tangent screw *t* it can be revolved slowly on the lower plate, moving the vernier along the divided circle, so that the instrument can easily be set at any given angle. The upper plate is attached to an accurately turned and slightly conical axis or spindle *Q*, Fig. 2, that extends down nearly to the tripod head. In transits of the most modern construction, this axis revolves within a socket that is controlled by the leveling screws S and about the upper portion of which revolves a socket that extends down from the lower plate, forming what is called a compound center. The centers, which control the entire instrument above the leveling screws, can be clamped against rotation by means of the clamp screw K', and the instrument can then be revolved slowly by means of the tangent screw t'. This clamp screw is called variously the lower clamp, clamp to the centers, or clamp to the lower plate, and the tangent screw is designated by corresponding terms. The centers are connected with plate O, sometimes called the lower leveling plate, by means of a hemispherical or ball-and-socket joint, shown at M. The centers and the entire instrument above them are supported in position by the four leveling screws, which serve also to level the instrument. The plate O screws on the tripod head. This plate and the leveling screws, considered together, are sometimes spoken of as the leveling head.

The graduated circle is numbered in various ways, three systems of numbering being employed. These may be described as the *azimuth system*, in which the figures extend from 0 continuously around the entire circle to 360; the *transit system*, in which the figures extend from 0 in opposite directions through the adjacent semicircles to 180 at the point diametrically opposite the zero point; and the *compass system*, in which the figures extend each way from two 0 points diametrically opposite each other through the adjacent quadrants to the 90° points.

There are usually two rows of figures extending around the graduated circle of a transit, each row being numbered according to one of the preceding systems. In some transits, both rows are of the azimuth system, extending in opposite directions around the circle. The different systems are also combined in various ways.

THE VERNIER

A *vernier* is an auxiliary scale used for measuring fractional parts of the smallest subdivisions of the main scale. The smallest fractional part of the main scale that can be read by means of the vernier is called the ***least reading of the vernier.*** If the least reading of the vernier is denoted by r, the smallest subdivision of the main scale by s, and the vernier is divided into n equal parts, then

$$r = \frac{s}{n}$$

For example, if the smallest division of a level rod is .01 ft., and the vernier is divided into 10 equal parts, then the least reading of the vernier is $r = .01 \div 10 = .001$ ft. In order to give this reading, the total length of the vernier must be $(n-1)s$. In the example just given the total length of the vernier must be $(10-1) \times .01 = .09$ ft. Such a rod and vernier are illustrated in Fig. 3.

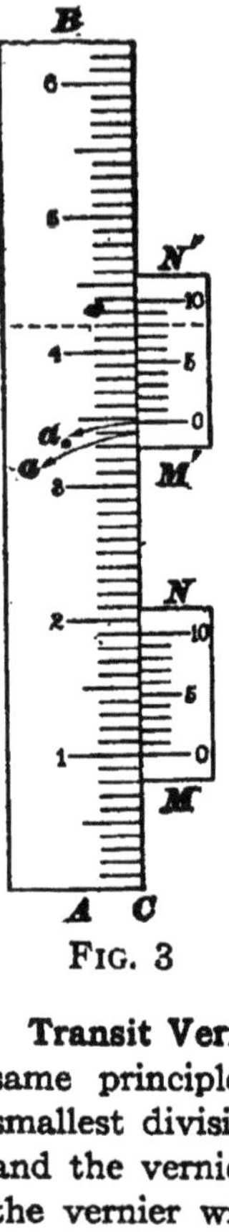

FIG. 3

To measure the length of a line, as Ca_0, Fig. 3, with a scale having a vernier, place the zero of the scale at the beginning of the line, as at C, and slide the vernier, as MN, up the scale until its zero coincides with the end of the line. In the example under consideration the position would be that of $M'N'$. Then the subdivision of the scale immediately preceding the end of the line will give the reading of the scale, in this case .34 ft. To this must be added the reading of the vernier. This is determined by the number of the division mark of the vernier that coincides with a division mark of the scale. In this case, this number is 8; the reading of the vernier is therefore $8 \times .001 = .008$. Hence, the length of Ca_0 is .348 ft.

Transit Verniers.—Transit verniers are constructed on the same principle as those used for measuring lines. If the smallest division of a horizontal circle of a transit is $\frac{1}{2}° = 30'$ and the vernier is divided into 30 parts, the least reading of the vernier will be $30' \div 30 = 1'$, and the vernier must cover on the circle a length equal to $\frac{1}{2}° \times (30-1) = 14°\ 30'$.

Fig. 4 shows part of the horizontal circle of a transit AB with the double verniers MN and MN'. The vernier MN is used when angles are turned to the left, that is, when the zero of the vernier slides in the direction *AB*, and the degrees are indicated by the upper figures (60, 70, 80, etc.) on the graduated circle. The vernier *MN'* is used when angles are turned to the right, and the degrees are indicated by the lower figures (90, 100, 110, etc.) on the graduated circle. Nearly all transits have two combinations of verniers similar to *NN'*, the zeros

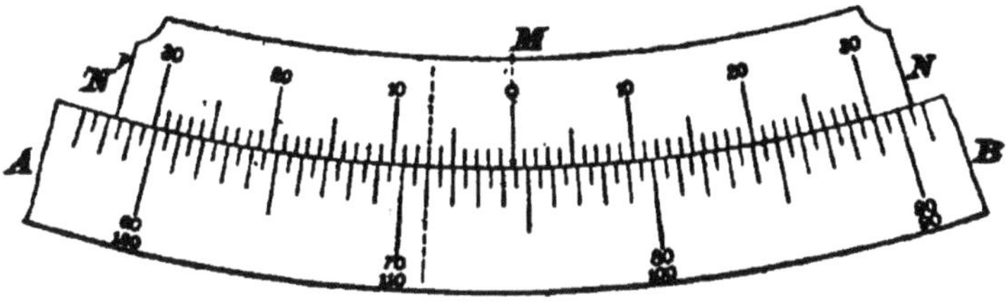

FIG. 4

of which are 180° apart. Each of these combinations, although it really consists of two verniers, is referred to as one vernier, one of them being called vernier *A* and the other vernier *B*. For very accurate work, both verniers are read, and if they do not agree, the mean of the two readings is taken as the true reading. The circle is divided into degrees and halves, and the vernier is divided into 30 equal parts covering 29 of the half-degree divisions of the circle; the vernier therefore reads to minutes.

Suppose that the center of the graduated circle is over the vertex of an angle to be measured; also, assume that its zero is on one of the sides, that the vernier has been slid to the left along the graduated circle until the other side of the angle passes through the zero mark of the vernier, and that the vernier has then the position shown in Fig. 4. Since the vernier has moved to the left, the side *MN* is to be read. The twenty-third mark of the vernier coincides with a division mark of the scale, and, as the least reading of the vernier is 1', its reading, in this case, is 23'. The reading of the scale, up to the division mark immediately preceding the zero of the vernier, is 74°. The reading of the instrument, or the measure of the angle is, therefore, $74°+23' = 74° 23'$. It can readily be seen that when the angle is measured from B toward A the reading of the instrument is 105° 37′.

The vertical circle, or arc, V, of the engineers' transit, Fig. 1, is often graduated to degrees and halves, and the vernier v', which is double, like the vernier of the horizontal circle, reads either to single minutes or to 5 minutes. If the vernier is attached to the standards, it is stationary; and instead of its sliding along the vertical arc, the vertical arc slides on it. Care should always be taken to read that side of the vernier whose numbers increase in the same direction as those by which degrees are measured on the graduated circle.

ADJUSTMENTS OF THE TRANSIT

When a transit is in perfect adjustment, it must, after being leveled, fulfil the following conditions:

1. The centers must revolve on a truly vertical axis, so that the plate levels will remain centered during a complete revolution.

2. The line of collimation—that is the line of sight—must be perpendicular to the transverse axis of the telescope, so that it will be in the same straight line when the telescope is plunged.

3. The axis of revolution (the transverse axis) of the telescope must be horizontal, and, therefore, perpendicular to the vertical axis of the instrument.

When a transit has a level and a vertical arc or circle attached to the telescope, it should fulfil the following additional conditions:

4. The line of collimation must be parallel to a line tangent to the tube of the telescope level at its middle point, so that the line of collimation will be horizontal when the bubble of the telescope level stands at the middle of its tube.

5. The vernier of the vertical arc or circle must read zero when the line of collimation is horizontal.

The adjustments establishing these conditions should be made in the order in which the conditions are stated. The best time for adjusting an instrument is on a cloudy day or in the early morning before the air has become heated and the sun dazzling. An open and nearly level space affording an unobstructed sight for at least 400 ft. from the transit in opposite directions should be chosen for making the adjustments. In setting up the instrument, the feet of the tripod should be planted firmly in solid ground that is not subject to jars from heavy machinery or other causes, so that its position will not be disturbed.

First Adjustment.—To make the axes of the plate levels perpendicular to the vertical axis of the instrument, so that when the bubbles are centered by the leveling screws the axis of the centers will be truly vertical and the plates will revolve in a horizontal plane. This adjustment is substantially the same as for the compass, and is performed as follows:

With the upper clamp set and the lower clamp loose, turn the instrument so that the plate levels l and l', Fig. 1, will be, respectively, parallel to the lines determined by the two pairs of leveling screws, and bring each bubble to the middle of its tube by means of the corresponding pair of leveling screws. Next, turn the instrument half way around; that is, revolve it in azimuth through 180°, so that each level will be in the reverse position with respect to the same pair of leveling screws. If the levels are in adjustment, the bubbles will remain in the centers of the tubes. If the bubbles do not remain so, but run to either end, bring them half way back to the middle of the tubes by means of the capstan-headed screws attached to the ends of the tubes, and the rest of the way back by the leveling screws. Then revolve the instrument again through 180° and observe the positions of the bubbles. Sometimes this adjustment is made by one trial, but it is usually necessary to repeat the operation.

Second Adjustment.—To make the line of sight perpendicular to the transverse axis of the telescope.

The manner of performing this adjustment is illustrated in Fig. 5. Set and level the instrument at a point A, and direct the telescope to some well-defined point B a few hundred feet distant. Both clamps being set, plunge the telescope and set another point, as a marking pin or a tack in the top of a stake, a few hundred feet away, on the opposite side of the instrument from B. If the line of sight is truly perpendicular

, and the
l the end
s are tau
eld will th
of the lat
The distanc
general, if
st be 5 a.
which case
35, li. As
one of the e
the other e
thirty-fifth

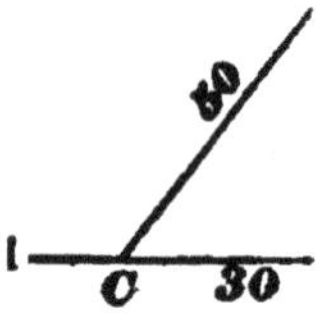

FIG.

To Determin
?, Fig. 3, be
gle DAE, m
ces AB and
gle DAE is c

XAMPLE.—I
t is the valu
LUTION.—S
eding equat

ce, $\frac{1}{2}DAE$ =
33° 28′.

[illegible] will be in the
[illegible] this is the
[illegible] a smooth
[illegible] tangent

[illegible] change the tele-
[illegible] as before,
[illegible] adjustment is
[illegible]

[illegible] plunging is
[illegible] is at E,
[illegible] sight must
[illegible] measure the
[illegible] same distance
[illegible] and set a
[illegible] DE.
[illegible] screws
[illegible] F, being
[illegible] that in
[illegible] move
[illegible] the telescope
[illegible] tighten
[illegible] a firm
[illegible] strain.
[illegible] the telescope
[illegible] the point B,
[illegible] EF. By
[illegible] sight again
[illegible] the adjust-
[illegible] the point C which
[illegible] midway between
[illegible] by reversing the
[illegible] the telescope
[illegible] necessary to

repeat the operation several times in order to obtain an exact adjustment.

Third Adjustment.—To make the transverse axis of the telescope perpendicular to the vertical axis of the instrument, so that when the instrument is leveled the transverse axis of the telescope will be horizontal.

Suspend a fine, smooth plumb-line from a rigid support at as high an elevation as convenient and at a distance from the instrument not exceeding the length of the line. The weight should be suspended in a pail of water, care being exercised that it does not touch the bottom of the pail and that the line is not exposed to wind. With both plate bubbles in the middle of their tubes, direct the line of sight to the upper end of the plumb-line; then, turning the telescope slowly downwards, notice whether the intersection of the cross-hairs exactly follows the line throughout its length. If it does follow it, the line of collimation revolves in a vertical plane. The plumb-line will usually vibrate slightly, but its mean position can be estimated by the eye. If the intersection of the cross-hairs does not coincide with the plumb-line throughout its length, but diverges to one side as it approaches the bottom of the line, the error must be corrected by raising or lowering one end of the transverse axis of the telescope, which is adjustable by means of screws placed in one of the standards. If the intersection of the cross-hairs diverges on the side of the plumb-line toward the adjustable end of the transverse axis, this end is to be lowered; if on the opposite side, it is to be raised.

This adjustment can also be tested and made in the following manner: Level the instrument, and direct the telescope to some well-defined point on a church spire or other high object, as the point *A*, Fig. 6. Having set both the upper and the lower clamp, depress the object end of the telescope and set a point in the line of sight on the ground at the base of the object; loosen the upper clamp, reverse the instrument in azimuth, plunge the telescope, sight again on the high point, again turn the telescope downwards, and notice whether or not the line

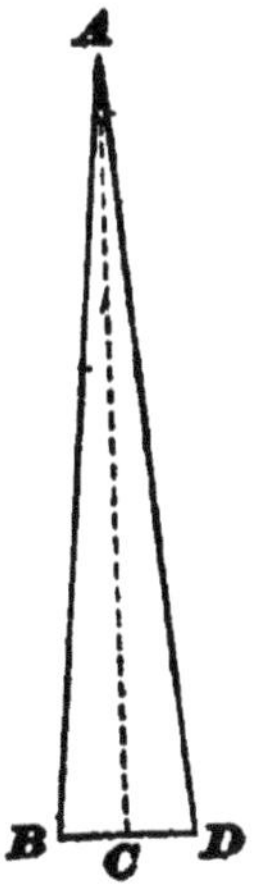

FIG. 6

of sight strikes the same point as before. If it does, the transverse axis of the telescope is horizontal. If the point first set is the point *B*, and the second line of sight passes through *D*, instead of *B*, the transverse axis is not horizontal, and must be adjusted. The adjustment is made by raising or lowering one end of the transverse axis (in this case the right-hand end would have to be lowered), and again repeating the test, until the points *B* and *D* coincide; that is, until the line of sight, when the telescope is depressed, strikes the same point, as *C*, both before and after reversal.

Fourth Adjustment.—To make the bubble of the telescope level stand in the middle of its tube when the line of sight is horizontal.

This adjustment makes the transit adapted to leveling work. It is the same as that of a regular level, and is described in connection with the level.

Fifth Adjustment.—To make the vernier of the vertical arc or circle read zero when the line of sight is horizontal.

To perform this adjustment, level the instrument and turn the telescope on its transverse axis until the bubble in the attached level is nearly in the middle of its tube; clamp the telescope, and center the bubble of the attached level exactly by means of the gradienter screw *g*, Fig. 1. If the vernier of the vertical limb does not read zero, set it so that it will read zero by means of the capstan-headed screws that control it.

This adjustment is not strictly necessary, provided the reading of the vernier when the telescope is horizontal is observed and noted. This reading is called the ***index error*** of the vertical circle or vernier and should be allowed for in reading vertical angles.

Adjustment of the Cross-Hairs.—For convenience in directing the telescope to a signal, it is desirable that the vertical cross-hair should be truly vertical, and the other truly horizontal. The two cross-hairs are attached to an adjustable diaphragm exactly at right angles to each other, so that when one is vertical the other is horizontal. In order to test the vertical cross-hair, sight on any sharply defined point, focusing the telescope perfectly and bringing the point exactly in range with either end of the vertical cross-hair. Then turn the telescope on its transverse axis slowly and notice whether the point sighted remains on the cross-hair throughout the motion. Should any deviation be discernible, loosen the capstan-headed screws that control the cross-hairs, and by the pressure of the hand, or by tapping lightly against the heads of the screws outside the telescope tube, rotate the cross-hairs very carefully in the direction opposite that in which they should apparently be rotated, until the point sighted remains on the cross-hair throughout the motion of the telescope. Then tighten the screws sufficiently to bring them to a firm bearing without straining them, and repeat the test.

This test should be applied before performing the third adjustment for the line of collimation. If the plate levels are in perfect adjustment, it can also be made by sighting at a plumb-line suspended at a suitable height and distance, with the plate levels centered perfectly, and observing whether the vertical cross-hair coincides exactly with the plumb-line.

TRANSIT FIELD WORK

To Prolong a Straight Line.—Let *AB*, Fig. 7, be a straight line whose position on the ground is fixed by stakes set at *A* and *B*, and let it be required to prolong the line to *C*. This

FIG. 7

can be done in two ways; namely, by foresight only, or by backsight and foresight, the latter method being commonly called backsight.

By Foresight.—The transit is set over the point at *A*, and the line of sight directed to a flag held at *B*; if the point *C* is to be set at a given distance from *B*, the chainmen measure the required distance, the head chainman being kept in line by the transitman. When the required distance has been measured, the point *C*, which evidently lies in the prolongation of *AB*, is marked by a stake or otherwise.

By Backsight.—Set the transit over the point at *B* and sight on a flag held at *A*. Plunge the telescope, which will then be directed along the prolongation of *AB*. Any required distance BC may then be measured from B in the direction indicated by the line of sight.

Measurement of Horizontal Angles.—The horizontal circle of the transit, like that of the compass, measures only horizontal angles; that is, angles between the horizontal projections of the lines of sight. Let AB and AC, Fig. 8, be two lines on the ground the angle between which it is desired to measure with the transit. Set up the instrument precisely over the vertex A, level it carefully, loosen the upper clamp, and turn the upper plate until the zero of the vernier to be read (say vernier A) nearly coincides with the zero of the graduated circle. Clamp the plates, and by turning the upper tangent screw bring the zero of the vernier exactly in line with that of the limb. This

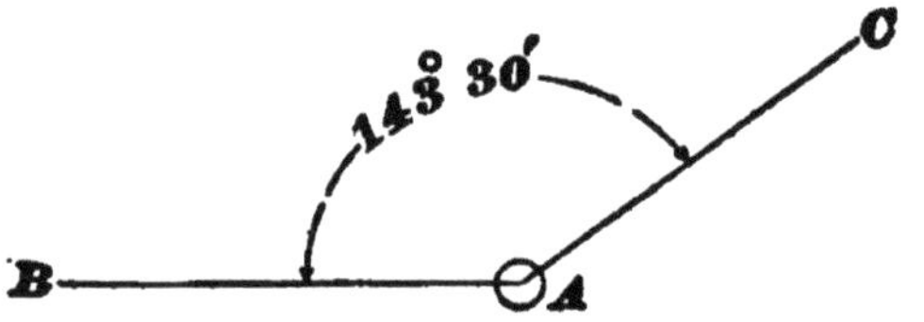

FIG. 8

operation is called *setting the vernier at zero.* Loosen the lower clamp (if it is not already loosened), and direct the telescope to a flag held at B. Next, loosen the upper clamp, and direct the telescope to a flag held at C. The arc of the graduated circle traversed by the zero point of the vernier will measure the angle BAC, whose value can be determined by reading the instrument; that is, by adding the reading of the vernier to that of the limb.

It is not always necessary nor convenient to set the vernier at zero before measuring an angle. The upper clamp being set, whatever the position of the vernier may be, the telescope is directed to B, as explained, and the reading of the instrument taken. The upper clamp is then loosened, the telescope directed to C, and the instrument read again. The difference between the two readings is the value of the angle.

TRAVERSING

In surveying, a *traverse* is a series of consecutive courses whose lengths and directions are determined by measurement. For determining the directions of the courses of a traverse, three methods are commonly employed, namely, *by bearings*, in which method the directions of the courses of the traverse are determined by their magnetic bearings; *by azimuths*, in which method the directions of the courses of the traverse are determined by their azimuths; and *by deflection angles*, or *by deflections*, in which method the relative directions of the courses of the traverse are determined by measuring the angle by which the direction of each course is deflected from the prolongation of the immediately preceding course.

Traversing by Azimuth.—The *azimuth* of a line is the angle that the line makes with the meridian. It is measured from 0° to 360°, either from the south in the direction west—north—east, or from the north in the direction east—south—west. Sometimes, a line that is neither a true nor a magnetic meridian is used as a line of reference from which azimuths are measured in the same manner as if the line were a meridian. Such a line of reference is called an *assumed meridian*, or simply a *meridian*.

When the directions of courses are given by their azimuths, a transit is used with its horizontal circle graduated from 0° to 360°. It often happens that, by the addition of certain angles, an azimuth greater than 360° is obtained. An azimuth greater than 360° is equal to the same azimuth diminished by 360°.

Process of Azimuth Traversing.—Referring to the illustration on page 52, suppose that *A* is a given point on a line *AF* previously surveyed, and that it is desired to connect this point with the point *E* by a traverse following the contour of the surface in such a manner as to give about the minimum rise and fall. The true bearing of *AF*, as previously determined, is N 42° 36′ W; therefore, its azimuth, counted from the north, is $360° - 42°\ 36' = 317°\ 24'$. The points *B*, *C*, and *D* are chosen in advance of the survey in such positions as will fulfil the required conditions as nearly as can be judged, each point being selected while the instrument is being moved forwards, set up, and oriented at the preceding point. The instrumental operations in running this traverse are as follows:

The transit is first set up at *A* and oriented by setting the vernier at 317° 24′ (azimuth of *AF*) and directing the telescope, with the upper plate clamped, along *AF*, the point *F* being marked by a flag. The lower clamp is then set, the vernier clamp is loosened, the telescope is turned in azimuth and directed to a flag held at *B*, and the vernier is read. The

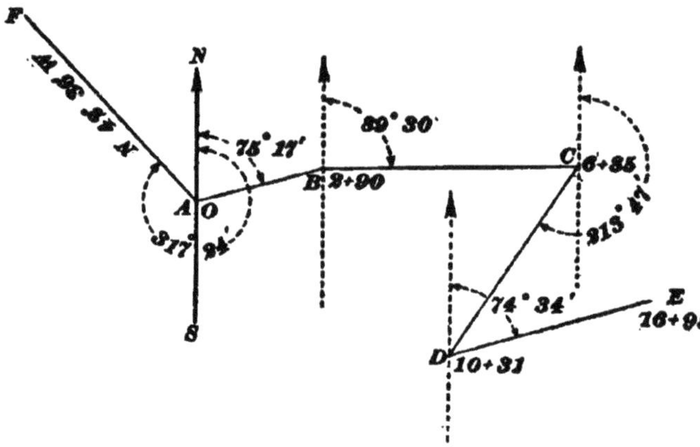

reading, which in this case is 75° 17′, is recorded as the azimuth of *AB*. As *A* is the initial point of the survey, complete information as to how the instrument is oriented should be described by means of a sketch or a written statement. As a check, the magnetic bearing of *AF* and that of the last line should be taken and recorded. Suppose the magnetic bearing of *AF* to be N 40° 10′ W; as the true bearing is N 42° 36′ W, the declination is 2° 26′ west, which should be noted.

The instrument is now moved forwards, set up at *B*, and oriented by making the reading of the vernier equal to the azimuth of *BA*, which is equal to that of *AB* plus 180°; that is, 75° 17′+180°=255° 17′. The upper clamp being set, the telescope is directed to *A*; the lower clamp is set, the upper clamp loosened, the telescope directed to *C*, and the vernier read. The reading is found to be 89° 30′ which is recorded as the azimuth of *BC*. The instrument is then moved to *C*, and the azimuth of *CD* is determined as explained for *BC*. This azimuth is found to be 213° 47′. The instrument is moved to *D* and oriented by backsighting on *C*. The forward azimuth of *CD* being 213° 47′, the back azimuth is 213° 47′+180°=393° 47′, or 393° 47′−360°=33° 47′. After setting the vernier at this reading and directing the telescope to *C*, the transit is oriented at *D*. The lower clamp is then set, the upper clamp is loosened, the telescope directed to *E*, and the vernier read again, the reading being the azimuth of *DE*.

The magnetic bearing of *DE* is now taken; suppose it to be N 77° 15′ E. As the declination is 2° 26′ west, the approximate true bearing of *DE*, as obtained from the compass, is

Station	Azimuth	True Bearing	Distances	Remarks
16+95	74° 34′	N 74° 49′ E		End of line.
10+31	213° 47′			
6+85	89° 30′			
2+90	75° 17′			
F, 64	317° 24′	N 42° 36′ W		
0				Sta. 0 is at Sta. 58+60 of surveyed line of O. & B. R. R. Oriented by forward azimuth on Inst. Point *F*, at Sta. 64 of same. True bearing N 42° 36′ W. Declination 2° 26′ west.

N 74° 49′ E. Since the line has an azimuth of 74° 34′, its true bearing is evidently N 74° 34′ E, which agrees with that given by the compass within the limit of accuracy of the latter instrument, with which angles are read to the nearest quarter of a degree. In a traverse consisting of many lines, it is advisable to take the magnetic bearing of every third or fourth line, and compare it with the true bearing obtained from the azimuth of the line.

The distance between *A* and *B* is measured when the transit is at *A*, the transitman keeping the head chainman in line; the distance between *B* and *C* when the transit is at *B*, etc.

Field Notes of an Azimuth Traverse.—The preceding notes are those of the azimuth traverse shown in the preceding illustration. The distances, which are obtained by merely subtracting the number of each instrument station from the number of the succeeding instrument station, are recorded in the fourth column. This is usually done in the office.

LATITUDE AND LONGITUDE

For the purposes of plotting and calculation, all the points of a survey are often located with reference to two coordinate axes perpendicular to each other, one being a north-and-south line, true or magnetic, and called a *reference meridian*, or *principal meridian;* the other, which is an east-and-west line, is called a *reference parallel of latitude*, or *principal parallel of latitude*. The distance of a point from the reference meridian is called the *longitude* of the point. It is east longitude or west longitude according as the point is east or west of the reference meridian. East longitudes are considered positive and west longitudes negative. The *latitude* of a point is the distance of the point from the reference parallel of latitude. It is a north latitude and considered positive when the point is north of the reference parallel; it is a south latitude and is negative when the point is south of the reference parallel. The algebraic difference obtained by subtracting the latitude of the beginning of a line, meaning the point from which the line is run, from the latitude of the other extremity of the line, is called the *latitude range* of the line. Likewise, the algebraic difference between the longitude of the end and the longitude of the beginning of the line is called the *longitude range* of the line. In Fig. 1, TT' and $G'G$ represent, respectively, a reference meridian and a principal parallel of latitude. The latitudes of the points P and Q are, respectively, HP and KQ; they are positive. The latitudes of the points P_1, Q_1, P_2, Q_2 are respectively H_1P_1, K_1Q_1, H_2P_2, and K_2Q_2; they are negative. The longitudes of P, Q, P_1, and Q_1 are, respectively, $H'P$, $K'Q$, $H_1'P_1$, and $K_1'Q_1$; they are positive. The longitudes of P_2 and Q_2 are, respectively, $H_2'P_2$ and $K_2'Q_2$; they are negative. If the line is run from P to Q, the latitude range of PQ is $KQ - HP = PD$, and is positive. Similarly, the longitude range of PQ is equal to $K'Q - H'P = DQ$. If run from Q to P, its latitude range would be $HP - KQ = -EQ = -PD$, and its longitude range $H'P - K'Q = -EP = -DQ$. The latitude range indicates how far the end of the line is north or south of the begin-

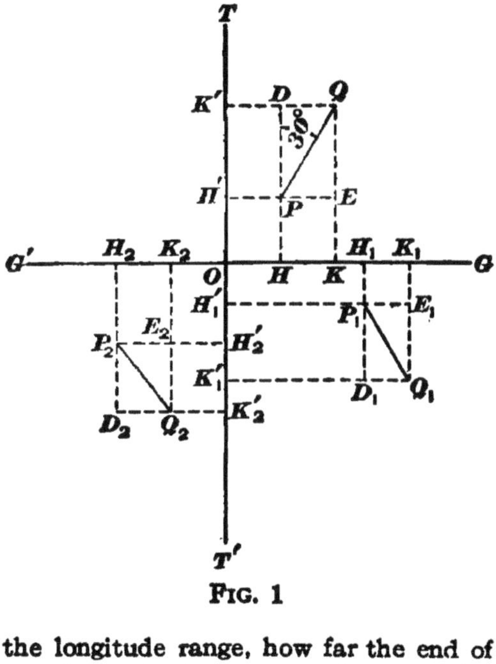

FIG. 1

ning; and the longitude range, how far the end of the line is east or west of the beginning, or of the meridian passing through the beginning. The latitude range is positive and is called a *north latitude range*, or a *northing*, whenever the line bears north; it is negative, and called a *south latitude range*, or a *southing*, whenever the line bears south. The longitude range is positive, and is called an *east longitude range*, or an *easting*, when the line bears east; it is negative, and called a *west longitude range*, or a *westing*, when the line bears west.

Thus, the latitude and the longitude range of PQ are, respectively, $+PD$ and $+DQ$; those of QP are $-QE$ and $-EP$. Likewise, the latitude range of P_1Q_1 is $-P_1D_1$, because the end of the line is south of the beginning. The longitude range is $+D_1Q_1$, because the end of the line is east of the beginning. These values may be verified by observing that the latitudes of P_1 and Q_1 are, respectively, $-H_1P_1$ and $-H_1D_1$, whose algebraic difference is $-H_1D_1-(-H_1P_1)=-H_1D_1+H_1P_1=-P_1D_1$; and that the longitudes of P_1 and Q_1 are, respectively, $+H_1'P_1$ and $+K_1'Q_1$, whose difference is equal to D_1Q_1.

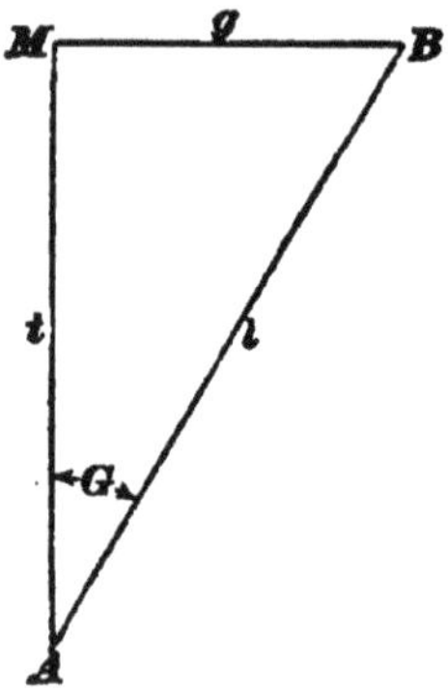

FIG. 2

General Formulas.—Let AB, Fig. 2, be a course whose length is l and whose bearing is G. In the right triangle AMB, in which AM is the direction of the meridian through A, the latitude range AM and the longitude range MB are denoted by t and g, respectively. According to trigonometry,

$$t=l\cos G \qquad (1)$$

$$g=l\sin G \qquad (2)$$

These formulas serve to compute the ranges when the length and bearing of the course have been measured. Special care should be taken to give t and g their proper signs, t being positive when G is north (that is, either northeast or northwest), and g being positive when G is east (that is, either northeast or southeast). When G is south (that is, either southeast or southwest), t is negative; and when G is west (that is, either northwest or southwest), g is negative.

If t and g are given, G is found by the formula

$$\tan G=\frac{g}{t} \qquad (3)$$

and l by either of the formulas following:

$$l=\frac{g}{\sin G} \qquad (4)$$

$$l=\sqrt{t^2+g^2} \qquad (5)$$

In applying formulas 3 and 5, the signs of t and g should be disregarded, both t and g being treated as positive.

EXAMPLE 1.—The length of a course is 896.7 ft. and its bearing is N 39° 15′ W; what are the ranges of the course?

SOLUTION.—Here $l = 896.7$ ft. and $G = 39° 15'$. Since the bearing is northwest, its latitude range is positive and its longitude range is negative. Applying formulas 1 and 2,

$$t = 896.7 \cos 39° 15' = 694.4 \text{ ft.}$$

$$g = -896.7 \sin 39° 15' = -567.4 \text{ ft.}$$

EXAMPLE 2.—The latitude range and the longitude range of a course are, respectively, −13.71 and −9.38 ch.; find the bearing and length of the course.

SOLUTION.—Since both ranges are negative, the course bears southwest. Applying formulas 3 and 4,

$$\tan G = \frac{9.38}{13.71}, \text{ whence } G = 34° 23', \text{ and}$$

$$l = \frac{9.38}{\sin 34° 23'} = 16.61 \text{ ch.}$$

When, instead of bearings, azimuths are measured, the same formulas hold good, only care must be taken to give the functions correct algebraic signs. When the azimuths are reckoned from the north, these formulas give both the numerical value and the algebraic sign of each range. This, however, is not the case when azimuths are reckoned from the south.

Platting by Latitudes and Longitudes.—To plat a traverse by latitudes and longitudes, pass reference lines through a convenient corner and figure the latitudes and longitudes of all the corners of the traverse. The courses are taken in the order in which they were run, the start being made at the initial point. The latitude or longitude of the end of the first course is equal to the corresponding range of that course; the latitude or longitude of the end of the second course is equal, respectively, to the latitude or the longitude of the end of the first plus the corresponding range of the second course; and, in general, the latitude or the longitude of the end of any course is equal, respectively, to the algebraic sum of the latitude or the longitude of the beginning of the course and the corresponding range of the course.

to the right for positive longitudes, and to the left for negative longitudes. The polygon *ABCDEFA* formed by joining the points, *A*, *B*, *C*, etc., is the required plat of the field.

Determination of Areas by Longitudes and Latitudes.—The *longitude of a course* is the longitude of its middle point. The *double longitude of a course* is twice its longitude, and is equal to the sum of the longitudes of the extremities of the course.

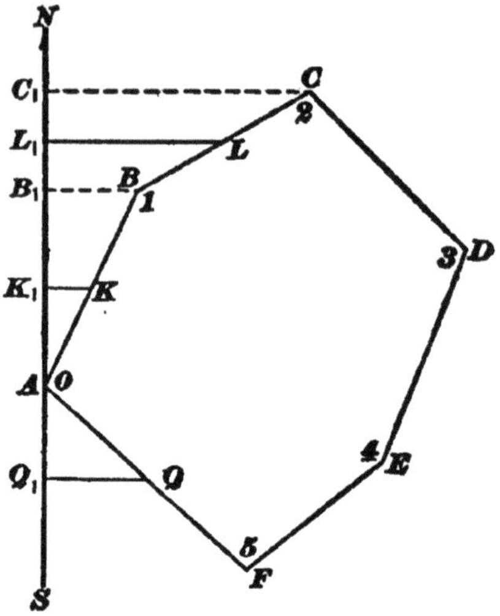

FIG. 4

In Fig. 4, *SN* is the reference meridian, *K*, *L*, and *Q* are the middle points, and K_1K, L_1L, and Q_1Q are the longitudes of the courses *AB*, *BC*, and *FA*.

The calculation of the area of a closed field requires that all the double longitudes be determined. This can be done by applying the following principle:

Principle.—*The double longitude of any course is equal to the algebraic sum of the double longitude of the preceding course, the longitude range of the preceding course, and the longitude range of the course considered.*

To apply this principle, note that the double longitude of the first course *AB* is equal to B_1B which is the longitude range of that course. As a check on the accuracy of the work, it should be noted that the double longitude of the last course is equal to its longitude range, but has the opposite algebraic sign.

After the double longitudes of all the courses have been calculated, the area of the field may be found by the following rule:

Rule.—*Multiply the latitude range of each course by the double longitude of the course, giving to the product its proper sign according to the signs of the factors. Add these products algebraically and divide the sum by 2.*

The following example shows the required calculation for determining the area of a closed field similar to the one shown in Fig. 4.

Courses	Longitude Ranges Chains	Double Longitudes	Latitude Ranges Chains	Double Areas +	Double Areas −
A B	+27.4	+ 27.4	+27.2	745.28	
B C	+63.2	+118.0	−23.8		2,808.40
C D	+13.1	+194.3	−37.5		7,286.25
D E	−33.1	+174.3	−33.3		5,804.19
E F	−50.1	+ 91.1	+24.1	2,195.51	
F A	−20.5	+ 20.5	+43.3	887.65	
				3,828.44	15,898.84
					3,828.44
					2)12,070.40
					6,035.20 sq. ch.
					= 603.52 A.

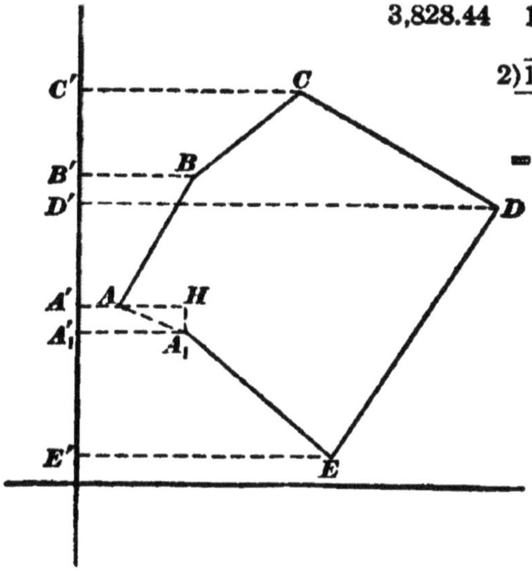

FIG. 5

Balancing the Survey of a Closed Field.—When a platted survey of a closed field does not close, as in Fig. 5, that is, when the point A_1, which is the end of the last line, does not coincide with the point A, which is the beginning of the first line, the line A_1A is called the *error of closure* and the ratio e of A_1A to the sum of all the courses is called the ***relative error of closure.*** Its value is

$$e=\sqrt{\left(\frac{S_t}{S_l}\right)^2+\left(\frac{S_g}{S_l}\right)^2},$$

in which S_t and S_g, denote, respectively, the algebraic sum of the latitude and of the longitude ranges and S_l is the sum of all the courses. In an ordinary compass survey, e should not exceed .002.

In order that a survey may close, it is necessary and sufficient that the algebraic sum of the latitude ranges and that

Courses	Bearings	Lengths Chains	Latitude Ranges N+	 S−	Longitude Ranges E+	 W−
AB	N 52° 00′ E	(10.62) 10.63	(6.57) 6.54		(8.34) 8.38	
BC	S 29° 45′ E	(4.08) 4.10		(3.55) 3.56	(2.01) 2.03	
CD	S 31° 45′ W	(7.68) 7.69		(6.51) 6.54		(4.08) 4.05
DA	N 61° 00′ W	(7.17) 7.13	(3.49) 3.46			(6.27) 6.24
		29.55 ($=S_l$)	10.00 −10.10	10.10	10.41 −10.29	10.29
			−.10 ($=S_t$)		+.12 ($=S_g$)	

of the longitude ranges should be equal to zero. When this is not the case, the ranges having the same sign as the algebraic sum must be shortened, and those of the opposite sign lengthened, until this condition is fulfilled. In a compass survey, the value of the correction on a longitude range is

$$c_g=\frac{S_g}{S_l}\times l$$

and on a latitude range, it is

$$c_t = \frac{S_t}{S_l} \times l$$

The altered length of the course is then

$$l_1 = \sqrt{t_1^2 + g_1^2}$$

In these formulas, S_g, S_t, and S_l have the same significance as in the formula for e; l is the length of the corresponding course; and t_1 and g_1 are the corrected latitude range and longitude range, respectively.

EXAMPLE.—The accompanying table contains the bearings and lengths of the courses of a compass survey. The lengths as measured, and the ranges, as calculated from the measured lengths and bearings, are printed horizontally opposite the letters denoting the corresponding corners. Above these numbers are placed in parentheses the corrected values of the lengths and ranges. Verify these corrected values and determine the relative error of closure.

SOLUTION.—First, determine the corrected latitude ranges. Here the sum of the courses, or S_l, is 29.55. The sum of the northings is 10.00, and that of the southings is -10.10. Therefore, the algebraic sum of the latitude ranges is $S_t = 10.00 + (-10.10) = -.10$. Applying the above formula

$$\frac{S_t}{S_l} = -\frac{.10}{29.55} = -\frac{10}{2,955} = -.003$$

Therefore,

c_t for $AB = 10.63 \times -.003 = -.03$
c_t for $BC = 4.10 \times -.003 = -.01$
c_t for $CD = 7.69 \times -.003 = -.02$ (See below)
c_t for $DA = 7.13 \times -.003 = -.02$

The sum of these corrections should be equal to S_t, or $-.10$, but it is only $-.08$. A correction of .01 therefore must be applied to two of the ranges. As the lengths of the third and fourth courses are nearly equal, 1 li. will be added arithmetically to the correction for CD and that for DA, writing c_t for $CD = -.03$, and c_t for $DA = -.03$. Subtracting algebraically the corrections just found from the corresponding latitude ranges, the corrected ranges are found to be

for AB, $6.54-(-.03)=6.54+.03=6.57$
for BC, $-3.56-(-.01)=-3.56+.01=-3.55$
for CD, $-6.54-(-.03)=-6.54+.03=-6.51$
for DA, $3.46-(-.03)=3.46+.03=3.49$

These are the corrected values placed in parentheses above the original values.

Second, determine the corrected longitude ranges. Here the sum of the eastings is 10.41, and that of the westings, -10.29. Therefore, $S_g=10.41-10.29=.12$, and

$$\frac{S_g}{S_l}=\frac{.12}{29.55}=.004$$

Therefore,

c_g for $AB=10.63\times.004=.04$
c_g for $BC=4.10\times.004=.02$
c_g for $CD=7.69\times.004=.03$
c_g for $DA=7.13\times.004=.03$

.12

The corrected longitude ranges are,

for AB, $8.38-.04=8.34$
for BC, $2.03-.02=2.01$
for CD, $-4.05-.03=-4.08$
for DA, $-6.24-.03=-6.27$

Third, determine the corrected lengths of the courses. Thus, applying the formula for l_1, page 63, and substituting the corrected ranges, the corrected length of

$$AB=\sqrt{6.57^2+8.34^2}=10.62$$
$$BC=\sqrt{3.55^2+2.01^2}=4.08$$
$$CD=\sqrt{6.51^2+4.08^2}=7.68$$
$$DA=\sqrt{3.49^2+6.27^2}=7.17$$

Fourth, determine the relative error of closure. Thus,

$e=\sqrt{.003^2+.004^2}=\sqrt{.000025}=.005$.

This error is 5 in 1,000, or 1 in 200, and is greater than would be allowed in any but exceedingly rough work.

The preceding method of balancing a closed survey is the one that is used for a compass survey, because the errors in the angular measurements are generally considerable.

In a transit survey in which the angular measurements, though sufficiently great to be considered, are small as compared with the error of closure, the formulas for the corrections of the ranges are

$$c_g = \frac{S_g}{S_r} \times r$$

and

$$c_t = \frac{S_t}{S_r} \times r$$

in which S_g and S_t have the same significance as before; r is the corresponding range to be corrected; and S_r is the arithmetical sum of the ranges of one kind, either latitude or longitude.

In a transit survey in which the angles are measured accurately the balancing is done by correcting the lengths of the sides, due consideration being given to the following principles:

Principle I.—*Measurements made either up or down a slope are likely to be too long as compared with measurements made under similar conditions on level ground.*

Principle II.—*Error in chaining is more likely to occur in lines measured over rough ground or under unfavorable conditions than in lines measured over smooth ground and under favorable conditions.*

These principles may serve as a guide in balancing a transit survey, an operation that must be done by trial, as no exact method has yet been devised.

Accuracy of Angular Measurements. The accuracy of the measurements of the angles of a closed survey can be checked by one of the following methods, depending on the method used in measuring the angles.

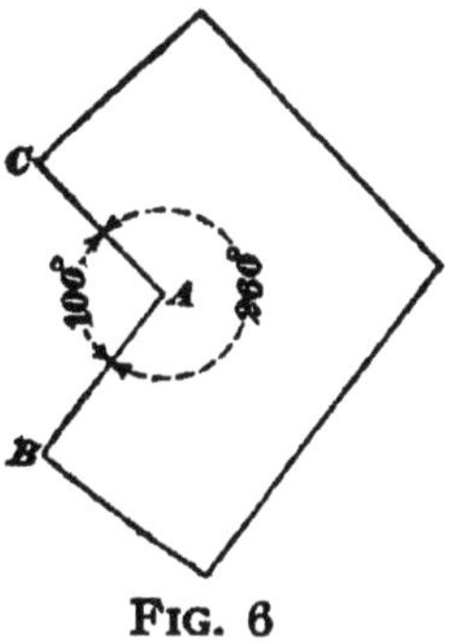

FIG. 6

1. When the angles are measured directly, the sum S of the interior angles of a polygon of n sides is given by the formula

$$S = 180° \times (n-2)$$

It should be borne in mind, in applying this formula, that reentrant angles, as that at A, Fig. 6, are greater than 180°. The angle A should be called 260°, not 100°. The sum of the measured angles should satisfy the formula within about 2 min. per angle.

2. When the deflection method is used each deflection angle, being the angle made by a side with the prolongation of the preceding, is an exterior angle of the polygon. The sum of all these angles should be equal to 360°, within the limits mentioned above.

SPECIAL PROBLEMS

SUPPLYING OMISSIONS

When, in surveying a closed field, omissions occur in the notes, they may in certain cases be supplied by computation. It must then be assumed that the remaining field notes are exactly correct; consequently, there are no means of balancing the work and all errors are thrown into the part or parts supplied. The following are the cases when it is possible to supply omissions by calculations:

1. When only one side is deficient, that is, when the bearing or the length, or both, are missing, the ranges of that course may be determined from the equations

$$t_x + S_t = 0$$

and

$$g_x + S_g = 0$$

in which t_x and g_x are, respectively, the latitude and the longitude range of the deficient side, and S_t and S_g are, respectively, the algebraic sums of the latitude ranges and longitude ranges of the known sides.

From these equations, $t_x = -S_t$ and $g_x = -S_g$. Then the bearing G_x and length l_x of the deficient side may be calculated by the formula

$$\tan G_x = \frac{g_x}{t_x}$$

and

$$l_x = \frac{g_x}{\sin G_x}$$

Since two angles correspond to a given tangent, in finding G_x two solutions are possible. The one to use may be determined by the signs of the ranges.

EXAMPLE.—The bearings and lengths of the first three courses of a survey are, respectively, N 32° 15′ E, 22 ch.; S 36° 30′ E, 10 ch.; and S 15° 45′ E, 5 ch. Determine the length and bearing of the fourth course, which closes the survey.

SOLUTION.—Let g_1, g_2, and g_3 be the longitude ranges, and t_1, t_2, and t_3 the latitude ranges of the known courses, which are as follows:

$$\begin{aligned} g_1 &= 22 \sin 32° 15' = 22 \times .53361 = 11.74 \\ g_2 &= 10 \sin 36° 30' = 10 \times .59482 = 5.95 \\ g_3 &= 5 \sin 15° 45' = 5 \times .27144 = \underline{1.36} \\ & \qquad 19.05 \text{ ch.} = S_g \end{aligned}$$

$$\begin{aligned} t_1 &= 22 \cos 32° 15' = 22 \times .84573 = 18.61 \\ t_2 &= -10 \cos 36° 30' = -10 \times .80386 = -8.04 \\ t_3 &= -5 \cos 15° 45' = -5 \times .96246 = \underline{-4.81} \\ & \qquad +5.76 = S_t \end{aligned}$$

Then, $g_x = -19.05$ and $t_x = -5.76$. Therefore, $\tan G_x = \frac{19.05}{5.76}$; whence, $G_x = 73° 11'$. The bearing is S 73° 11′ W. Also, as both ranges are negative,

$$l_x = \frac{19.05}{\sin 73° 11^1} = 19.9 \text{ ch.}$$

2. When the lengths of two sides are missing, let l_x and l_y be these lengths of the deficient sides, and G_x and G_y their corresponding bearings. Then,

$$l_y = \frac{S_g \cos G_x - S_t \sin G_x}{\sin G_x \cos G_y - \cos G_x \sin G_y}$$

and

$$l_x = -\frac{S_g + l_y \sin G_y}{\sin G_x}$$

EXAMPLE.—In a six-sided field, the lengths and bearings of four sides are N 30° 36′ E, 314 ft.; N 89° 35′ E, 406.0 ft.; S 32° 14′ E, 212.0 ft.; and N 26° 15′ W, 196.2 ft. The bearings of the other two sides are S 57° 46′ W and N 79° 47′ W. Determine their lengths.

SOLUTION.—By calculation it is found that $S_t=238.00$ and $S_g=636.63$. Taking G_x as S 57° 46′ W and G_y as N 79° 47′ W and substituting in the formulas,

$$l_y=\frac{636.63\ (-\cos 57°\ 46')-238\ (-\sin 57°\ 46')}{(-\sin 57°\ 46')\cos 79°\ 47'-(-\cos 57°\ 46')\ (-\sin 79°\ 47')}$$

$$=\frac{-339.56+201.32}{-.15003-.52491}=204.8\text{ ft.}$$

and $$l_x=-\frac{636.63+204.8\ (-\sin 79°\ 47')}{-\sin 57°46'}=514.3\text{ ft}$$

3. When the bearings of two sides are missing, let

$$M=\frac{l_y^2-l_x^2+S_t^2+S_g^2}{2l_y}$$

Then, $$\cos G_y=\frac{-S_tM\pm\left[-S_g\sqrt{S_t^2-M^2+S_g^2}\right]}{S_t^2+S_g^2},$$

from which G_y is found, thus reducing the remainder of the problem to case 1.

EXAMPLE.—The bearings and lengths of two sides of a field are N 52° 00′ E, 10.63 ch., and S 29° 45′ E, 4.10 ch. The bearings of the other two sides are to be determined, their lengths being 7.69 ch. and 7.13 ch., respectively.

SOLUTION.—By calculation, $S_t=6.54-3.56=2.98$ and $S_g=8.38+2.03=10.41$. Then,

$$M=\frac{7.13^2-7.69^2+2.98^2+10.41^2}{2\times 7.13}=7.75$$

and

$$\cos G_y=\frac{-2.98\times 7.75\pm\left[-10.41\sqrt{2.98^2-7.75^2+10.41^2}\right]}{2.98^2+10.41^2}$$

$$=-.8682,\text{ or }.47425$$

Suppose that it can be seen from a sketch that the bearing G_y is northwest. Then the cosine will be positive, and the angle corresponding to .47425 is the correct bearing; that is, G_y = N 61° 41′ W. Then, applying the method illustrated in case 1,

$$\tan G_x=\frac{-10.41-7.13\ (-\sin 61°\ 41')}{-2.98-7.13\cos 61°\ 41'}$$

$$=\frac{-4.13}{-6.36}=.64937=\tan 33°$$

As both ranges are negative, the bearing is S 33° W.

4. When the length l_x of one side and the bearing G_y of another are missing, l_x is determined by the formula:

$$l_x = -S_g \sin G_x - S_t \cos G_x$$
$$\sqrt{l_y^2 - S_t^2 - S_g^2 + (S_g \sin G_x + S_t \cos G_x)^2}$$

When l_x has been determined, the unknown bearing G_y is found as in case 1.

NOTE.—In the two preceding cases, two sets of results will generally be obtained. The problems are therefore indeterminate. However, if the notes contain a sketch showing the shape of the tract, both sets may be plotted and the correct figure identified.

EXAMPLE.—Two sides of a four-sided field have the bearings and lengths N 77° 24′ W, 32 ch., and N 38° 49′ E, 14 ch. The other two sides are deficient, one having the length 32.52 ch., bearing unknown, and the other the bearing S 18° 15′ W, length unknown.

SOLUTION.—In this example the required values are l_x and G_y. By calculation $S_g = -22.45$ and $S_t = 17.89$. Then, substituting known values in the formula, $l_x = 28.2$ ch. or -8.3 ch.

As the second value of l_x is negative, it shows that in this case only one solution is possible.

The required bearing G_y is now determined as in case 1. Thus $g_y = 22.45 + 28.2 \sin 18°15' = 31.28$, and $t_y = -17.89 + 28.2 \cos 18°15' = 8.89$. Then, $\tan G_y = \frac{31.28}{8.89} = \tan 74°08'$, and the bearing is N 74°08′ E.

In applying these formulas, careful attention should be given to the algebraic signs of the functions and of the ranges, which signs depend on the bearings. For northeast and northwest bearings, the latitude ranges and the cosines are +, and for southeast and southwest bearings, the cosines are −; the longitude ranges and the sines are + for northeast and southeast bearings, and − for northwest and southwest bearings.

PROBLEMS ON DIVISION OF LAND

Problem I.—*To divide a trapezoid into two parts, whose areas shall be proportional to two given numbers, by a line parallel to the bases.*

Let it be required to divide the trapezoid $ABCD$, Fig. 1, into two parts whose areas S_1 and S_2 are to be in the ratio $\frac{m}{n}$. In solving this problem, it may be necessary to find the length x of the dividing line EF, the distances AE and ED, and the altitudes

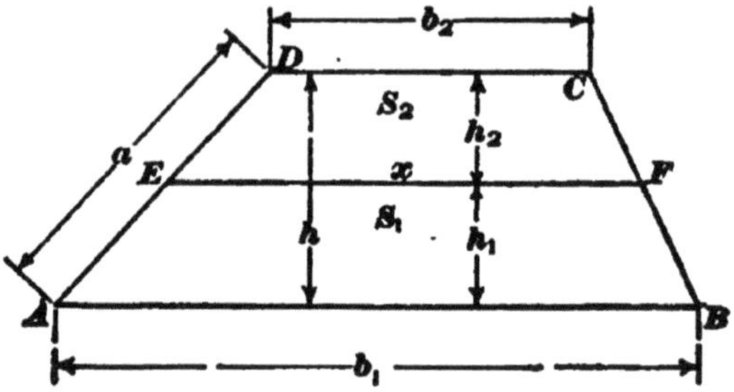

FIG. 1

h_1 and h_2. The following formulas are used, the notation being shown in the illustration:

$$x = \sqrt{\frac{mb_2^2 + nb_1^2}{m+n}}$$

$$DE = \frac{a(x-b_2)}{b_1-b_2}$$

$$AE = a - DE, \text{ or } \frac{a(b_1-x)}{b_1-b_2}$$

$$h_1 = \frac{h(b_1-x)}{b_1-b_2}$$

and

$$h_2 = \frac{h(x-b_2)}{b_1-b_2}$$

These formulas can be applied to a triangular tract, by taking the upper base b_2 as zero; then, $mb_2^2 = 0$.

EXAMPLE.—Suppose that the trapezoid $ABCD$, Fig. 1, represents a tract of land in which $DC = 50$ ch., $AB = 100$ ch., $AD = 47.50$ ch., and $h = 35$ ch., and that the tract is to be so divided by the line EF that the parts will be as 3 and 2, respectively, that is, $\frac{m}{n} = \frac{3}{2}$. Required, EF and DE.

SOLUTION.—By substituting the given values in the formulas,

$$EF=\sqrt{\frac{3\times 50^2+2\times 100^2}{5}}=\sqrt{5,500}=74.16 \text{ ch.}$$

and

$$DE=\frac{47.50\times(74.16-50)}{100-50}=22.95 \text{ ch.}$$

Problem II.—*To cut off a given area by a line starting from a given point on the boundary of a polygonal field.*

Let $ABCDEF$, Fig. 2, be a field from which it is required to cut off S acres by a line run through a given point G in the boundary. Draw a line GD from G to one of the opposite angles of the plat in such a position as to cut off an area nearly equal to the required area. Calculate the length and bearing of GD by the method given under *Supplying Omissions.* Calculate the area $GBCD$, which will be called S_1. Find the difference between the required area S and the calculated area S_1. If S is greater than S_1, an additional area S' must be found; let GDH be this area. Then, area $GDH=S-S_1=S'$. In the triangle GDH, the side GD and the angle D' are known. Then,

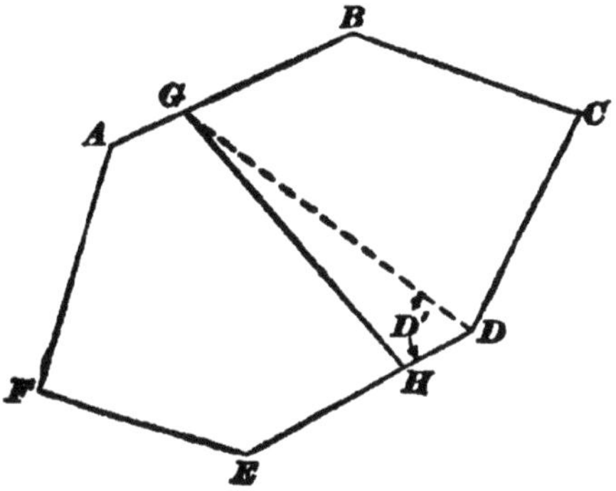

FIG. 2

$$DH=\frac{2\,S'}{GD \sin D'}$$

If the required area S is less than S_1, the process is substantially the same, except that the required distance should be calculated and measured from D along the line DC.

EXAMPLE.—In Fig. 2, assume that the length of the line GD is 8.93 ch., that the angle GDH is 61°, and that the area of $GBCD$ is 3.58 A. What must be the distance of the point H from the point D, in order that the line GH will cut off 5 A.; that is, in order that the area of the figure $GBCDH$ will be 5 A.?

SOLUTION.—The area S' of GDH is equal to $5-3.58=1.42$ A., or 14.2 sq. ch. Substituting in the formula,

$$DH=\frac{2\times 14.2}{8.93 \sin 61^\circ}=3.64 \text{ ch.}$$

PROBLEMS ON INACCESSIBLE LINES

Problem I.—*To determine the length of a line, AB, whose one end A is accessible and the other end B, is visible but not accessible.*

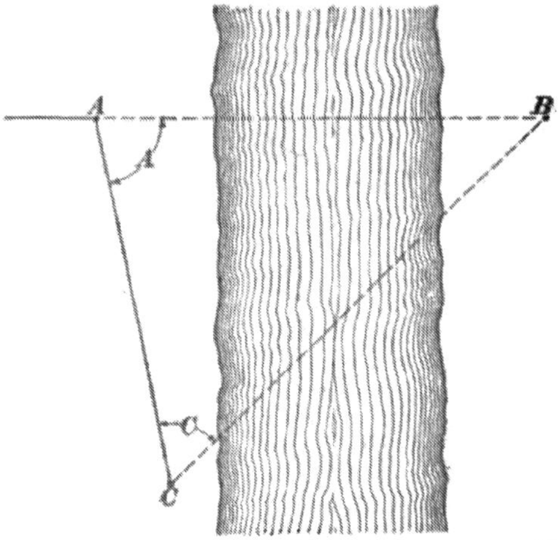

FIG. 3

Set the transit at A, Fig. 3, and turn off an angle BAC, which, if practicable, should be made equal to 90°. Measure along AC a distance of about 300 or 400 ft., and measure the angle C. Then, if $A=90°$, $AB=AC \tan C$. For any other angle A, $AB=\frac{AC \sin C}{\sin B}$.

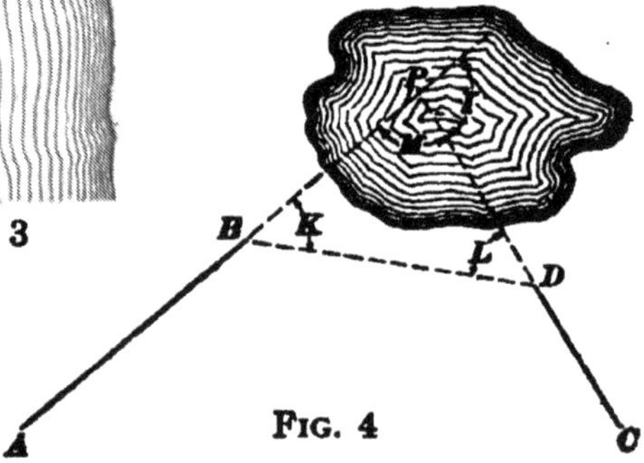

FIG. 4

Problem II.—*To determine the angle between two lines AB and CD, whose point of intersection P is inaccessible; also, the distances BP and DP.*

This problem is of frequent occurrence in railroad work, the two given lines being the center lines of two tracks that are to be connected by a curve.

Measure the distance BD, Fig. 4, and the angles K and L. Then, $M=180°-(K+L)$, $I=K+L$, $BP=\frac{BD \sin L}{\sin M}$, and $DP=\frac{BD \sin K}{\sin M}$.

Problem III.—*To determine the length of a line both ends of which are inaccessible.*

Let AB, Fig. 5, be the line, the ends A and B of which are inaccessible. Select two points P, Q from which both ends of the line can be seen, and at a distance from each other of about 300 or 400 ft. Measure the line PQ, and the angles K, L, M, and N. Then, from triangle APQ,

$$AP=\frac{PQ \sin M}{\sin R}$$

in which $R=180°-(K+L)-M$. From triangle BPQ,

$$BP=\frac{PQ \sin (M+N)}{\sin S}$$

in which $S=180°-L-(M+N)$. Then, from triangle ABP,

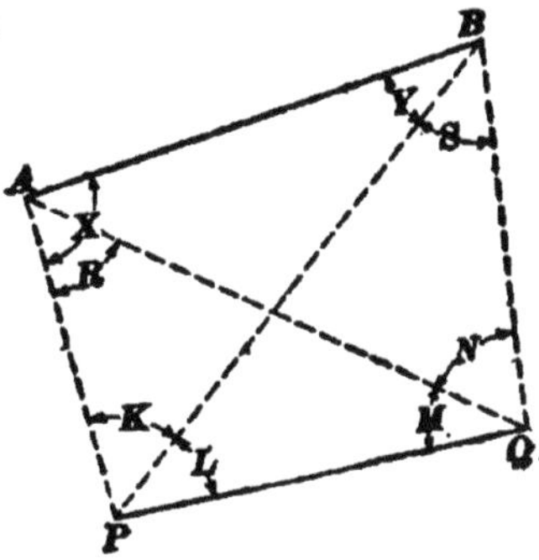

FIG. 5

$$\tan \tfrac{1}{2}(X-Y)=\frac{BP-AP}{BP+AP}\cot \tfrac{1}{2}K$$

Finally,

$$AB=\frac{(BP-AP)\cos \tfrac{1}{2}K}{\sin \tfrac{1}{2}(X-Y)}$$

EXAMPLE.—If, in Fig. 5, the distance PQ is 400 ft., and the angles, as measured, are $K=37° 10'$, $L=36° 30'$, $M=52° 15'$, $N=32° 55'$, what is the distance AB?

SOLUTION.—In the triangle APQ, $R=180°-(37° 10'+36° 30'+52° 15')=54° 05'$, and

$$AP=\frac{400 \sin 52° 15'}{\sin 54° 05'}=390.53 \text{ ft.}$$

In the triangle BPQ, $S=180°-(36° 30'+52° 15'+32° 55')=58° 20'$, $M+N=52° 15'+32° 55'=85° 10'$, and

$$BP=\frac{400 \sin 85° 10'}{\sin 58° 20'}=468.30 \text{ ft.}$$

Also, $K=37° 10'$, $\tfrac{1}{2}K=18° 35'$, and

$$\tan \tfrac{1}{2}(X-Y)=\frac{(468.30-390.53)}{468.30+390.53}\cot 18° 35'$$

whence, $\tfrac{1}{2}(X-Y)=15° 04'$, and therefore

$$AB=\frac{(468.30-390.53)\cos 18° 35'}{\sin 15° 04'}=283.58 \text{ ft.}$$

LEVELING

SPIRIT LEVELING

Leveling is the process of determining the relative elevations of a series of points. There are three general methods of determining elevations, namely, gravity leveling, commonly called spirit leveling, and also designated as direct leveling; trigonometric leveling, also called indirect leveling; and barometric leveling.

The most highly developed form of the spirit level is the *engineers' level.* It consists essentially of a telescope, having a very accurate spirit level attached, mounted on a tripod and controlled by leveling screws in such a manner that the line of sight can be made truly horizontal. There are two general classes of engineers' levels, namely, the *wye level,* also written *Y level,* in which the telescope rests in Y-shaped supports from which it can be removed, and the *dumpy level,* in which the telescope is fixed. The wye level is much the more popular with American engineers because of the facility with which it can be adjusted, while the dumpy level is favored in Europe.

THE WYE LEVEL

An engineers' wye level is shown in Fig. 1. The telescope AB rests in the Y-shaped supports Y, in which it is held firmly by semicircular clasps, commonly called clips; these are hinged at one end, and passing over the telescope are held at the other end by small pins. The lower ends of the wyes pass through the ends of the horizontal bar CD, sometimes called the level bar, and are adjustable vertically by means of the capstan-pattern nuts shown at C and D, which bear against the upper and lower surfaces of the bar. The bar CD is attached rigidly to the center or spindle, which turns in the socket V, permitting the telescope to be revolved in a horizontal plane. The spindle can be clamped by the screw K and the telescope then revolved slowly by means of the tangent screw t, which operates against a short projecting arm having a spring bearing against its opposite side. The position of the socket *V* is controlled by the four leveling screws *S*, which, together with the lower leveling plate *M*, and the tripod *P*, are substantially the same as in a transit, except that a level does not commonly have a shifting center.

The telescope is in every respect similar to that of the transit except that it is longer, and having no horizontal axis, it cannot be revolved in a vertical plane.

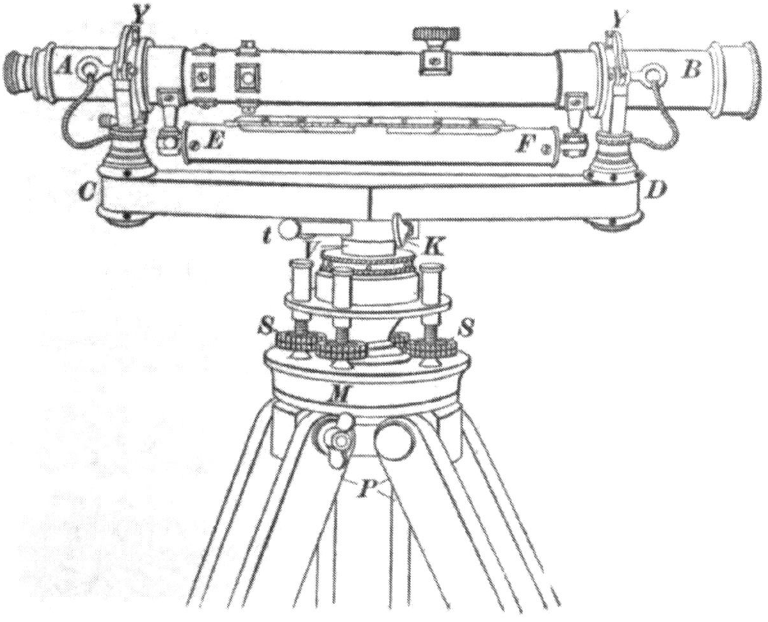

FIG. 1

The spirit level *EF* is also similar to that attached to the telescope of a transit, but in a leveling instrument, it is usually more accurate and sensitive. It consists of a hermetically sealed glass tube, curved slightly in a manner corresponding to the short upper arc of a large vertical circle, having the upper portion of its inner surface on a longitudinal section ground truly to the arc, and so nearly filled with alcohol, or a mixture of alcohol and ether, as to leave only a small bubble of air. Alcohol is used extensively for the levels of surveying instruments, but is rather slow acting. Ether, though more sensitive and quick acting, is affected too greatly by changes of temperature to be used in surveying instruments. A mixture of alcohol and ether gives excellent results. Since the air bubble rises to the highest point of the inner surface of the level tube in which it is confined, and since the upper portion of the inner surface of the tube is ground truly to the arc of a circle in the plane of its longitudinal section, it follows that a tangent to this arc at the center of the bubble is a horizontal line. A line tangent to the inner upper surface of the bubble tube at its center is called the *axis of the bubble*, or *axis of the level tube*. When the bubble is in the center of the tube, this line will be tangent to the center of the bubble, and consequently, will be a horizontal line. Hence, the axis of the level tube is horizontal when the bubble is in the center of the tube.

Adjustments of the Wye Level.—There are three adjustments of the wye level, as follows:

1. To make the line of sight, or line of collimation, parallel to the axis of the collars, or rings, on the telescope by which it rests in the wyes.

2. To make the axis of the level tube bubble parallel to the axis of the collars, and, consequently, parallel to the line of collimation.

3. To make the axis of the level tube perpendicular to the vertical axis of the instrument, so that when the instrument is leveled up the bubble will remain centered while the telescope is revolved horizontally.

First Adjustment.—Plant the tripod firmly; choose some distant and clearly defined point, the more distant the better, so long as it is distinctly visible and sharply defined. Remove the pins from the clips, clamp the spindle, and by means of the tangent screw and leveling screws bring the intersection of the cross-hairs to coincide exactly with the point sighted. Revolve or turn the telescope in the wyes through one-half a revolution, that is, until it is bottom side up. If the intersection of the cross-hairs is still on the point of sight, it shows that the line of sight coincides with the axis of the collars. But if, when the telescope is turned bottom side up, the line of sight defined by the intersection of the cross-hairs is no longer on the point, move the cross-hairs by means of the capstan-headed adjusting screws so as to correct one-half the apparent error, being careful to move them in the opposite direction to which it would appear they should be moved. The apparent error shown by reversing the telescope is double the real error, as is illustrated in Fig. 2.

Suppose that with the instrument at *A* the line of sight given by the intersection of the cross-hairs is directed to the point *B*, and that when the telescope has been revolved or turned upside down in the wyes, the line of sight strikes the point *C*; then the distance *BC* will be double the real error,

FIG. 2

and the true line of sight will be at *D*, half way between *B* and *C*. Sometimes both the horizontal and the vertical cross-hairs are out of adjustment, in which case they should be moved alternately until their intersection will coincide with the same point throughout a complete revolution of the telescope.

Second Adjustment.—The second adjustment consists of two parts, one lateral and the other vertical.

To adjust the level tube laterally, level up the instrument, remove the pins from the wyes, and open the clips; place the telescope over a pair of leveling screws and clamp the spindle. Bring the bubble exactly to the middle of the tube by means of the leveling screws and revolve the telescope in the wyes, first in one direction and then in the other, through about an eighth of a revolution. If the bubble runs toward one end of the tube when in the first position and toward the other end when in the second, it shows that the longitudinal axis of the bubble tube and the line of collimation, or longitudinal axis, of the telescope do not lie in the same plane. To correct the error, bring the bubble nearly to the center by means of the capstan-headed adjusting screws at one end of the level tube, which regulate its lateral movement, and repeat the operation until the bubble will remain centered during the partial revolution of the telescope.

To adjust the level tube vertically, center the bubble accurately, take the telescope out of the wyes, turn it end for end, and replace it in the wyes very carefully so as not to disturb their position. If the bubble remains in the center of the tube, the adjustment is perfect. If the bubble runs to one end, bring it half way back by means of the capstan-pattern adjusting nuts at one end of the level tube, by which it can be raised or lowered, and then bring it to the middle of the tube by means of the leveling screws. Repeat the operation until the bubble will remain truly centered when the telescope is reversed in the wyes.

Third Adjustment.—Level up the instrument, using each pair of leveling screws. Having centered the bubble carefully with the telescope over one pair of leveling screws, reverse the telescope or turn it end for end over the same pair of leveling screws. If the bubble runs toward one end, bring it half way back by means of the capstan-pattern nuts at the end of the level bar; then center it perfectly with the leveling screws. Repeat the operation over each pair of leveling screws alternately until the bubble will remain perfectly centered throughout an entire horizontal revolution of the telescope.

Adjustment of the Wye-Level Cross-Hairs.—Besides the preceding adjustments, it is convenient in leveling to have the horizontal cross-hair truly horizontal so as to be able to sight with any portion of it. To test this, sight upon any sharply defined point, focusing the telescope perfectly and bringing the point exactly in range with the horizontal cross-hair near either end; that is, near the right-hand or left-hand edge of the field of view. Then, revolve the telescope slowly on its vertical axis and notice whether or not the point sighted is cut exactly the same by the cross-hair throughout its entire length. If any deviation is discernible, it should be corrected by carefully rotating the cross-hairs in a direction opposite to that in which it appears they should be rotated, until the horizontal cross-hair will cut the point exactly the same throughout its length when the telescope is revolved slowly on the vertical axis of the instrument.

THE DUMPY LEVEL

An engineers' dumpy level of American make is shown in the accompanying illustration. In its general construction it is similar to the wye level. The essential difference is that in the dumpy level the telescope *AB* is attached rigidly to the horizontal level bar *CD*, and the level tube *EF* is attached to the

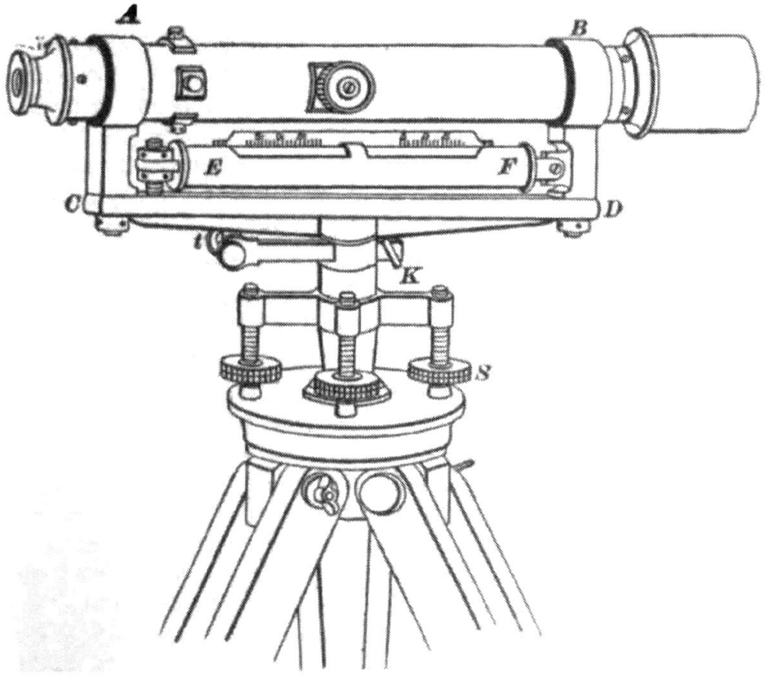

level bar and is adjustable at one end and in a vertical direction only, while the other end is attached permanently by a hinge. Since the telescope cannot be revolved in its supports, there is no necessity for the lateral adjustment of the level tube.

Adjustments of the Dumpy Level.—There are two adjustments of the dumpy level, namely:

1. To make the axis of the level tube perpendicular to the vertical axis of rotation, so that the bubble will stand in the center of its scale when the telescope is revolved.

2. To make the line of sight parallel to the axis of the level tube, so that the line of sight will be horizontal when the level bubble stands in the center of its scale.

First Adjustment.—Plant the tripod firmly and level up the instrument, using each pair of leveling screws. With the telescope over one pair of leveling screws, center the bubble accurately, then reverse the telescope end for end over the same pair of leveling screws. If the bubble runs toward either end, bring it half way back by means of the capstan nuts at one end of the level tube; then center it with the leveling screws. Repeat the operation over each pair of leveling screws alternately until the bubble will remain centered perfectly throughout a complete revolution of the telescope.

Second Adjustment.—The second adjustment is effected by establishing a horizontal line and adjusting the cross-hairs to agree with it while the bubble is at the center of the tube. To establish this line, drive two pegs into the ground several hundred feet apart and determine the true difference in elevation of these pegs. This can be accomplished even with an unadjusted instrument by setting it up at a place having the same distance from each peg and then taking rod readings and subtracting them. Next, set up the instrument over one peg with its center at a distance from the peg horizontally equal to about one-half the length of the telescope; bring the level bubble to the center of the tube, and with the leveling rod measure the exact height of the intersection of the cross-hairs above the peg. To determine this height on the rod, hold the graduated face of the rod about a half-inch from the eye end of the telescope, and by looking into the object end of the telescope bring the point of a pencil in the center of the small field of view on the face of the rod. Set the target at this height, plus or minus the difference in the elevations of the pegs, according as the rod reading on the distant peg was more or less than on the peg over which the instrument is set; then direct the telescope toward the rod held on the distant peg and adjust the cross-hairs so that the horizontal cross-hair will exactly bisect the target when the level bubble stands in the middle of its scale.

GENERAL PROPERTIES OF LEVELS

Sensitiveness of Level Bubble.—The sensitiveness or delicacy of the level bubble is indicated by the angle through which the line of sight must move in order to cause the bubble to move over one division of it. The smaller the angle, the more sensitive will be the bubble. The tangent of this angle can be determined by setting up the instrument and taking two rod readings at a distance of, say, 400 ft. from the station. Take one reading when the center of the bubble is exactly at a division mark of its scale, and then by means of the leveling screws, tip the instrument just sufficiently to cause the bubble to move one division of its scale and note again the reading of the rod. If the difference of the two rod readings is r, the distance of the rod from the station is d, and the angle through which the line of sight has been moved is a, then

$$\tan a = \frac{r}{d}$$

Magnifying Power and Definition.—The magnifying power of a telescope is the measure of its capacity to enlarge the apparent size of an object. It is commonly expressed by the number of times greater any linear dimension of an object appears when viewed through the telescope than when viewed with the naked eye, and is commonly spoken of as the number of *diameters* of magnifying power.

The magnifying power of a telescope can be determined approximately in the following manner: Cut out a white card exactly .1 ft. in width and attach it to a leveling rod so as to cover exactly one of the tenth divisions; set up the rod at a distance of, say, 25 ft., direct the telescope toward the rod, and focus it perfectly. Then, by observing the rod with both eyes, but with one eye looking through the telescope, note the number of divisions on the rod, as viewed with the naked eye, that appear to be covered by the white card, as viewed through the telescope. This will be, approximately, the number of diameters of magnifying power of the telescope. It is well to repeat the observation with the other eye looking through the telescope.

The *definition* of a telescope indicates the degree of clearness and sharpness of outline with which objects can be seen through it. In a general way, magnifying power and definition are opposed; that is, for the same size, a low-power telescope will have better definition than a high-power telescope, provided the excellence of the optical construction is the same in each case.

It is well to note here, that for telescopes of the same length, the inverting telescope gives considerably higher magnifying power, better definition, better light, and a much more brilliant image than the erecting telescope. A well-constructed erecting telescope 18 in. long may have a magnifying power of 30 diameters, and an inverting telescope of the same length has a power of about 40 diameters.

Care of Level.—The level should not be exposed to the burning rays of the sun, to rapid changes of temperature, to unequal temperatures on its different parts, or to dust, and should not be used in rainy weather when possible to avoid it. Changes of temperature disturb the adjustments, dust is injurious to the bearings and the lenses, and moisture obscures the lenses and is otherwise injurious to the instrument. Where it is impossible to avoid working in the rain, wipe the lenses frequently and carefully with a soft linen cloth, and after returning to the office or camp, wipe very carefully and thoroughly, finishing with a piece of dry chamois skin, and place in a moderately warm, dry place, so that every particle of moisture will be removed. When carrying a level on its tripod in open country, the spindle should always be clamped slightly to prevent the wearing of the centers by swinging, and the instrument should be carried with the object end of the telescope down. When working in a wooded country where underbrush is dense, the level should be carried with the spindle unclamped, so that the telescope will turn freely on the spindle and yield readily to any pressure. A blow that would inflict no injury upon an unclamped instrument might seriously damage one while clamped rigidly.

Leveling Rods.—There are two classes of leveling rods, namely, (1) rods on which the graduations are sufficiently distinct to be read directly by the leveler, and called *self-reading rods*, and (2) rods on which the graduations are small and which have a sliding target brought into the line of sight by signals from the leveler. For ordinary work, the type first mentioned is preferred by engineers, the target rod being used where very accurate work is required.

It is very important that the rod should be held truly vertical when sighted at. Different devices are employed for this purpose, and for work requiring great accuracy, such as bridge foundations, a rod level that fits closely to the angle of the rod and carries two small spirit levels is used to plumb it accurately. For ordinary work, however, this is not required. The leveler can plumb the rod across the line of sight by observing whether it coincides with the vertical cross-hair of the instrument, and he can obtain good results by making the rodman slowly tip the rod backwards and forwards in the direction of the line of sight and then taking the shortest reading.

FIELD WORK IN LEVELING

Example in Direct Leveling.—The principles of direct leveling are illustrated in the accompanying illustration.

Let *A* be the starting point, which has a known elevation of 20 ft. The instrument is set at *B*, leveled up and sighted to a rod held at *A*. The target being set, the reading, 8.42 ft., called a *backsight*, is the distance that the point where the line of sight cuts the rod is above the point *A*, and is to be added to the elevation of the point *A*; $20.00+8.42=28.42$ is called the *height of instrument* and is designated by *H. I.* The instrument being turned in the opposite direction, a point *C* is chosen, which must be below the line of sight. This point is called a *turning point*, and is designated by the abbreviation *T. P.* Drive a peg at *C*, or take for a turning point a rock or some other permanent object upon which the rod is held. The first reading on a turning point is a foresight, and is to be subtracted from the height of instrument at *B* to find the elevation of the point *C*. Let the rod reading be 1.20 ft. Then, $28.42-1.20 = 27.22$ ft., is the elevation of the point *C*. The leveler carries the instrument to *D*, which should be of such a height above *C* that, when leveled up, the line of sight will cut the rod near the top. The backsight to *C* gives a reading of 11.56 ft., which, added to 27.22 ft., the elevation of *C*, gives 38.78 ft., the height of instrument at *D*. The

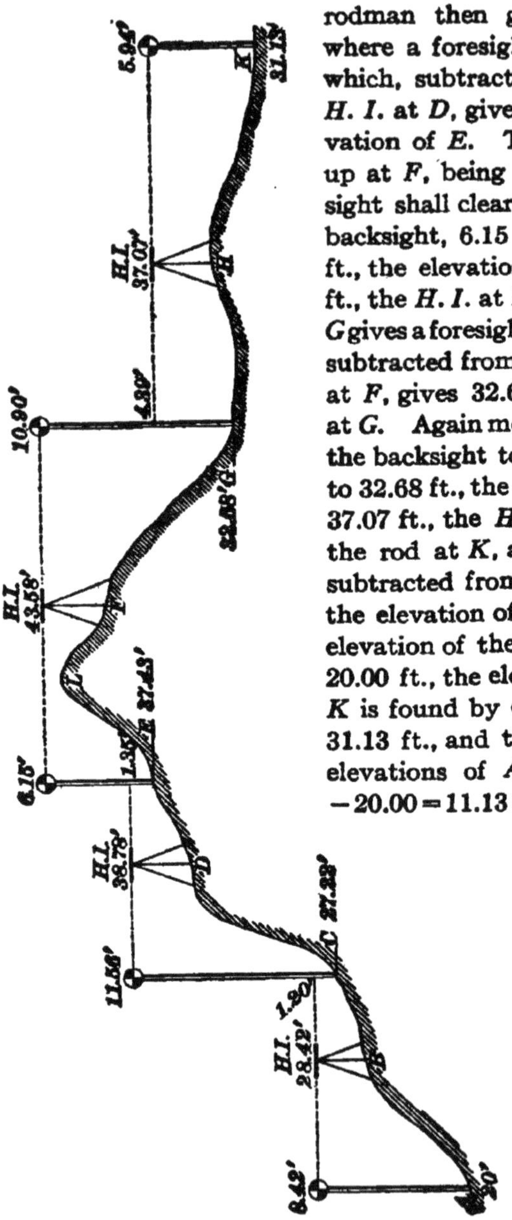

rodman then goes to *E*, a point where a foresight reading is 1.35, which, subtracted from 38.78, the *H. I.* at *D*, gives 37.43 ft., the elevation of *E*. The level is then set up at *F*, being careful that line of sight shall clear the hill at *L*. The backsight, 6.15 ft., added to 37.43 ft., the elevation of *E*, gives 43.58 ft., the *H. I.* at *F*. The rod held at *G* gives a foresight of 10.90 ft., which, subtracted from 43.58 ft., the H. I. at *F*, gives 32.68 ft., the elevation at *G*. Again moving the level to *H*, the backsight to *G* of 4.39 ft. added to 32.68 ft., the elevation of *G*, gives 37.07 ft., the *H. I.* at *H*. Holding the rod at *K*, a foresight of 5.94, subtracted from 37.07, gives 31.13, the elevation of the point *K*. The elevation of the starting point *A* is 20.00 ft., the elevation of the point *K* is found by direct leveling to be 31.13 ft., and the difference in the elevations of *A* and *K* is 31.13 − 20.00 = 11.13 ft.; that is, the point *K* is 11.13 ft. higher than the point *A*.

At each setting of the level, foresight readings can be taken on a number of points, before taking a foresight on a turning point, preparatory to moving the level to a new position. The elevation of any point will be equal to the *H. I.* minus the foresight reading.

A turning point is a point where the rod is held for a foresight, and after the level has been moved to a new position, for a backsight. The backsights are (+) readings, and are to be added; the foresights are (−) readings, and are to be subtracted. A point for a foresight having been determined, the rodman drives a peg firmly in the ground and holds the rod upon it. After the instrument is moved, set up, and a backsight taken, the peg is pulled up and carried in the pocket until another turning point is called for.

Balancing Backsights and Foresights.—The most valuable and reliable safeguard against errors in leveling is obtained by equal backsights and foresights on turning points. They should usually be equal in pairs; that is, each pair of sights on turning points, one backsight and one foresight, should be of approximately equal lengths. Should any inequality of length occur in one pair of sights, it should be balanced up in the next pair, or as soon as possible. For example, should the foresight in one pair of sights be longer than the backsight, then in the next pair of sights the backsight should be made correspondingly longer than the foresight. The sights should be balanced as perfectly as possible between bench marks. It is not necessary to measure the lengths of the sights accurately; they can be determined closely enough by counting steps in walking. A man of ordinary stature, when walking naturally, will average about 40 steps in each 100 ft. of distance, usually a somewhat less number on smooth and level ground, and a greater number where the ground is rough or sloping, either ascending or descending.

Keeping Level Notes.—Many forms on which to keep level notes are used. The distinguishing feature of one of the best, which is here shown, is a single column for all rod readings. The backsights being additive and the foresights subtractive readings, they are distinguished from other rod readings by the signs + and −.

Checking Level Notes.—A well-known method of checking level notes provides for checking the elevations of turning points and heights of instrument only, which is sufficient, as

LEVEL NOTES

Station	Rod Reading	Ht. Instru-ment	Eleva-tion	Grade	Cut	Fill	Remarks	
B. M.	+ 5.61	105.61	100.00				On root of white oak	
0	6.1		99.5				stump 10′ L. Sta. 0	
1	7.3		98.3					
2	8.4		97.2					
3	9.2		96.4					
T. P.	−10.22		95.39					
	+ 5.41	100.80						
4	6.3		94.5					
5	4.2		96.6					
5+50	11.5		89.3				Spring Brook	
T. P.	− 2.52		98.28					
	+11.57	109.85						
6	6.2		103.7					
7	8.5		101.4					
8	10.1		99.8					
T. P.	−11.53		98.32					

all other elevations are deduced from them. The method depends on the fact that all backsights are additive (+) quantities, and all foresights are subtractive (−) quantities. The accompanying level notes are checked as follows: The elevation of the bench mark at station 0 is 100.00 ft., to which all backsights, or + readings, are to be added and from this sum all foresights, or − readings, are to be subtracted. The sum of the backsights, with elevation of bench mark at Sta. 0, is 122.59. Sum of foresights on turning points is 24.27, and difference is 98.32 ft., the elevation of the last turning point. When a page of level notes is filled, the notes should be checked and a check-mark placed at the last height of instrument or elevation checked. When the work of staking out or cross-sectioning is being done, the levels should be checked at each bench mark on the line. After each day's work, the leveler must check on the nearest bench mark.

Profiles.—A *profile* represents a longitudinal section of the line of survey. In it all abrupt changes in elevation are clearly outlined. Vertical and horizontal measurements are usually represented to different scales, to render irregularities of surface more distinct through exaggeration. For railroad work, profiles are commonly made to the following scales: horizontal, 400 ft. = 1 in.; vertical, 20 ft. = 1 in.

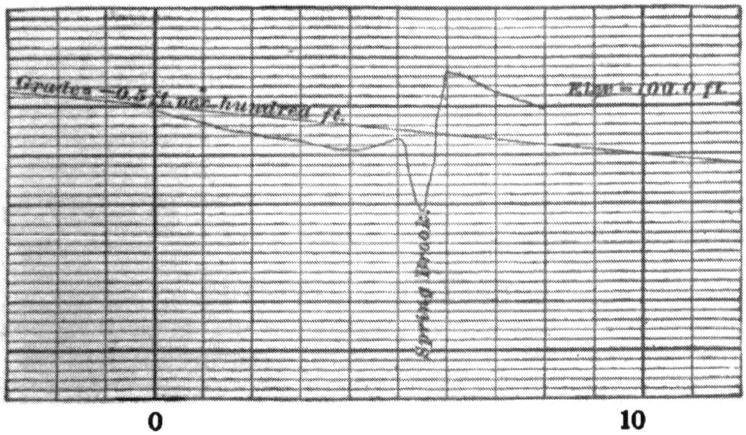

A section of profile paper is shown in the accompanying diagram. Every fifth horizontal line and every tenth vertical line is heavy. By the aid of these heavy lines, distances and elevations are quickly and correctly estimated and the work of platting greatly facilitated. The elevations given in the preceding notes are platted in the accompanying diagram. The elevation of some horizontal line is assumed. This elevation is, of course, referred to the datum plane, and is the base from which the other elevations are estimated. Every tenth station number is written at the bottom of the sheet under the heavy vertical lines.

Grade Lines.—The principal use of a profile is to enable the engineer to establish a *grade line*; that is, a line showing the slope of the road on which the amounts of excavation and embankment depend. The *rate* of a grade line is measured by the vertical rise or fall in each hundred feet of its length, and is designated by the term *per cent.*, abbreviated %. Thus, a grade line that rises or falls 1 ft. in each hundred feet of its length is called an ascending or a descending 1 % grade, and is written + 1 or − 1 per hundred. A rise or fall of ½ ft. in each hundred feet is called a .5% grade, and is written + .5 or − .5 per hundred. The grade line having been decided on, it is drawn in red ink, and the rate of grade is written on the line.

Example.—The elevation of station 20 is 140 ft.; between stations 20 and 100 there is an ascending grade of .75%. What is the elevation of the grade at station 71?

Solution.—To obtain the elevation of the grade at station 71, add to the elevation of the grade at station 20, or 140 ft., the total rise in grade between stations 20 and 71. The distance is 71−20=51 stations. The total rise is, therefore, .75 ft.×51=38.25 ft.; 140 ft.+38.25 ft.=178.25 ft., the elevation of grade at station 71.

ACCURACY IN LEVELING

Curvature and Refraction.—Owing to the spherical form of the earth, the difference in elevation, as shown by the rod reading, between the line of sight and the point on which the rod is held is not equal to the difference in elevation between the cross-hairs and the point, the rod reading being in excess of the true difference in elevation.

Let this excess be denoted by e_c, the radius of the earth (about 20,900,000 ft.) by r, and the horizontal distance between the instrument station and the leveling point by d; then,

$$e_c = \frac{d^2}{2r}$$

Another source of error in leveling, due to atmospheric refraction, tends to lessen the error due to curvature. Its value e_r can be figured from the formula

$$e_r = .071\,\frac{d^2}{r}$$

The combined error due to curvature and refraction is equal to

$$e = e_c - e_r = \frac{3d^2}{7\,r}$$

The errors due to curvature and refraction are very small for a single sight of ordinary length, and their cumulation may be eliminated by balancing backsights and foresights.

Degree of Accuracy Required in Spirit Leveling.—If M denotes the length of a leveling circuit and E the permissible error of closure, in feet—that is the permissible divergence between the elevation of a point as obtained at the beginning of the circuit and the elevation of the same point as obtained when ending the circuit—then, for very accurate surveys,

$$E = .012\,\sqrt{M} \text{ to } .029\,\sqrt{M}$$

For good average work of ordinary character,

$$E = .05\,\sqrt{M}$$

For preliminary railroad surveys,

$$E = .1\,\sqrt{M}$$

EXAMPLE.—Determine the error permissible in making the preliminary survey for a railroad 100 mi. long.

SOLUTION.—By substituting a value of 100 for M in the proper equation,

$$E = .1\,\sqrt{100} = 1.0 \text{ ft.}$$

TRIGONOMETRIC LEVELING

Trigonometric leveling is the process of determining the relative elevations of two points, trigonometrically; that is, by solving a triangle of which the unknown difference in elevation is one side, the other necessary data having been measured.

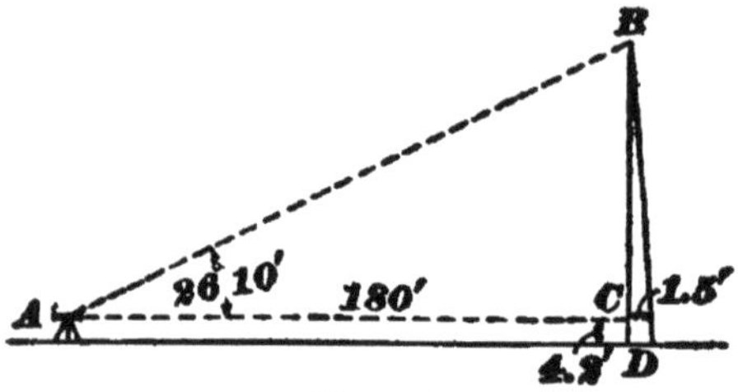

FIG. 1

Problem I.—*To determine the height of a vertical flagstaff.*

Let DB, Fig. 1, represent the flagstaff, the height of which is to be determined. Set a transit up at A, and then find the intersection of the line of sight of the telescope, when perfectly horizontal, with the flagstaff at C. Let this distance be found by measurement to be 180 ft. Then measure the vertical angle CAB; measure also CD, the height of the instrument over D, and the diameter of the flagpole at C. Let these measurements be respectively, $CAB = 26° 10'$ and $CD = 4.2$ ft. and let the diameter of the flagstaff at $C = 1.5$ ft. Then, the vertical height of B over the line AC is $\left(180+\frac{1.5}{2}\right) \times \tan 26° 10' = 88.81$ ft., and the total height $BD = 88.81 + 4.2 = 93.01$ ft.

Problem II.—*To determine the elevation of an inaccessible point.*

Let it be required to determine the elevation of the inaccessible point B over A, Fig. 2, and let the point D also be inaccessible. Set the transit up at any point, as A, and measure the vertical angle a. Select a point C' in the vertical plane ABD; move to it the instrument, and measure the angle c; then measure the horizontal distance m. Also, determine y, the height of A over C'; then,

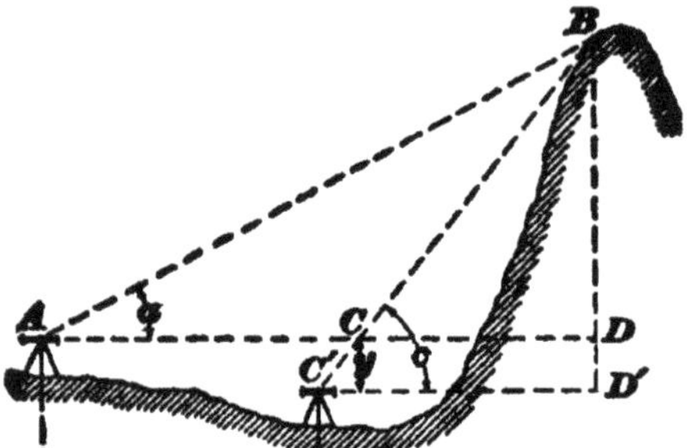

FIG. 2

$$BD = \frac{m + y \cot c}{\cot a - \cot c}$$

If C' is higher than A, y will be taken as minus, and the quantity $y \cot c$ will be negative.

If convenient, select the point C' in the same horizontal plane as A, Fig. 3; then, $y=0$, $y \cot c$ is also zero, and

$$BD = \frac{m}{\cot a - \cot c}$$

EXAMPLE.—If in Fig. 2, the angle $a = 17° 37'$, the angle $c = 31° 24'$, the horizontal distance m between the two positions of the instrument is 300 ft., and its position at C' is 2.5 ft. higher than its position at A, what is the elevation of the point B above the horizontal line AD?

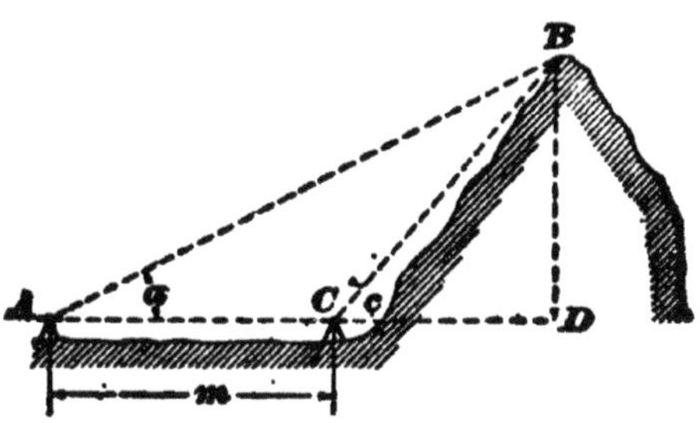

FIG. 3

SOLUTION.—Substituting known values in the proper formula and giving y the minus sign, since the point C' is above the point A,

$$BD = \frac{300 - 2.5 \times \cot 31° 24'}{\cot 17° 37' - \cot 31° 24'} = \frac{300 - 4.09565}{3.14922 - 1.63826} = 195.84 \text{ ft.}$$

BAROMETRIC LEVELING

The variation in air pressure at different altitudes, as observed by a barometer is made the basis for measuring differences in elevations. As mercury barometers are not readily portable, aneroid barometers are substituted. These barometers are adjusted to agree with the mercurial barometer at a temperature of 32° F. at the sea level in latitude 45°. Observations at the two stations whose difference in elevation is required should be made as nearly simultaneous as possible, because temperature and atmospheric conditions are constantly changing.

HEIGHTS CORRESPONDING TO BAROMETER READINGS

Arranged for Temperature of 50° F.

Height Feet	Aneroid or Corrected Barometer Inches	Height Feet	Aneroid or Corrected Barometer Inches	Height Feet	Aneroid or Corrected Barometer Inches
0	31.000	2,050	28.754	4,100	26.671
50	30.943	2,100	28.701	4,150	26.622
100	30.886	2,150	28.649	4,200	26.573
150	30.830	2,200	28.596	4,250	26.524
200	30.773	2,250	28.544	4,300	26.476
250	30.717	2,300	28.491	4,350	26.427
300	30.661	2,350	28.439	4,400	26.379
350	30.604	2,400	28.387	4,450	26.330
400	30.548	2,450	28.335	4,500	26.282
450	30.492	2,500	28.283	4,550	26.234
500	30.436	2,550	28.231	4,600	26.186
550	30.381	2,600	28.180	4,650	26.138
600	30.325	2,650	28.128	4,700	26.090
650	30.269	2,700	28.076	4,750	26.042
700	30.214	2,750	28.025	4,800	25.994
750	30.159	2,800	27.973	4,850	25.947
800	30.103	2,850	27.922	4,900	25.899
850	30.048	2,900	27.871	4,950	25.852
900	29.993	2,950	27.820	5,000	25.804
950	29.938	3,000	27.769	5,050	25.757
1,000	29.883	3,050	27.718	5,100	25.710
1,050	29.828	3,100	27.667	5,150	25.663
1,100	29.774	3,150	27.616	5,200	25.616
1,150	29.719	3,200	27.566	5,250	25.569
1,200	29.665	3,250	27.515	5,300	25.522
1,250	29.610	3,300	27.465	5,350	25.475
1,300	29.556	3,350	27.415	5,400	25.428
1,350	29.502	3,400	27.364	5,450	25.382
1,400	29.448	3,450	27.314	5,500	25.335
1,450	29.394	3,500	27.264	5,550	25.289
1,500	29.340	3,550	27.214	5,600	25.242
1,550	29.286	3,600	27.164	5,650	25.196
1,600	29.233	3,650	27.115	5,700	25.150
1,650	29.179	3,700	27.065	5,750	25.104
1,700	29.126	3,750	27.015	5,800	25.058
1,750	29.072	3,800	26.966	5,850	25.012
1,800	29.019	3,850	26.916	5,900	24.966
1,850	28.966	3,900	26.867	5,950	24.920
1,900	28.913	3,950	26.818	6,000	24.875
1,950	28.860	4,000	26.769		
2,000	28.807	4,050	26.720		

Let z = difference in elevation of the two stations, in feet;

h = the reading, in inches, of the barometer at the lower station;

H = the reading, in inches, of the barometer at the higher station;

t and T = temperatures of the air at the two stations.

Then,

$$z = 60{,}384.3\ (\log h - \log H)\left(1 + \frac{t + T - 64}{900}\right)$$

EXAMPLE.—Suppose that the barometer at the lower station reads 26.25 in. with the temperature at 72° F. and that at the upper station it reads 24.95 in. with the temperature at 46° F. What is the difference in elevation?

SOLUTION.—Substituting known values in the preceding formula,

$$z = 60{,}384.3\ (\log 26.25 - \log 24.95)\left(1 + \frac{72 + 46 - 64}{900}\right)$$

or $$z = 60{,}384.3 \times .02206 \times 1.06 = 1{,}412 \text{ ft.}$$

The accompanying table was compiled from the preceding formula for a mean temperature of 50° F.; that is, for $\frac{T+t}{2} = 50°$ F. Therefore, for this condition, the heights corresponding to the barometer readings may be taken directly from the table. If the heights at the upper and lower stations as taken from the table are denoted by H and h, respectively, the difference in elevation is

$$z = H - h$$

When the mean temperature is more or less than 50° F., the result, as obtained by means of the table, must be multiplied by the factor $\left(\frac{T+t}{1{,}000} + .9\right)$. Then,

$$z = (H - h)\left(\frac{T + t}{1{,}000} + .9\right)$$

STADIA AND PLANE-TABLE SURVEYING

STADIA SURVEYING

Stadia surveying is the process of determining distances by observing through a telescope (usually that of a plane table or a transit) the intercept on a graduated rod. The intercept is formed by two horizontal cross-hairs, which are carried on the same reticle as the regular cross-hair and are equidistant from it. The intercept bears a certain relation to the distance of the instrument from the rod. The instrument is also provided with a vertical circle, so that the vertical angle that an inclined sight makes with a level line may be measured. This angle serves for determining horizontal distances, as well as for figuring the relative elevation between the instrument point and the point where the rod is held. When the line of sight is nearly level, the distance d of the instrument from the rod can be determined by the formula:

$$d = sR + i,$$

in which R denotes the stadia reading or the intercept between the stadia wires, and s and i are called, respectively, the stadia constant and the instrument constant. Their values are usually determined by the instrument maker. The instrument constant varies from about .75 to 1.33 ft. in different transits, according to the size and power of their telescopes. Its value is usually marked on a card attached to the inside of the instrument box.

The stadia constant is customarily made equal to 100; so that, in a horizontal line of sight, the stadia wire will intercept a distance of 1 ft. on a rod whose distance from the instrument is 100 ft. plus the instrument constant. Thus, if the stadia wires intercept a distance of 8.37 ft. on the rod, the distance from the rod to the transit would be 837 ft. plus the instrument constant. For ordinary topographical work, especially for long distances, it is sufficiently close to take for the distance 100 times the length intercepted on the rod, the instrument constant being disregarded; but, for more accurate work, the constant usually taken is 1 ft.

To verify the constants, a line from 400 to 800 ft. is run on level ground and careful rod readings are taken at intervals of 50 ft. Let R_2 and R_1 be two stadia readings taken at the respective distances d_2 and d_1; then,

$$s=\frac{d_2-d_1}{R_2-R_1}$$

and

$$i=\frac{d_1R_2-d_2R_1}{R_2-R_1}$$

Several pairs of readings and their corresponding distances are substituted in these formulas, and the mean of all the resulting values of s and i is calculated.

EXAMPLE.—Determine the stadia and the instrument constant from the following data:

Distance Measured *Feet*	*Rod Reading* *Feet*
50	.488
100	.988
200	1.988
300	2.991
400	3.986

SOLUTION.—Take 50 ft. for the value of d_1 and 100 ft.; 200 ft., etc. successively for the values of d_2, and apply the preceding formulas for s and i. For the first pair of observations:

$$s=\frac{100-50}{.988-.488}=100.000$$

and

$$i=\frac{50\times.988-.488\times100}{.988-.488}=1.200 \text{ ft.}$$

The other values are figured in a similar manner and the whole is tabulated as follows:

	s	*i*
	100.000	1.200
	100.000	1.200
	99.880	1.258
	100.057	1.172
	4) 399.937	4) 4.830
means	99.984 = s	1.208 = i

Inclined Sights.—When the line of sight is inclined, the rod is held vertical and the vertical angle that the line of sight makes with a horizontal is measured. Denoting this angle by V and using the previous notation,

$$d=(sR \cos V+i) \cos V$$

When V is less than 3°, the angle is not considered and formula on page 94 is used.

Vertical Distances.—For finding differences in elevation the following formula is used:

$$v=\tfrac{1}{2}sR \sin 2V+i \sin V$$

In this formula, v is the difference in elevation between the center of the instrument and the point of intersection of the line of sight with the rod.

To determine the difference in elevation between the point on which the rod is held and the point over which the instrument is set, add to the value of v, as obtained from the formula, the height of the instrument, and from the result subtract the reading of the middle cross-hair. To avoid these calculations, the middle cross-hair may be made to intersect the rod at a point whose height above the ground is equal to that of the instrument. The result obtained from the formula is then the required difference in elevation.

The stadia point is higher or lower than the instrument point according as the angle V is one of elevation or depression.

EXAMPLE.—The length intercepted on the rod is 7 ft., and the vertical angle when the line of sight intersects the rod at a height equal to the height of the instrument is 18° 23′. If the stadia constant is 100 and the instrument constant 1 ft., (*a*) what is the horizontal distance of the rod from the center of the instrument? (*b*) what is the difference of elevation between the center of transit and the point where the line of sight intersects the rod, as indicated by the center cross-hair?

SOLUTION.—(*a*) Here $s=100$, $R=7$, $i=1$, and $\cos V=\cos 18° 23'=.94897$. Substituting these values in the formula for d,

$$d=(100\times7\times.94897+1)\times.94897=631.3 \text{ ft.}$$

(*b*) Here $\sin V=\sin 18° 23'=.31537$, and $\sin 2V=\sin 36° 46'=.59856$. Substituting these values and those given above in the formula for v,

$$v=\tfrac{1}{2}\times100\times7\times.59856+1\times.31537=209.8 \text{ ft.}$$

STADIA-SURVEYING NOTES

Sta.	Azimuth	Stadia	Vert. Angle	Hor. Dist.	Elev.	⊡ 1 = Old corner stone. Elev. assumed as 100.0.
	Readings fr	om ⊡ 1 Ele	v. = 100.0			Property line from ⊡ 1 to A
A	230° 0′	6.18	0° 0′			Corner post.
B	100° 24′	4.47	−18° 10′			Artesian well.
⊡ 2	125° 14′	12.10	−11° 45′			
	Readings fr	om ⊡ 2 Ele	v. =			
⊡ 1	305° 14′	12.11	+11° 44′			

Form of Stadia Notes.—A regular transit book is used for keeping notes in stadia surveying, its arrangement being shown herewith. The letters A and B in the first column signify the points where stadia readings were taken, and the marks ▣ designate instrument stations. The vertical angles are prefixed with + or −, according as they are angles of elevation or depression. The columns headed Hor. Dist. and Elev. are filled out in the office. The notes to the right of the double line are made on the right-hand page of the actual notebook.

Stadia Reduction Tables.—The work of reducing the notes in stadia surveying is conveniently done by means of the accompanying tables. In these tables are shown the horizontal distances and differences of elevation for various vertical angles, for the stadia constant 100 and for the rod reading 1. Thus, in the column headed Hor. Dist. is given the value of $100 \cos^2 V$ or d_1, and at the bottom of the page the value $i \cos V$ or i_d for $i = .75$, 1.00 or 1.25 may be found. From this,

$$d = d_1R + i_d$$

Similarly in the column headed Diff. Elev. are given values of $\frac{100 \sin 2V}{2}$, or v_1, and at the bottom are found values of $i \sin V$ or i_v. From this,

$$v = v_1R + i_v$$

EXAMPLE.—The stadia rod reading is 3.96 ft., the vertical angle is 10° 26′; $s = 100$, and $i = 1.00$. Find d and v.

SOLUTION.—From the table d_1 for 10° 26′ = 96.72, and $i_d = .98$. Hence, $d = 96.72 \times 3.96 + .98 = 383.99$ ft. Likewise, $v_1 = 17.81$ and $i_v = .18$. Finally, $v = 17.81 \times 3.96 + .18 = 70.71$.

STADIA REDUCTION TABLE

Minutes	0°		1°		2°		3°	
	Hor. Dist.	Diff. Elev.	Hor. Dist.	Diff. Elev.	Hor. Dist.	Diff. Elev.	Hor. Dist.	Diff. Elev.
0	100.00	.00	99.97	1.74	99.88	3.49	99.73	5.23
2	100.00	.06	99.97	1.80	99.87	3.55	99.72	5.28
4	100.00	.12	99.97	1.86	99.87	3.60	99.71	5.34
6	100.00	.17	99.96	1.92	99.87	3.66	99.71	5.40
8	100.00	.23	99.96	1.98	99.86	3.72	99.70	5.46
10	100.00	.29	99.96	2.04	99.86	3.78	99.69	5.52
12	100.00	.35	99.96	2.09	99.85	3.84	99.69	5.57
14	100.00	.41	99.95	2.15	99.85	3.89	99.68	5.63
16	100.00	.47	99.95	2.21	99.84	3.95	99.68	5.69
18	100.00	.52	99.95	2.27	99.84	4.01	99.67	5.75
20	100.00	.58	99.95	2.33	99.83	4.07	99.66	5.80
22	100.00	.64	99.94	2.38	99.83	4.13	99.66	5.86
24	100.00	.70	99.94	2.44	99.82	4.18	99.65	5.92
26	99.99	.76	99.94	2.50	99.82	4.24	99.64	5.98
28	99.99	.81	99.93	2.56	99.81	4.30	99.63	6.04
30	99.99	.87	99.93	2.62	99.81	4.36	99.63	6.09
32	99.99	.93	99.93	2.67	99.80	4.42	99.62	6.15
34	99.99	.99	99.93	2.73	99.80	4.47	99.61	6.21
36	99.99	1.05	99.92	2.79	99.79	4.53	99.61	6.27
38	99.99	1.11	99.92	2.85	99.79	4.59	99.60	6.32
40	99.99	1.16	99.92	2.91	99.78	4.65	99.59	6.38
42	99.99	1.22	99.91	2.97	99.78	4.71	99.58	6.44
44	99.98	1.28	99.91	3.02	99.77	4.76	99.58	6.50
46	99.98	1.34	99.90	3.08	99.77	4.82	99.57	6.56
48	99.98	1.40	99.90	3.14	99.76	4.88	99.56	6.61
50	99.98	1.45	99.90	3.20	99.76	4.94	99.55	6.67
52	99.98	1.51	99.89	3.26	99.75	4.99	99.55	6.73
54	99.98	1.57	99.89	3.31	99.74	5.05	99.54	6.79
56	99.97	1.63	99.89	3.37	99.74	5.11	99.53	6.84
58	99.97	1.69	99.88	3.43	99.73	5.17	99.52	6.90
60	99.97	1.74	99.88	3.49	99.73	5.23	99.51	6.96
$i = .75$	.75	.01	.75	.02	.75	.03	.75	.05
$i = 1.00$	1.00	.01	1.00	.03	1.00	.04	1.00	.06
$i = 1.25$	1.25	.02	1.25	.03	1.25	.05	1.25	.08

TABLE—*(Continued)*

Minutes	4°		5°		6°		7°	
	Hor. Dist.	Diff. Elev.	Hor. Dist.	Diff. Elev.	Hor. Dist.	Diff. Elev.	Hor. Dist.	Diff. Elev.
0	99.51	6.96	99.24	8.68	98.91	10.40	98.51	12.10
2	99.51	7.02	99.23	8.74	98.90	10.45	98.50	12.15
4	99.50	7.07	99.22	8.80	98.88	10.51	98.49	12.21
6	99.49	7.13	99.21	8.85	98.87	10.57	98.47	12.27
8	99.48	7.19	99.20	8.91	98.86	10.62	98.46	12.32
10	99.47	7.25	99.19	8.97	98.85	10.68	98.44	12.38
12	99.46	7.30	99.18	9.03	98.83	10.74	98.43	12.43
14	99.46	7.36	99.17	9.08	98.82	10.79	98.41	12.49
16	99.45	7.42	99.16	9.14	98.81	10.85	98.40	12.55
18	99.44	7.48	99.15	9.20	98.80	10.91	98.39	12.60
20	99.43	7.53	99.14	9.25	98.78	10.96	98.37	12.66
22	99.42	7.59	99.13	9.31	98.77	11.02	98.36	12.72
24	99.41	7.65	99.11	9.37	98.76	11.08	98.34	12.77
26	99.40	7.71	99.10	9.43	98.74	11.13	98.33	12.83
28	99.39	7.76	99.09	9.48	98.73	11.19	98.31	12.88
30	99.38	7.82	99.08	9.54	98.72	11.25	98.30	12.94
32	99.38	7.88	99.07	9.60	98.71	11.30	98.28	13.00
34	99.37	7.94	99.06	9.65	98.69	11.36	98.27	13.05
36	99.36	7.99	99.05	9.71	98.68	11.42	98.25	13.11
38	99.35	8.05	99.04	9.77	98.67	11.47	98.24	13.17
40	99.34	8.11	99.03	9.83	98.65	11.53	98.22	13.22
42	99.33	8.17	99.01	9.88	98.64	11.59	98.20	13.28
44	99.32	8.22	99.00	9.94	98.63	11.64	98.19	13.33
46	99.31	8.28	98.99	10.00	98.61	11.70	98.17	13.39
48	99.30	8.34	98.98	10.05	98.60	11.76	98.16	13.45
50	99.29	8.40	98.97	10.11	98.58	11.81	98.14	13.50
52	99.28	8.45	98.96	10.17	98.57	11.87	98.13	13.56
54	99.27	8.51	98.94	10.22	98.56	11.93	98.11	13.61
56	99.26	8.57	98.93	10.28	98.54	11.98	98.10	13.67
58	99.25	8.63	98.92	10.34	98.53	12.04	98.08	13.73
60	99.24	8.68	98.91	10.40	98.51	12.10	98.06	13.78
$i = .75$	.75	.06	.75	.07	.75	.08	.74	.10
$i = 1.00$	1.00	.08	1.00	.10	.99	.11	.99	.13
$i = 1.25$	1.25	.10	1.24	.12	1.24	.14	1.24	.16

TABLE—*(Continued)*

Minutes	8°		9°		10°		11°	
	Hor. Dist.	Diff. Elev.	Hor. Dist.	Diff. Elev.	Hor. Dist.	Diff. Elev.	Hor. Dist.	Diff. Elev.
0	98.06	13.78	97.55	15.45	96.98	17.10	96.36	18.73
2	98.05	13.84	97.53	15.51	96.96	17.16	96.34	18.78
4	98.03	13.89	97.52	15.56	96.94	17.21	96.32	18.84
6	98.01	13.95	97.50	15.62	96.92	17.26	96.29	18.89
8	98.00	14.01	97.48	15.67	96.90	17.32	96.27	18.95
10	97.98	14.06	97.46	15.73	96.88	17.37	96.25	19.00
12	97.97	14.12	97.44	15.78	96.86	17.43	96.23	19.05
14	97.95	14.17	97.43	15.84	96.84	17.48	96.21	19.11
16	97.93	14.23	97.41	15.89	96.82	17.54	96.18	19.16
18	97.92	14.28	97.39	15.95	96.80	17.59	96.16	19.21
20	97.90	14.34	97.37	16.00	96.78	17.65	96.14	19.27
22	97.88	14.40	97.35	16.06	96.76	17.70	96.12	19.32
24	97.87	14.45	97.33	16.11	96.74	17.76	96.09	19.38
26	97.85	14.51	97.31	16.17	96.72	17.81	96.07	19.43
28	97.83	14.56	97.29	16.22	96.70	17.86	96.05	19.48
30	97.82	14.62	97.28	16.28	96.68	17.92	96.03	19.54
32	97.80	14.67	97.26	16.33	96.66	17.97	96.00	19.59
34	97.78	14.73	97.24	16.39	96.64	18.03	95.98	19.64
36	97.76	14.79	97.22	16.44	96.62	18.08	95.96	19.70
38	97.75	14.84	97.20	16.50	96.60	18.14	95.93	19.75
40	97.73	14.90	97.18	16.55	96.57	18.19	95.91	19.80
42	97.71	14.95	97.16	16.61	96.55	18.24	95.89	19.86
44	97.69	15.01	97.14	16.66	96.53	18.30	95.86	19.91
46	97.68	15.06	97.12	16.72	96.51	18.35	95.84	19.96
48	97.66	15.12	97.10	16.77	96.49	18.41	95.82	20.02
50	97.64	15.17	97.08	16.83	96.47	18.46	95.79	20.07
52	97.62	15.23	97.06	16.88	96.45	18.51	95.77	20.12
54	97.61	15.28	97.04	16.94	96.42	18.57	95.75	20.18
56	97.59	15.34	97.02	16.99	96.40	18.62	95.72	20.23
58	97.57	15.40	97.00	17.05	96.38	18.68	95.70	20.28
60	97.55	15.45	96.98	17.10	96.36	18.73	95.68	20.34
$i = .75$	.74	.11	.74	.12	.74	.14	.73	.15
$i = 1.00$	.99	.15	.99	.17	.98	.18	.98	.20
$i = 1.25$	1.24	.18	1.23	.21	1.23	.23	1.22	.25

TABLE—*(Continued)*

Minutes	12°		13°		14°		15°	
	Hor. Dist.	Diff. Elev.	Hor Dist.	Diff. Elev.	Hor. Dist.	Diff. Elev.	Hor. Dist.	Diff. Elev.
0	95.68	20.34	94.94	21.92	94.15	23.47	93.30	25.00
2	95.65	20.39	94.91	21.97	94.12	23.52	93.27	25.05
4	95.63	20.44	94.89	22.02	94.09	23.58	93.24	25.10
6	95.61	20.50	94.86	22.08	94.07	23.63	93.21	25.15
8	95.58	20.55	94.84	22.13	94.04	23.68	93.18	25.20
10	95.56	20.60	94.81	22.18	94.01	23.73	93.16	25.25
12	95.53	20.66	94.79	22.23	93.98	23.78	93.13	25.30
14	95.51	20.71	94.76	22.28	93.95	23.83	93.10	25.35
16	95.49	20.76	94.73	22.34	93.93	23.88	93.07	25.40
18	95.46	20.81	94.71	22.39	93.90	23.93	93.04	25.45
20	95.44	20.87	94.68	22.44	93.87	23.99	93.01	25.50
22	95.41	20.92	94.66	22.49	93.84	24.04	92.98	25.55
24	95.39	20.97	94.63	22.54	93.82	24.09	92.95	25.60
26	95.36	21.03	94.60	22.60	93.79	24.14	92.92	25.65
28	95.34	21.08	94.58	22.65	93.76	24.19	92.89	25.70
30	95.32	21.13	94.55	22.70	93.73	24.24	92.86	25.75
32	95.29	21.18	94.52	22.75	93.70	24.29	92.83	25.80
34	95.27	21.24	94.50	22.80	93.67	24.34	92.80	25.85
36	95.24	21.29	94.47	22.85	93.65	24.39	92.77	25.90
38	95.22	21.34	94.44	22.91	93.62	24.44	92.74	25.95
40	95.19	21.39	94.42	22.96	93.59	24.49	92.71	26.00
42	95.17	21.45	94.39	23.01	93.56	24.55	92.68	26.05
44	95.14	21.50	94.36	23.06	93.53	24.60	92.65	26.10
46	95.12	21.55	94.34	23.11	93.50	24.65	92.62	26.15
48	95.09	21.60	94.31	23.16	93.47	24.70	92.59	26.20
50	95.07	21.66	94.28	23.22	93.45	24.75	92.56	26.25
52	95.04	21.71	94.26	23.27	93.42	24.80	92.53	26.30
54	95.02	21.76	94.23	23.32	93.39	24.85	92.49	26.35
56	94.99	21.81	94.20	23.37	93.36	24.90	92.46	26.40
58	94.97	21.87	94.17	23.42	93.33	24.95	92.43	26.45
60	94.94	21.92	94.15	23.47	93.30	25.00	92.40	26.50
$i=.75$	.73	.16	.73	.18	.73	.19	.72	.20
$i=1.00$	.98	.22	.97	.23	.97	.25	.96	.27
$i=1.25$	1.22	.27	1.22	.29	1.21	.31	1.20	.33

TABLE—(*Continued*)

Minutes	16°		17°		18°		19°	
	Hor. Dist.	Diff. Elev.	Hor. Dist.	Diff. Elev.	Hor. Dist.	Diff. Elev.	Hor. Dist.	Diff. Elev.
0	92.40	26.50	91.45	27.96	90.45	29.39	89.40	30.78
2	92.37	26.55	91.42	28.01	90.42	29.44	89.36	30.83
4	92.34	26.59	91.39	28.06	90.38	29.48	89.33	30.87
6	92.31	26.64	91.35	28.10	90.35	29.53	89.29	30.92
8	92.28	26.69	91.32	28.15	90.31	29.58	89.26	30.97
10	92.25	26.74	91.29	28.20	90.28	29.62	89.22	31.01
12	92.22	26.79	91.26	28.25	90.24	29.67	89.18	31.06
14	92.19	26.84	91 22	28.30	90.21	29.72	89.15	31.10
16	92.15	26.89	91.19	28.34	90.18	29.76	89.11	31.15
18	92.12	26.94	91.16	28.39	90.14	29.81	89.08	31.19
20	92.09	26.99	91.12	28.44	90.11	29.86	89.04	31.24
22	92.06	27.04	91.09	28.49	90.07	29.90	89.00	31.28
24	92.03	27.09	91.06	28.54	90.04	29.95	88.97	31.33
26	92.00	27.13	91.02	28.58	90.00	30.00	88.93	31.38
28	91.97	27.18	90.99	28.63	89.97	30 04	88.89	31.42
30	91.93	27.23	90.96	28.68	89.93	30.09	88.86	31.47
32	91.90	27.28	90.92	28.73	89.90	30.14	88.82	31.51
34	91.87	27.33	90.89	28.77	89.86	30.18	88.78	31.56
36	91.84	27.38	90.86	28.82	89.83	30.23	88.75	31.60
38	91.81	27.43	90.82	28.87	89.79	30.28	88.71	31.65
40	91.77	27.48	90.79	28.92	89.76	30.32	88.67	31.69
42	91.74	27.52	90.76	28.96	89.72	30.37	88.64	31.74
44	91.71	27.57	90.72	29.01	89.69	30.41	88.60	31.78
46	91.68	27.62	90.69	29.06	89.65	30.46	88.56	31.83
48	91.65	27.67	90.66	29.11	89.61	30.51	88.53	31.87
50	91.61	27.72	90.62	29.15	89.58	30.55	88.49	31.92
52	91.58	27.77	90.59	29.20	89.54	30.60	88.45	31.96
54	91.55	27.81	90.55	29.25	89.51	30.65	88.41	32.01
56	91.52	27.86	90.52	29.30	89.47	30.69	88.38	32.05
58	91.48	27.91	90.49	29.34	89.44	30.74	88.34	32.09
60	91.45	27.96	90.45	29.39	89.40	30.78	88.30	32.14
$i = .75$	.72	.21	.72	.23	.71	.24	.71	.25
$i = 1.00$	.96	.28	.95	.30	.95	.32	.94	.33
$i = 1.25$	1.20	.36	1.19	.38	1.19	.40	1.18	.42

TABLE—(*Continued*)

Minutes	20°		21°		22°		23°	
	Hor. Dist.	Diff. Elev.	Hor. Dist.	Diff. Elev.	Hor. Dist.	Diff. Elev.	Hor. Dist.	Diff. Elev.
0	88.30	32.14	87.16	33.46	85.97	34.73	84.73	35.97
2	88.26	32.18	87.12	33.50	85.93	34.77	84.69	36.01
4	88.23	32.23	87.08	33.54	85.89	34.82	84.65	36.05
6	88.19	32.27	87.04	33.59	85.85	34.86	84.61	36.09
8	88.15	32.32	87.00	33.63	85.80	34.90	84.57	36.13
10	88.11	32.36	86.96	33.67	85.76	34.94	84.52	36.17
12	88.08	32.41	86.92	33.72	85.72	34.98	84.48	36.21
14	88.04	32.45	86.88	33.76	85.68	35.02	84.44	36.25
16	88.00	32.49	86.84	33.80	85.64	35.07	84.40	36.29
18	87.96	32.54	86.80	33.84	85.60	35.11	84.35	36.33
20	87.93	32.58	86.77	33.89	85.56	35.15	84.31	36.37
22	87.89	32.63	86.73	33.93	85.52	35.19	84.27	36.41
24	87.85	32.67	86.69	33.97	85.48	35.23	84.23	36.45
26	87.81	32.72	86.65	34.01	85.44	35.27	84.18	36.49
28	87.77	32.76	86.61	34.06	85.40	35.31	84.14	36.53
30	87.74	32.80	86.57	34.10	85.36	35.36	84.10	36.57
32	87.70	32.85	86.53	34.14	85.31	35.40	84.06	36.61
34	87.66	32.89	86.49	34.18	85.27	35.44	84.01	36.65
36	87.62	32.93	86.45	34.23	85.23	35.48	83.97	36.69
38	87.58	32.98	86.41	34.27	85.19	35.52	83.93	36.73
40	87.54	33.02	86.37	34.31	85.15	35.56	83.89	36.77
42	87.51	33.07	86.33	34.35	85.11	35.60	83.84	36.80
44	87.47	33.11	86.29	34.40	85.07	35.64	83.80	36.84
46	87.43	33.15	86.25	34.44	85.02	35.68	83.76	36.88
48	87.39	33.20	86.21	34.48	84.98	35.72	83.72	36.92
50	87.35	33.24	86.17	34.52	84.94	35.76	83.67	36.96
52	87.31	33.28	86.13	34.57	84.90	35.80	83.63	37.00
54	87.27	33.33	86.09	34.61	84.86	35.85	83.59	37.04
56	87.24	33.37	86.05	34.65	84.82	35.89	83.54	37.08
58	87.20	33.41	86.01	34.69	84.77	35.93	83.50	37.12
60	87.16	33.46	85.97	34.73	84.73	35.97	83.46	37.16
$i=$ .75	.70	.26	.70	.27	.69	.29	.69	.30
$i=$ 1.00	.94	.35	.93	.37	.92	.38	.92	.40
$i=$ 1.25	1.17	.44	1.16	.46	1.15	.48	1.15	.50

TABLE—(*Continued*)

Minutes	24°		25°		26°		27°	
	Hor. Dist.	Diff. Elev.	Hor. Dist.	Diff. Elev.	Hor. Dist.	Diff. Elev.	Hor. Dist.	Diff. Elev.
0	83.46	37.16	82.14	38.30	80.78	39.40	79.39	40.45
2	83.41	37.20	82.09	38.34	80.74	39.44	79.34	40.49
4	83.37	37.23	82.05	38.38	80.69	39.47	79.30	40.52
6	83.33	37.27	82.01	38.41	80.65	39.51	79.25	40.55
8	83.28	37.31	81.96	38.45	80.60	39.54	79.20	40.59
10	83.24	37.35	81.92	38.49	80.55	39.58	79.15	40.62
12	83.20	37.39	81.87	38.53	80.51	39.61	79.11	40.66
14	83.15	37.43	81.83	38.56	80.46	39.65	79.06	40.69
16	83.11	37.47	81.78	38.60	80.41	39.69	79.01	40.72
18	83.07	37.51	81.74	38.64	80.37	39.72	78.96	40.76
20	83.02	37.54	81.69	38.67	80.32	39.76	78.92	40.79
22	82.98	37.58	81.65	38.71	80.28	39.79	78.87	40.82
24	82.93	37.62	81.60	38.75	80.23	39.83	78.82	40.86
26	82.89	37.66	81.56	38.78	80.18	39.86	78.77	40.89
28	82.85	37.70	81.51	38.82	80.14	39.90	78.73	40.92
30	82.80	37.74	81.47	38.86	80.09	39.93	78.68	40.96
32	82.76	37.77	81.42	38.89	80.04	39.97	78.63	40 99
34	82.72	37.81	81.38	38.93	80.00	40.00	78.58	41.02
36	82.67	37.85	81.33	38.97	79.95	40.04	78.54	41.06
38	82.63	37.89	81.28	39.00	79.90	40.07	78.49	41.09
40	82.58	37.93	81.24	39.04	79.86	40.11	78.44	41.12
42	82.54	37.96	81.19	39.08	79.81	40.14	78.39	41.16
44	82.49	38.00	81.15	39.11	79.76	40.18	78.34	41.19
46	82.45	38.04	81.10	39.15	79.72	40.21	78.30	41.22
48	82.41	38.08	81.06	39.18	79.67	40.24	78.25	41.26
50	82.36	38.11	81.01	39.22	79.62	40.28	78.20	41.29
52	82.32	38.15	80.97	39.26	79.58	40.31	78.15	41.32
54	82.27	38.19	80.92	39.29	79.53	40.35	78.10	41.35
56	82.23	38.23	80.87	39.33	79.48	40.38	78.06	41.39
58	82.18	38.26	80.83	39.36	79.44	40.42	78.01	41.42
60	82.14	38.30	80.78	39.40	79.39	40.45	77.96	41.45
$i=$ .75	.68	.31	.68	.32	.67	.33	.67	.35
$i=$ 1.00	.91	.41	.90	.43	.89	.45	.89	.46
$i=$ 1.25	1.14	.52	1.13	.54	1.12	.56	1.11	.58

TABLE—*(Continued)*

Minutes	28°		29°		30°	
	Hor. Dist.	Diff. Elev.	Hor. Dist.	Diff. Elev.	Hor. Dist.	Diff. Elev.
0	77.96	41.45	76.50	42.40	75.00	43.30
2	77.91	41.48	76.45	42.43	74.95	43.33
4	77.86	41.52	76.40	42.46	74.90	43.36
6	77.81	41.55	76.35	42.49	74.85	43.39
8	77.77	41.58	76.30	42.53	74.80	43.42
10	77.72	41.61	76.25	42.56	74.75	43.45
12	77.67	41.65	76.20	42.59	74.70	43.47
14	77.62	41.68	76.15	42.62	74.65	43.50
16	77.57	41.71	76.10	42.65	74.60	43.53
18	77.52	41.74	76.05	42.68	74.55	43.56
20	77.48	41.77	76.00	42.71	74.49	43.59
22	77.42	41.81	75.95	42.74	74.44	43.62
24	77.38	41.84	75.90	42.77	74.39	43.65
26	77.33	41.87	75.85	42.80	74.34	43.67
28	77.28	41.90	75.80	42.83	74.29	43.70
30	77.23	41.93	75.75	42.86	74.24	43.73
32	77.18	41.97	75.70	42.89	74.19	43.76
34	77.13	42.00	75.65	42.92	74.14	43.79
36	77.09	42.03	75.60	42.95	74.09	43.82
38	77.04	42.06	75.55	42.98	74.04	43.84
40	76.99	42.09	75.50	43.01	73.99	43.87
42	76.94	42.12	75.45	43.04	73.93	43.90
44	76.89	42.15	75.40	43.07	73.88	43.93
46	76.84	42.19	75.35	43.10	73.83	43.95
48	76.79	42.22	75.30	43.13	73.78	43.98
50	76.74	42.25	75.25	43.16	73.73	44.01
52	76.69	42.28	75.20	43.18	73.68	44.04
54	76.64	42.31	75.15	43.21	73.63	44.07
56	76.59	42.34	75.10	43.24	73.58	44.09
58	76.55	42.37	75.05	43.27	73.52	44.12
60	76.50	42.40	75.00	43.30	73.47	44.15
$i = .75$	.66	.36	.65	.37	.65	.38
$i = 1.00$	.88	.48	.87	.49	.86	.51
$i = 1.25$	1.10	.60	1.09	.62	1.08	.63

PLANE-TABLE SURVEYING

Plane Table.—Fig. 1 shows a Johnson plane table, which is the one most generally used in private work. Its essential parts are: (1) a drawing board mounted on a tripod, with contrivances for leveling the board and for turning it horizontally, called the *movement*, and (2) an instrument for sighting and transferring the line of sight to the paper on the board,

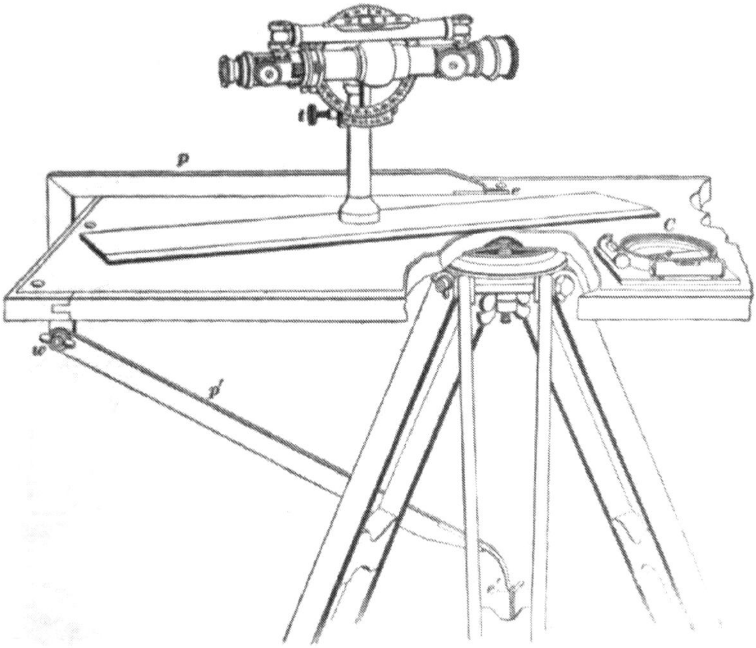

FIG. 1

called the *alidade*. The latter consists of a telescope provided with a level tube, a vertical circle, and stadia wires. The telescope is carried by an upright resting on a metal ruler. The vertical plane in which the line of sight of the telescope is moving is parallel to the edge of the ruler. The *declinator* C is a compass mounted on a base whose edges are parallel to the line joining the zero marks of the compass; it serves for determining the magnetic meridian on the drawing. The plumbing arm *e p p′e′* serves for suspending a plumb-bob, so that it will be directly under a point *e* on the paper representing the point determined on the ground by the plumb-bob.

The plane table is used for preparing topographical maps. In the survey of larger areas, reference lines forming a network of triangles are run with a transit and platted on the drawing of the plane table; the vertexes of these triangles, called

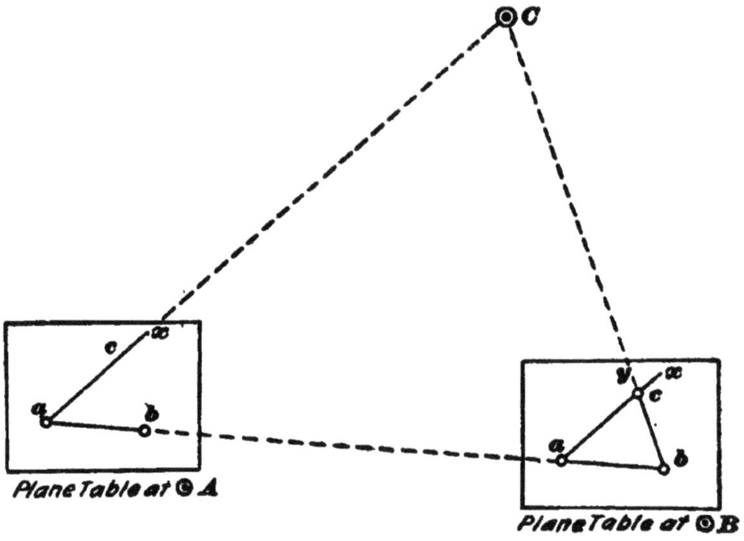

FIG. 2

triangulation stations, are used for determining other points of the survey by means of the plane table.

Orienting.—When the plane table set up over a point has each line platted on it parallel to the corresponding line in the field, it is said to be *oriented*.

Let *ab*, Fig. 2, be the platted position of the line *AB* on the ground, and assume that the plane table is to be oriented at *A*. First, orient the table approximately by the eye and at the same time, by means of the plumbing arm, bring the point *a* over *A*. Then level the table and, with the edge of the alidade ruler along the line *ab*, move the table horizontally until the telescope is accurately directed to *B*. The table is then clamped and another point, as *c*, may be platted by directing the telescope to *C* and at the same time having the edge of the alidade ruler in contact with the point *a*; the line *ax* is then drawn and the distance *AC*, measured by stadia or otherwise, is laid off to scale, giving the point *c*.

Plotting by Intersection.—After the line *ax* in the preceding example has been drawn, the point *c* can be located without measuring the distance *ac*. This is done by moving the table

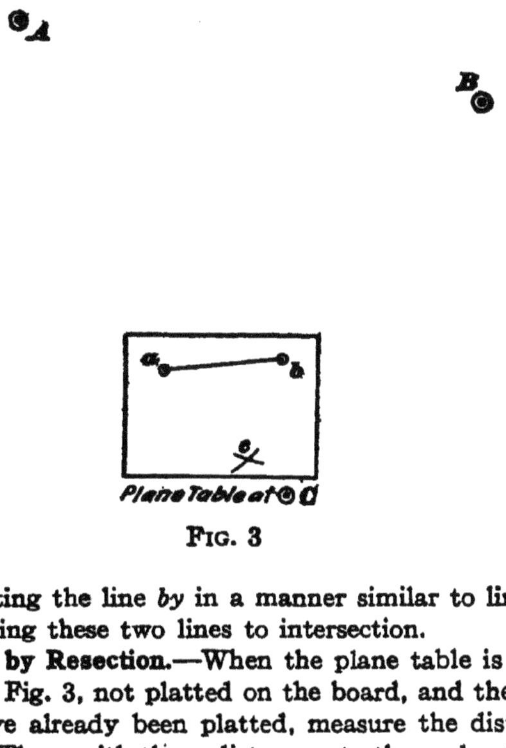

FIG. 3

to *B*, platting the line *by* in a manner similar to line *ax*, and then bringing these two lines to intersection.

Platting by Resection.—When the plane table is set up on a point *C*, Fig. 3, not platted on the board, and the points *A* and *B* have already been platted, measure the distances *CA* and *CB*. Then, with these distances, to the scale of the map, as radii, swing arcs from *a* and *b* as centers. The point of intersection of these arcs is the platted position of the point *C*, and the table can then be oriented in the usual manner.

The Three-Point Problem.—Let the plane table be set over a point *P*, Fig. 4, not platted on the board, from which three points *A*, *B*, and *C* platted at *a*, *b*, and *c*, respectively, are visible, but whose distances from *P* cannot conveniently be measured. To plat this point fasten a piece of tracing cloth over the plane-table paper; orient the table approximately with the eye, and select on the tracing cloth a point *p′* approximately corresponding to the true position of *p* with regard to *a*, *b*, and *c*, plat the lines *p′c′*, *p′b′*, and *p′a′* as if *p′* were the correct point *p*. Then unfasten the tracing cloth and

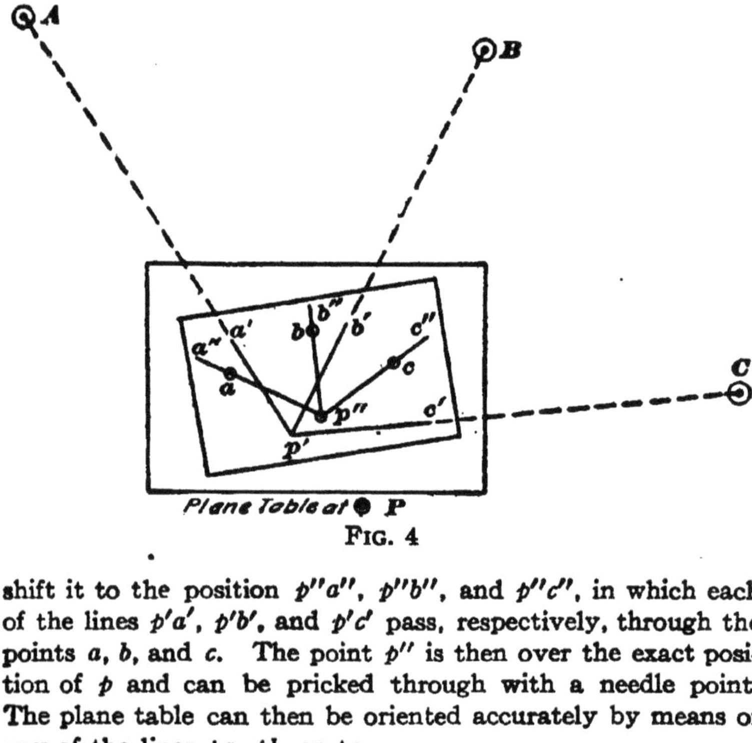

FIG. 4

shift it to the position *p″a″*, *p″b″*, and *p″c″*, in which each of the lines *p′a′*, *p′b′*, and *p′c′* pass, respectively, through the points *a*, *b*, and *c*. The point *p″* is then over the exact position of *p* and can be pricked through with a needle point. The plane table can then be oriented accurately by means of any of the lines *pa*, *pb*, or *pc*.

The Two-Point Problem.—When only two points *A* and *B*, Fig. 5, platted at *a* and *b* are visible, but inaccessible, the platted position of a third point *C* may be determined by establishing through it a line parallel to *A B* and orienting the table by means of that line. The field work is as follows: First, set up the plane table at D, Fig. 5; orient it approximately by the eye, and plat the point d and the lines dc', da', and db'. Then move the table to C and orient it with reference to the line CD by placing the edge of the ruler on the line $c'd$ and directing the telescope to station D. Through any point c'' on the line $c'd$ plat the lines of sight to B and A, the intersections of which with da' and db' give, respectively, the points a'' and b''. The line $a''b''$ is then parallel to AB. Now place the edge of the ruler on the

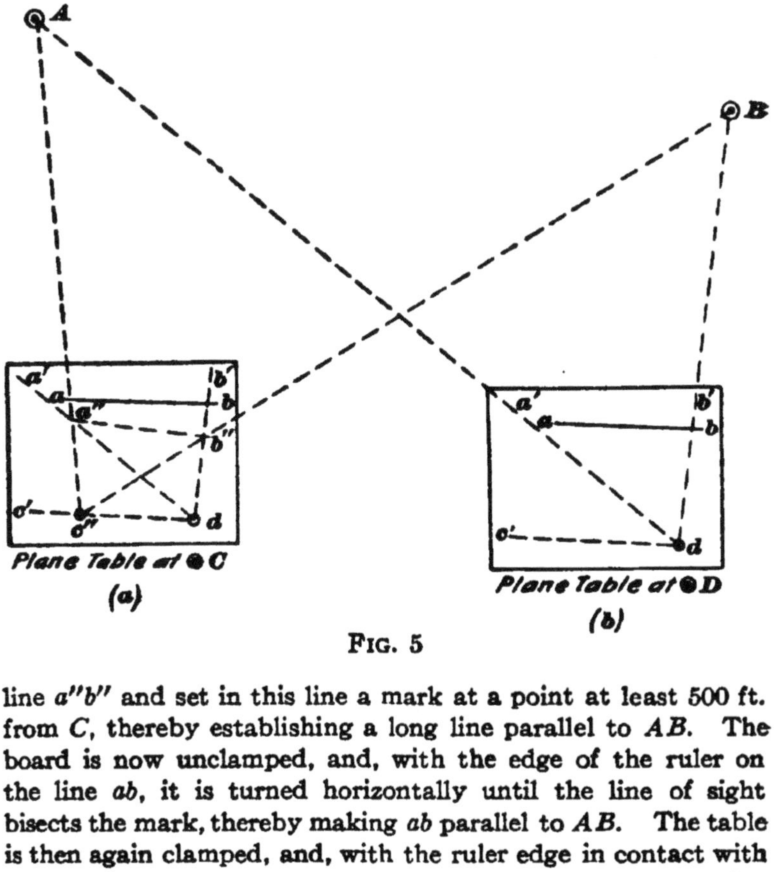

FIG. 5

line $a''b''$ and set in this line a mark at a point at least 500 ft. from C, thereby establishing a long line parallel to AB. The board is now unclamped, and, with the edge of the ruler on the line ab, it is turned horizontally until the line of sight bisects the mark, thereby making ab parallel to AB. The table is then again clamped, and, with the ruler edge in contact with a and b in turn, the telescope is directed to the points A and B and the lines ca and cb are drawn. The intersection of these lines will give the platted position of the point c.

TOPOGRAPHIC SURVEYING

METHODS EMPLOYED

In a topographic survey, the relative elevations or depressions of points and objects are determined in addition to their positions. Three methods, differing with regard to the instruments used, are employed in making topographic surveys; These are the *transit method*, the *stadia method*, and the *plane-table method*.

TRANSIT METHOD

The transit method is well adapted to surveys for the location of railroads and to similar surveys that relate to lines rather than to areas, and in which the topography is required to cover only comparatively narrow strips of country contiguous to the lines. In such surveys, the entire process is based on the line of the survey, which is usually alined with a transit and measured with a chain or tape. Along with the survey, a line of levels is run with a leveling instrument and at suitable intervals, generally every 100 ft., cross-sections are taken at right angles to the line. For the latter purpose the hand level and the clinometer are often used.

Hand Level and Clinometer.—The hand level, also called the Locke level, is shown in Fig. 1. The bubble of the level tube *C* can be seen through the opening *D* by means of a

FIG. 1

reflecting prism. A cross-hair placed in the main tube *AB* serves to fix the object observed, and when this hair at the same time bisects the bubble the line of sight is horizontal.

By means of a clinometer, one form of which is shown in Fig. 2, the angle that a slope makes with a horizontal can be measured. The bar *ab* is placed on any sloping surface, and the arm *mn* is raised until the bubble *t* is at the center of the

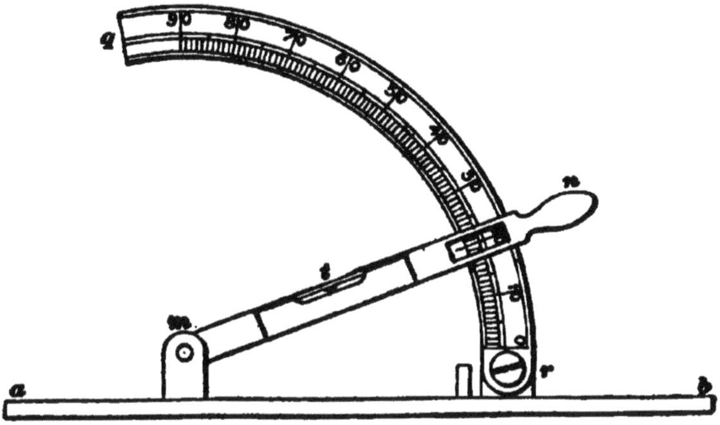

FIG. 2

level tube; the arm will be horizontal and its reading on the graduated quadrant *qr* will be the required angle.

The Abney level, shown in Fig. 3, is a combination of a hand level and a clinometer. The spirit level is movable in a vertical

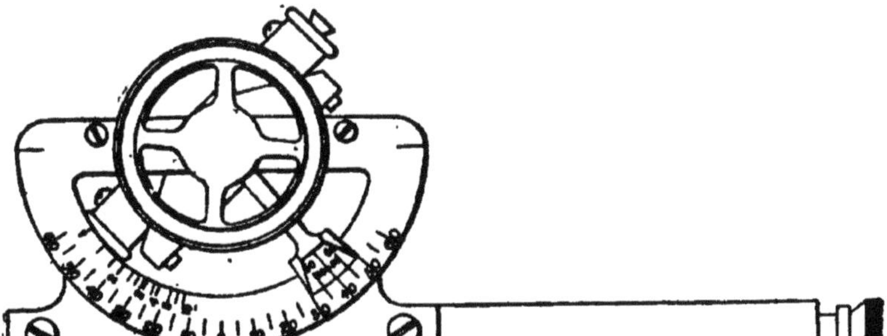

FIG. 3

plane, so that when the main tube is given any inclination the level can be turned to a horizontal position and the angle of inclination determined. When the spirit level is set parallel to the main tube it can be used as a hand level.

If the horizontal distance of a slope is h and the angle of slope a, the difference in elevation between the top and the bottom of the slope, e, is

$$e = h \tan a$$

Also,

$$h = e \cot a$$

Example of Cross-Sectioning With Hand Level.—Fig. 4 represents the right slope at Station 108 of a railroad survey. The topographer, having determined that his eye is 5.1 ft. above the ground, stands at the station and the rodman holds the rod at B, where the slope changes. The topographer, by

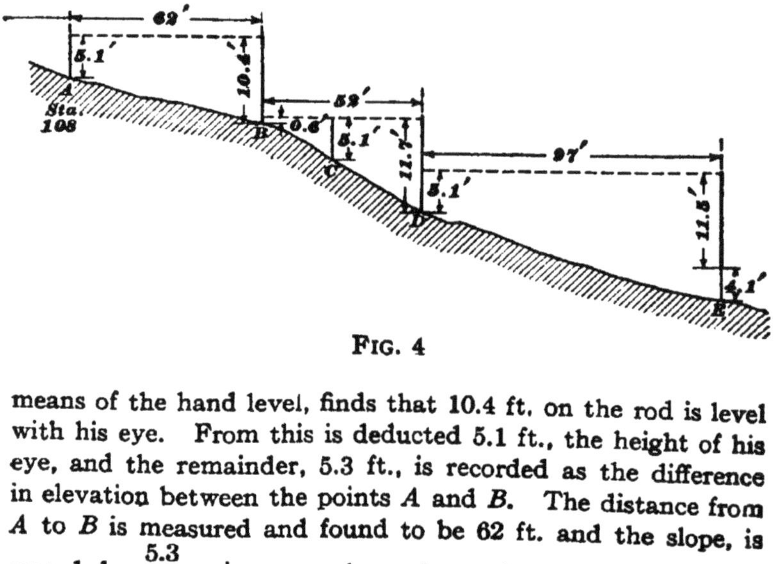

FIG. 4

means of the hand level, finds that 10.4 ft. on the rod is level with his eye. From this is deducted 5.1 ft., the height of his eye, and the remainder, 5.3 ft., is recorded as the difference in elevation between the points A and B. The distance from A to B is measured and found to be 62 ft. and the slope, is recorded $-\frac{5.3}{62}$, minus meaning a descending slope.

The topographer then proceeds down the slope to C, where his eye is about level with the bottom of the rod at B. The rod reading on B is .6 ft. The rodman proceeds to D, where the slope again changes. The topographer turns around at C and obtains the rod reading on D, which is 11.7. The difference of these rod readings, $11.7 - .6 = 11.1$, is the difference in elevation between *B* and *D*. Since the elevation of point *C* is not desired, its location is not recorded. The distance from *B* to *D* is found to be 52 ft., and the second slope is recorded $-\frac{11.1}{52}$.

The topographer moves forwards to the point *D*, and the rodman holds the rod at *E*, the foot of the slope. The top of the rod is below the level line of sight from the topographer's eye, so the rodman "shins the rod," holding it against his body sufficiently high to be intersected by the level line of sight. The rod reading is found to be 11.5 ft. The rodman then measures with the rod the distance from the ground to the point on his body to which the bottom of the rod was raised. This distance, 4.1 ft., is called out to the topographer, who adds it to the rod reading and then deducts the height of the eye. The distance from *D* to *E* is found to be 97 ft. This slope is recorded $-\frac{10.5}{97}$.

STADIA METHOD

In the stadia method, points are located by means of a transit for the azimuths. The transit is equipped with a level on the telescope, a vertical arc or circle, and stadia wires. The distances and the differences of elevation are determined by stadia measurement. This method is adapted to all kinds of surveys in which a great degree of accuracy is not required. It is the best method of making a general topographical survey of considerable extent, and is especially convenient for preliminary railroad location surveys. The stadia method was officially adopted by the United States Lake Survey in 1864.

PLANE-TABLE METHOD

In the plane-table method, points located by the plane table are at once platted on the map, which is thus prepared in the field without the intermediate process of reading and recording angles and distances. This method is well adapted to mapping, especially for filling in the details after the principal lines of a survey have been determined by other means

It has been used extensively for this purpose by the United States Coast and Geodetic Survey and the United States Geological Survey. It is also adapted for smaller surveys, such as that of a park, in which it is desired to locate numer-

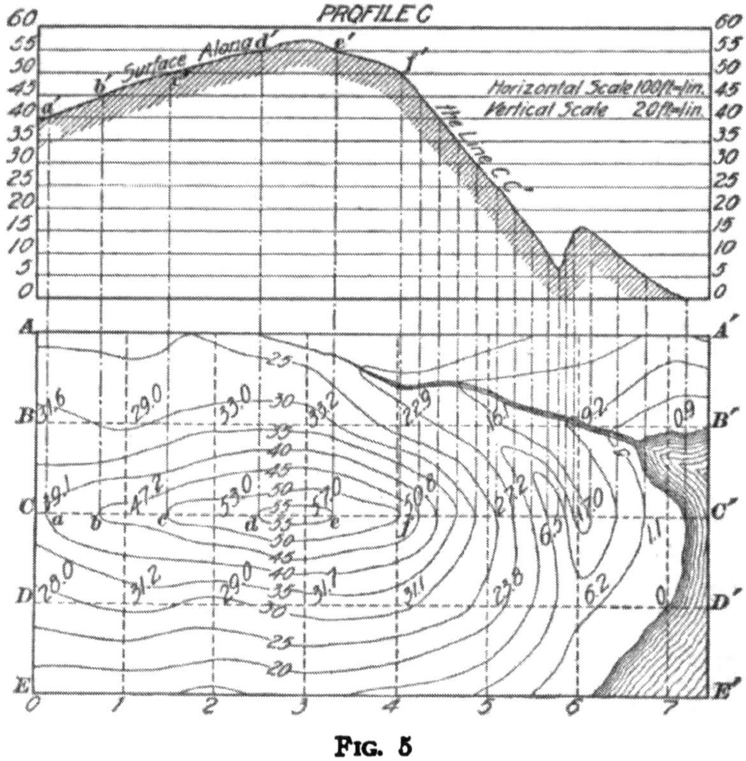

FIG. 5

ous objects within a small area, and in surveys for rough maps, the time for making which is limited and in which only some of the principal points are located accurately, the other features being sketched in by eye.

CONTOURS

Contour curves are lines joining points of equal elevations. Fig. 5 illustrates part of a contour map of a survey. The tract was divided by the lines *AA'*, *BB'*, *CC'*, *1*, *2*, etc. 100 ft. apart into squares of uniform size. Levels were taken at all points of intersection and at any intermediate points where the slope changes abruptly, and the positions of the contour points were determined as follows: Take for instance the line *CC'*. The elevation of Station *C-0* is 39.1 ft., and that of Station *C-1*, is 47.2 ft., giving a rise equal to $47.2-39.1=8.1$ ft. from the former station to the latter. Since the horizontal distance between the stations is 100 ft., the rate of slope is equal to $\frac{100}{8.1}$ or 12.3 ft. horizontal for 1 ft. rise. The contour interval is taken at 5 ft., and, consequently, the elevation of each contour is some multiple of 5 ft. The first contour above Station *C-0* is contour *40*, and to locate this contour a rise of $40.0-39.1=.9$ ft. above this station must be made. Since the rate of slope is 12.3 ft. horizontal for 1 ft. rise, the horizontal distance from Station *0* on this line to contour *40* is equal to $12.3\times.9=11.1$ ft. The rise from contour *40* to contour *45* is 5 ft. As the rate of slope continues the same, contour *45* will intersect line *C* at a distance of $12.3\times5=61.5$ ft. from contour *40*, or $61.5+11.1=72.6$ ft. from Station *0*.

From an inspection of the elevation, it is evident that contour *50* must occur between Stations *1* and *2*, since the elevation of the former is 47.2 ft., and that of the latter is 53 ft. The rise from Station *1* to Station *2* is equal to $53.0-47.2=5.8$ ft. Since the horizontal distance giving this rise is 100 ft., the rate of slope is equal to $\frac{100}{5.8}=17.2$ ft. horizontal for 1 ft. rise. To locate contour *50*, a rise of $50.0-47.2=2.8$ ft. must be made, and since the rate of slope is 17.2 ft. horizontal for 1 ft. rise, contour *50* will intersect line *C* at a distance of $17.2\times2.8=48.2$ ft. from Station *1*.

In a similar manner are determined the other points on the

line *CC'*, as well as those on the lines *AA'*, *BB'*, *1*, *2*, etc. The points having the same elevation are then joined by continuous lines, forming the contour lines.

The upper part of the figure shows a vertical section, or a profile, on the line *CC'* of the contour map. The horizontal lines *0–0*, *5–5*, etc. correspond to the elevations of the contour lines. The points *a*, *b*, *c*, *d*, etc. are projected on the lines of corresponding elevations on the profile, giving the points *a'*, *b'*, *c'*, etc. These points are then joined by a continuous line representing the surface of the ground along the line *CC'*.

MAPPING

Conventional Signs.—Fig. 1 shows the manner of representing the shore line of a body of water. Fig. 2 shows a rocky and abrupt shore, the irregular dotted surfaces surrounded by shore lines representing sandbars and the dotted outlines beyond the shore line shoals or submerged rocks. Fig. 3

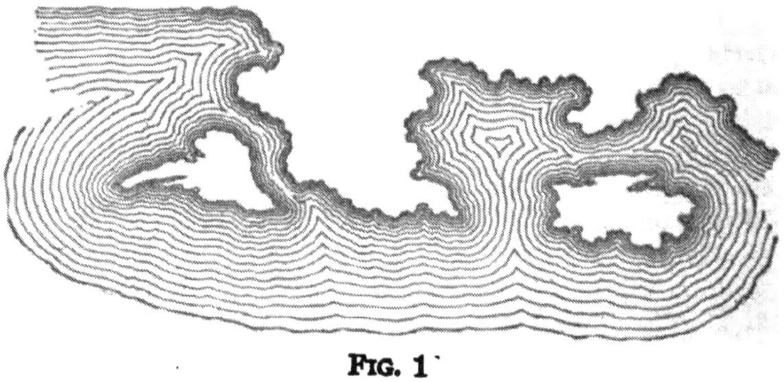

FIG. 1

shows how to represent a sandy shore, the irregular dotted surfaces inland from the shore line representing sand dunes. Fig. 4 shows the manner of representing the shore lines of rivers; for small brooks and creeks, one line is used. Fig. 5 shows the manner of representing grass; Fig. 6, cultivated land; Fig. 7, orchard; Figs. 8 and 9, woods; Fig. 10, clearings;

Fig. 11, underbrush; Fig. 12, swamps; Fig. 13, fresh-water ponds and marshes; Fig. 14, salt-water ponds and marshes; and Fig. 15, rice dikes and ditches.

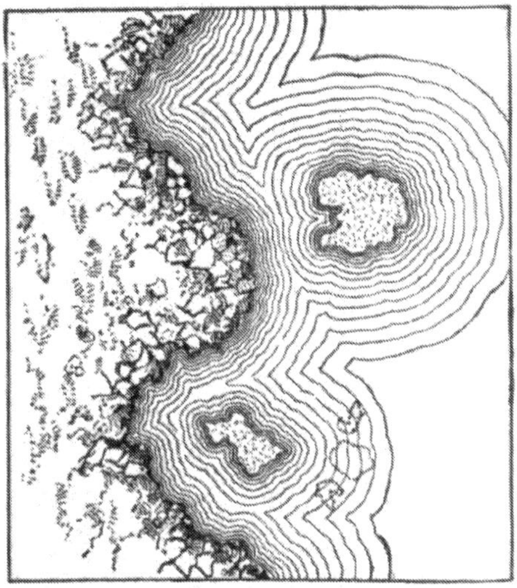

FIG. 2

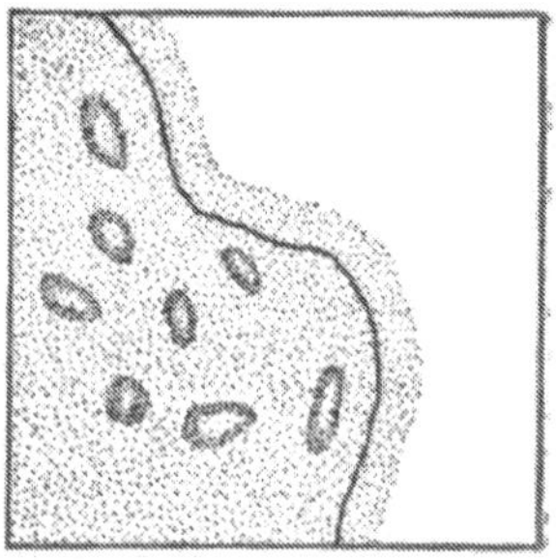

FIG. 3

Platting Angles.—In platting a traverse requiring great accuracy, as, for example, a difficult railroad location, the method of latitudes and longitudes given under Angular Surveying

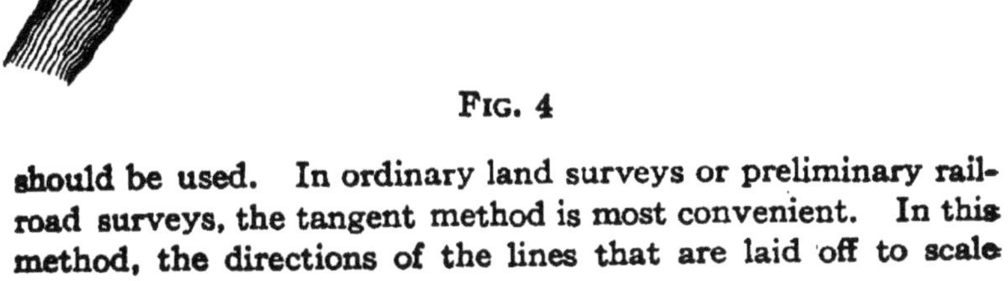

FIG. 4

should be used. In ordinary land surveys or preliminary railroad surveys, the tangent method is most convenient. In this method, the directions of the lines that are laid off to scale

FIG. 5

FIG. 6

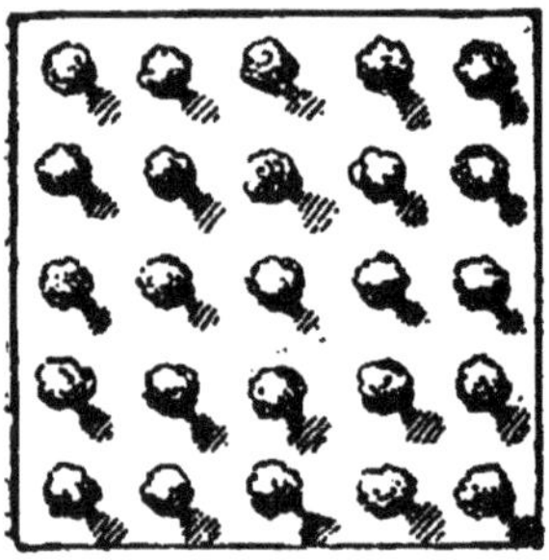

FIG. 7

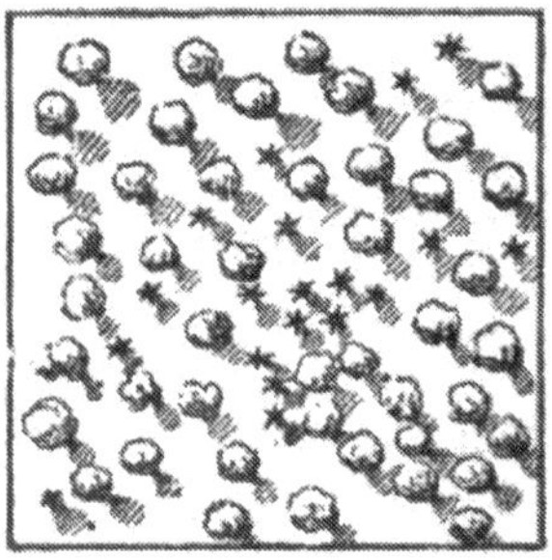

FIG. 8

FIG. 9

FIG. 10

FIG. 11

FIG. 12

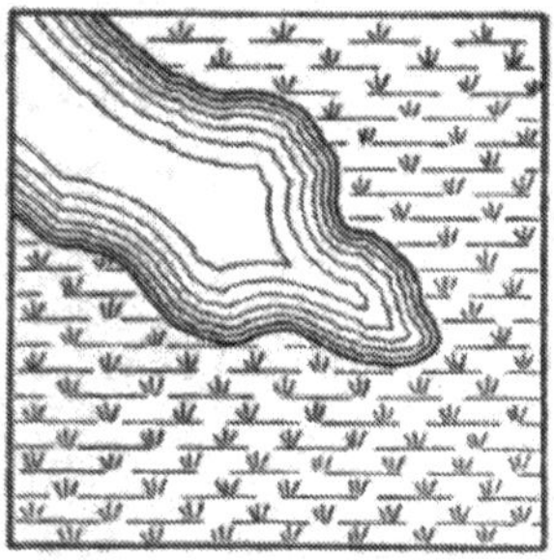

FIG. 13

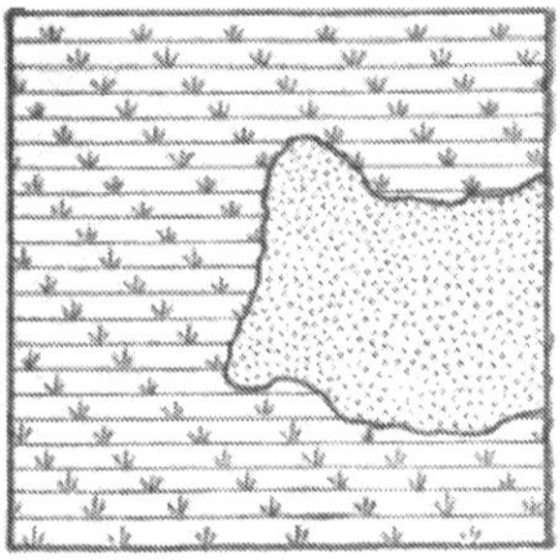

FIG. 14

FIG. 15

are determined by means of the tangents of the angles they are making with each other. Let *BA*, Fig. 16, be a line of a traverse from which an angle of 30° 15′ is to be turned to the left at *B*. Lay off *BC* equal to any convenient distance, say 10 in., and draw *CC′* perpendicular to the left of *BA*. From a table of natural functions the tangent of 30° 15′ is taken as .58318, which, multiplied by the distance *BC*—in this case 10—gives the length *CC′*, which is 5.83 in. This length is laid off as *CC′* and the line *BC′* gives the desired direction.

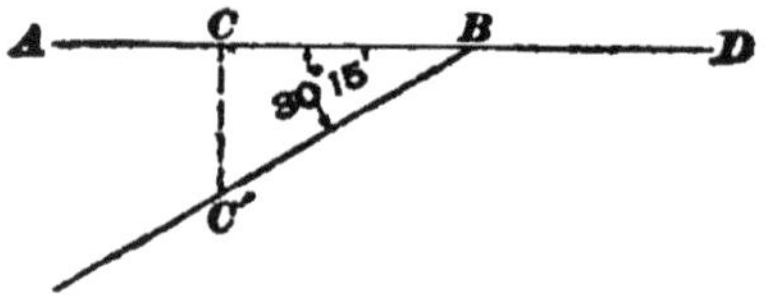

FIG. 16

To plat an obtuse angle as *DBC′*, Fig. 16, turned off to the right of *BD*, produce *DB* and construct the supplement *ABC′*, which is 180°−*DBC′*, as before.

HYDROGRAPHIC SURVEYING

SURVEY OF THE OUTLINE OF A BODY OF WATER

Hydrographic surveying is the process of surveying a body of water with a view of obtaining the outline of its shore, the

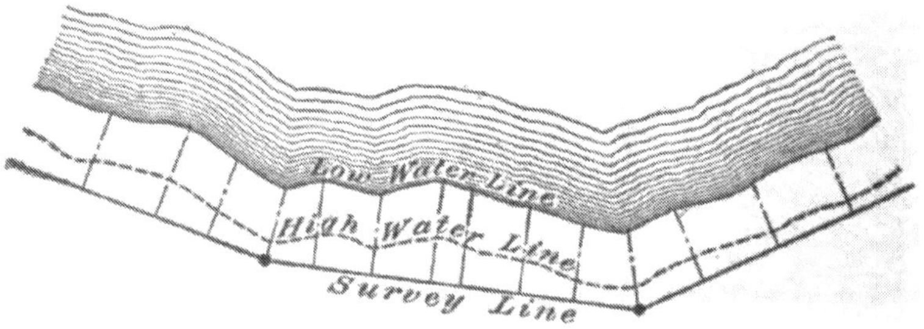

FIG. 1

topography of its bottom, and the volume of the body of water.

The outline of a body of water is determined by means of a traverse and offsets from the line of survey, as shown in Fig. 1, or by triangulation, an example of which method is shown in Fig. 2. A carefully measured base line, as *1-3*, is selected, and the angles of all the triangles are measured. From the triangle *1-2-3*, in which the side *1-3* and the angles are known, the sides *1-2* and *2-3* are computed by trigonometry. Then, in the triangle *2-3-4*, the side *2-3* and the angles are known and the other sides are calculated; and so on with the other triangles. In the last triangle, a side, as *5-7*, may be measured as a check on the work.

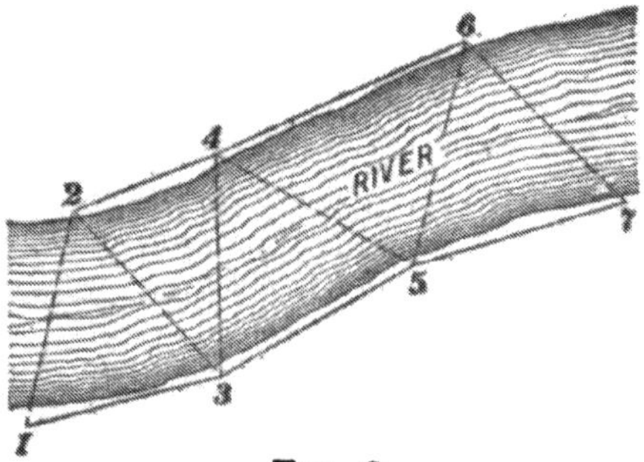

FIG. 2

SOUNDINGS

The configuration of the bottom of a body of water is determined by means of soundings. For depths of 18 ft. or less, *sounding poles* are used. The lower portion of one form of sounding pole is shown in Fig. 3. It is made of white pine 3 to 3½ in. in diameter at the bottom and 2 to 2½ in. at the upper end. It is fitted with a disk-shaped iron shoe, which prevents the rod from sinking into soft mud. The bottom of the shoe is sometimes hollowed out for the purpose of bringing up samples of materials.

FIG. 3

For depths greater than 18 ft., a *lead line* is used in making soundings. It consists of twisted hemp or closely plaited linen, about ⅜ in. in diameter, to the end of which is secured a weight called the *lead*. One form of lead L having the cross-section S is shown in Fig. 4. It is molded around an iron rod R to which small cross-bars are attached to prevent the lead from slipping. At the bottom is a cup C covered with a washer W, which prevents samples of material from being washed out while the lead is being drawn to the surface.

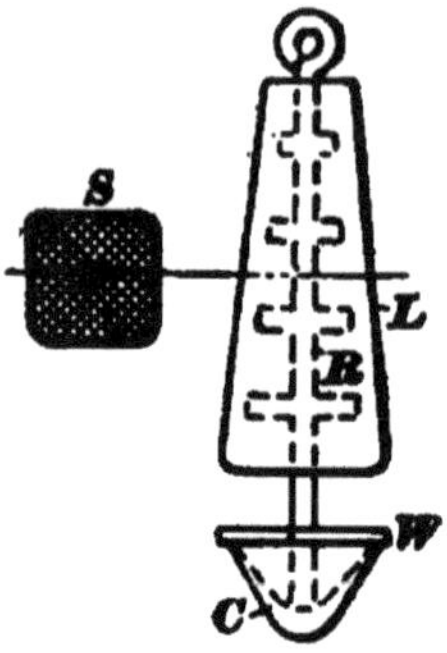

FIG. 4

Methods of Locating Soundings. Soundings are usually made on well-defined lines called *ranges*. The position of each sounding is located by various methods, depending on local conditions, the degree of accuracy required, etc. The most important methods are as follows:

1. *By Time Intervals.*—The soundings are made at stated intervals of time from a boat moving at uniform speed along a range. The distance between the end soundings being known, the position of each sounding can be determined by proportion.

2. *By One Angle Measured on the Shore.* The ranges are fixed with regard to a shore base line AB, Fig. 5, and the position of a sounding as C is found by the intersection of the range line with the line AC, the angle of which with AB is measured with a transit located at A.

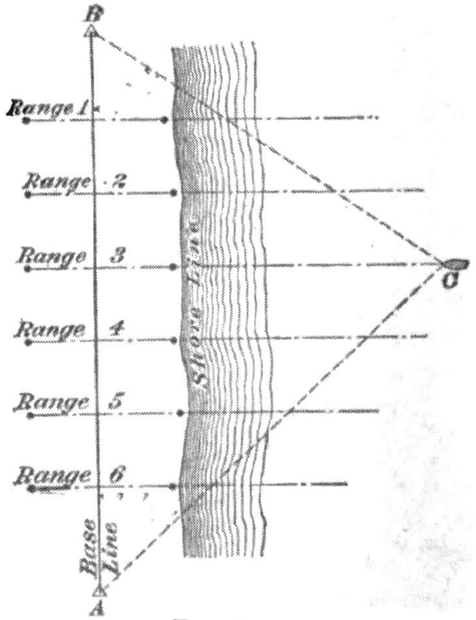

FIG. 5

3. *By Two Angles Measured Simultaneously on Shore.*—A transit is also placed at B, Fig. 5, and the angle CBA is measured simultaneously with the angle CAB, the position of C being determined by the intersection of the lines AC and BC. The ranges

need not be very accurately located; but if they are located accurately, they afford a means of checking the accuracy of the angular measurements.

4. *By Transit and Stadia.*—In calm and smooth water, the distance AC, Fig. 5, may be determined by observing a

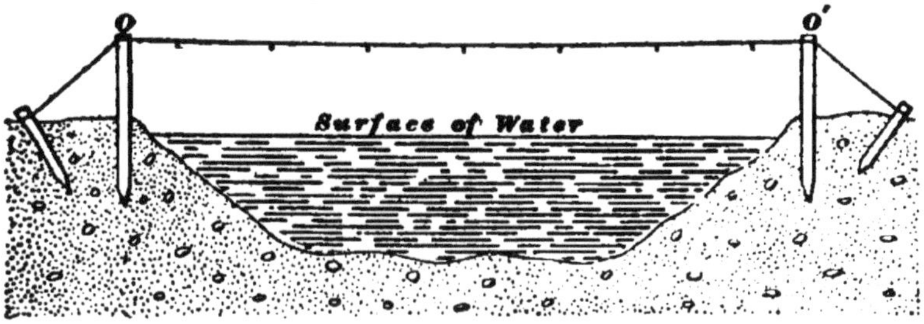

FIG. 6

stadia rod held in the boat at the same time that the angle CAB is measured.

5. *By Stretching a Rope from Bank to Bank of a Narrow River or Channel.*—The points where soundings are taken are marked by tin tags secured to the rope, as shown in Fig. 6.

FIG. 7

6. *By Two Angles Measured in a Sounding Boat With Two Sextants.*—Three prominent objects E, C, and D, Fig. 7, such as church spires, lighthouses, etc., are located by determining the distances $EC = a$, and $CD = b$ and by measuring the angle W. A sounding, as F, is then located by simultaneously measuring the angles A and B with two sextants. Then, the angles X and Y can be obtained by the formula

$$\cot X = \frac{a \sin B}{b \sin S \sin A} + \cot S,$$

in which $S = 360° - (W + A + B)$,

and $Y = S - X$

After X and Y are found the distances FE and FD can be figured by trigonometry.

Example.—Given $a=850$ ft., $b=760$ ft., $W=150°$, $A=41°\ 30'$, and $B=35°\ 30'$. What are the values of EF and DF?

Solution.—Substituting the given values, $S=360°-227°=133°$ and $\cot S=-\cot(180°-S)=-\cot 47°$.

Substituting known values in the preceding formula,

$$\cot X=\frac{850 \sin 35°\ 30'}{760 \sin 133° \sin 41°\ 30'}-\cot 47°$$

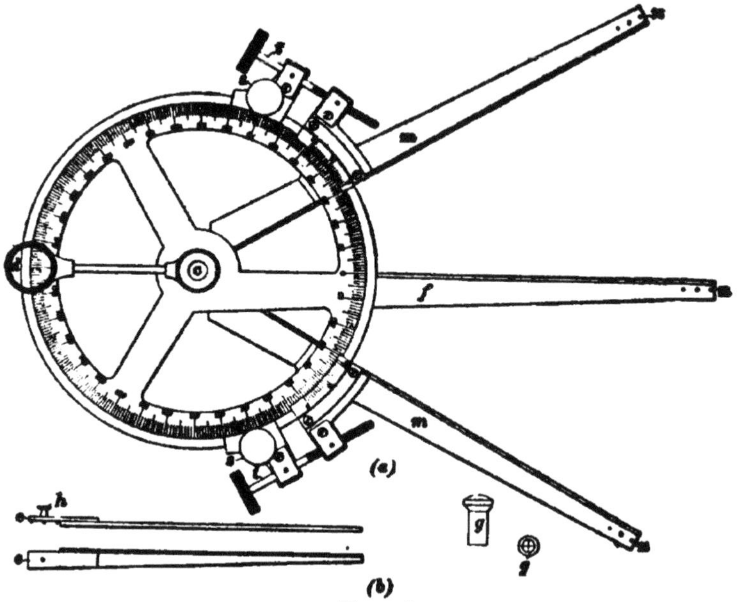

Fig. 8

$X=67°\ 49'$; whence, $Y=133°-67°\ 49'=65°\ 11'$.

In the triangle FCE, $ECF=180°-(41°\ 30'+67°\ 49')=70°\ 41'$. Therefore,

$$EF=\frac{850 \sin 70°\ 41'}{\sin 41°\ 30'}=1{,}211 \text{ ft.}$$

In the triangle DCF, $DCF=180°-(35°\ 30'+65°\ 11')=79°\ 19'$. Therefore,

$$DF=\frac{760 \sin 79°\ 19'}{\sin 35°\ 30'}=1{,}286 \text{ ft.}$$

Three-Arm Protractor.—The positions of soundings made by method 6 can most conveniently be platted by means of the three-arm protractor, Fig. 8. The arm *f* is fixed and its beveled edge is in line with the center and the zero point of the graduated circle. The arms *m* are movable, and their beveled edges also pass through the center of the circle. To determine the position of a sounding *F*, Fig. 7, when the positions of *E*, *C*, and *D*, are platted, set the movable arms of the protractor to form the measured angles *A* and *B* with *f*; then, with the beveled edge of *f* passing through *C*, slide the instrument around on the paper until the beveled edges of the arms *m* pass through *E* and *D*; the center of the circle *c* will then be over the point *F*.

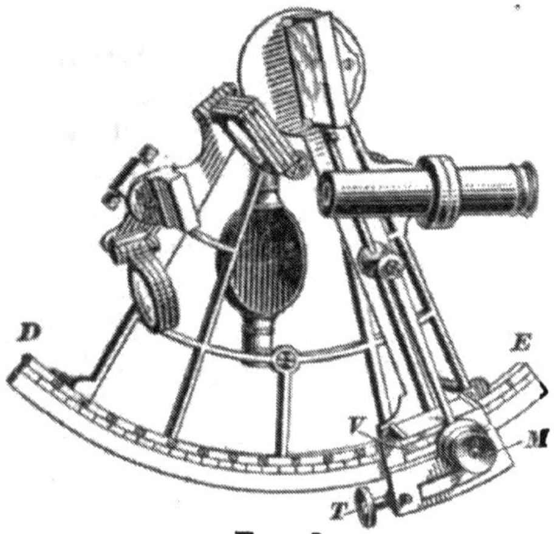

FIG. 9

The Sextant. — The sextant is a hand instrument for measuring angles. With it angles can be rapidly measured while in a boat when in motion. A sextant is illustrated in Fig. 9, and its essential parts are diagrammatically shown in Fig. 10. It has two fixed arms *BD* and *BE*, Fig. 10, to which are attached a telescope *T* and a horizon glass *A*, one-half

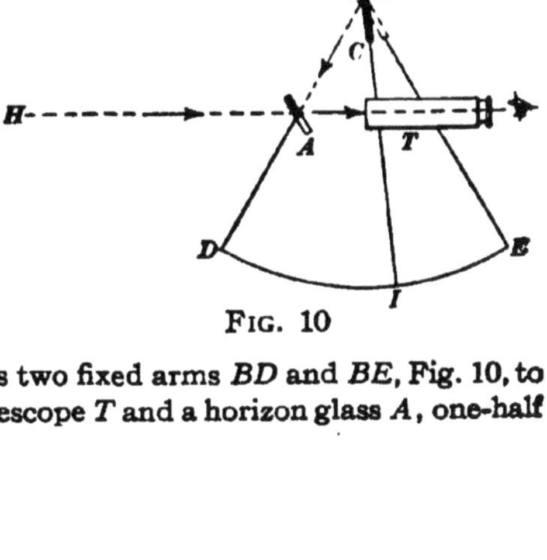

FIG. 10

of which is of transparent glass and the other half a mirror. The movable arm, or index arm, *BI*, revolves around the point *B*. It is fitted with a vernier at *I* and carries an index mirror *B'C*. The rays of light from an object *S* reflect from the index mirror to the mirror at *A*, and from this mirror to the telescope, through which *S* can be seen. To measure the angle between the lines of sight *SB* and *HA*, direct the telescope to *H*, which can be seen through the transparent half of the horizon glass, and revolve the index arm, by using the clamp *M* and tangent screw *T*, until the reflected image of *S* coincides with *H*. When in this position, the angle *EBI* equals one-half of the required angle, but since the arc *ED* has each half degree marked as a whole degree the angle can be read directly from the arc by means of the vernier *V*.

Adjustments of Sextant.—There are four adjustments of the sextant, as follows:

1. To make the plane of the index glass perpendicular to the plane of the limb.

2. To make the plane of the horizon glass perpendicular to the plane of the limb.

3. To make the line of collimation of the telescope parallel to the plane of the limb.

4. To make the planes of the mirrors parallel when the index reading is zero.

First Adjustment.—Place the index bar near the middle of the limb; with the eye near the plane of the limb, observe whether the limb as seen directly and its image as reflected in the index glass form a smooth continuous curve; if they do, the glass is perpendicular to the plane of the limb and the adjustment is correct. But if the reflected limb appears to be above that part of the limb seen directly, the glass leans forwards; if it appears to be below, it leans backwards. In either case it is made perpendicular to the plane of the limb by means of the adjusting screws at its base.

Second Adjustment.—Look through the telescope and horizon glass toward a star or other well-defined distant object. Move the index bar slowly until the reflected image passes over the image seen directly. If these images coincide, the horizon glass is perpendicular to the plane of the limb. If they do not coincide, the horizon glass is adjusted by an adjusting screw placed under, behind, or beside the glass, according to the construction of the instrument.

Third Adjustment.—Place the sextant in a horizontal position on a table or other support, and direct the telescope at some well-defined point or mark about 20 ft. away. Place two small blocks of equal height on the limb, one near each extremity. These blocks should be of exactly equal height, so that a line of sight over their tops will be parallel to the plane of the limb, and should be at the same height above the limb as the center of the telescope. Sight over the tops of the two blocks in the direction of the point or mark sighted through the telescope, and note whether the line of sight intersects the mark. If it does not, but falls above or below the mark, the telescope is not parallel to the limb. It can be made parallel to the limb by means of the screws in the collar that holds the telescope. This adjustment, however, is not usually made unless the error is considerable, because a slight lack of parallelism between the line of sight and the plane of the limb does not appreciably affect the angular measurements on the limb.

Fourth Adjustment.—Set the index at zero, look through the telescope toward a star and note whether the direct and reflected images of the star coincide. If they do, the adjustment is correct. If they do not, move the index bar until they do coincide, and clamp it in this position. The reading of the index when in this position is called the ***index error***. This error can be corrected by means of screws at the back of the index glass, which cause it to revolve about an axis perpendicular to the plane of the limb. To make the correction, set the index bar at zero and, by turning the screws, revolve the index glass until the two images exactly coincide. This adjustment will usually disturb the previous adjustment of the index glass, and, as a rule, it is not made unless the index error is greater than 3 min.

When the index error is less than 3 min., it is usually applied as a correction to all observations. If the error is *off* the arc, that is, if the index is to the right of the zero mark, it is additive, or plus, and should be added to all readings. If the error is *on* the arc, that is, if the index is to the left of the zero mark, the error is subtractive, or minus, and should be subtracted from all readings.

EXAMPLE.—The angular distance between two objects, as measured with a sextant, reads on the vernier 35° 36′ 30″. What is the true angular distance if the index error of the sextant is: (*a*) +1′ 20″; (*b*) −1′ 40″?

SOLUTION.—(*a*) Since the vernier reading is 35° 36′ 30″ and the index error is +1′ 20″, the true angular distance is equal to 35° 36′ 30″+1′ 20″ = 35° 37′ 50″.

(*b*) Since in this case the index error is −1′ 40″, the true angular distance is equal to 35° 36′ 30″ − 1′ 40″ = 35° 34′ 50″.

VOLUME OF A RESERVOIR

By means of the platted soundings a contour map is prepared in the manner explained under the heading Topographic

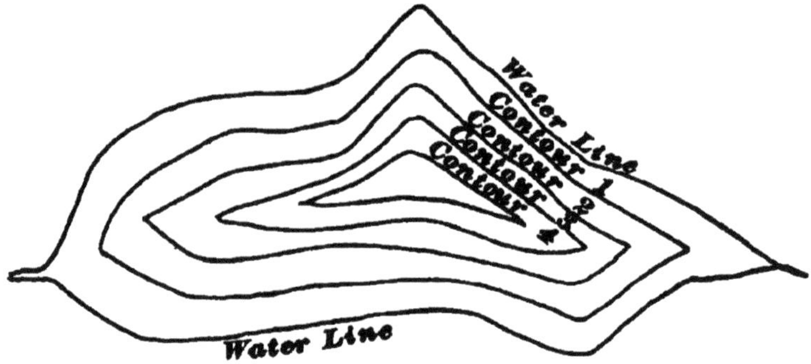

Surveying; the outline of the reservoir being the surface contour. The contour interval is fixed according to the slopes of the valley and the degree of accuracy required. The volume of water included between two plane surfaces passing through two adjacent contours is that of a prismoid whose bases are those surfaces included by the contour lines and whose height is the contour interval. The sum of the volumes of the several prismoids will be the volume in the reservoir. When the number of prismoids is even, the following expression which is based on the prismoidal formula, will give the total volume V.

$$V=\frac{h}{3}(A_0+4\Sigma A_1+2\Sigma A_2+A_n),$$

in which h = contour interval;

A_0 = area included by surface contour;

A_n = area included by lowest contour;

ΣA_1 = sum of areas of odd-numbered contours;

ΣA_2 = sum of areas of even-numbered contours.

EXAMPLE.—Let, in the accompanying illustration, $h=5$ ft.; $A_0=13{,}350$ sq. ft., $A_1=8{,}100$ sq. ft., $A_2=4{,}280$ sq. ft., $A_3=1{,}925$ sq. ft., and $A_4=520$ sq. ft. Find the volume V.

SOLUTION.—By substituting the given values in the formula, $V=\frac{5}{3}(13{,}350+4\times 8{,}100+4\times 1{,}925+2\times 4{,}280+520)=104{,}217$ cu. ft.

When there is an odd number of prismoids, the last prismoid may be computed separately by multiplying one-half the sum of its end areas by the contour interval.

CITY SURVEYING

LINEAR MEASUREMENTS

The surveying work to be done by a city often requires a great degree of precision, necessitating the employment of special methods and instruments.

Corrections for Temperature.—The steel tape is the standard instrument for city work. The usual lengths are 50 and 100 ft. When a high degree of precision is required, corrections for temperature, pull, and sag of the tape are necessary. For such work, the temperature at which the tape is exactly its graduated length should be determined by a test in a responsible testing laboratory, such as the Bureau of Standards in Washington, which for a small charge will furnish the constants of temperatures and pull for any tape.

Let this temperature be t_0, and let a line of the true length l_0 be measured with a tape at a temperature t. The correction for temperature is then equal to

$$c(t-t_0)l,$$

in which c is the coefficient of expansion of the tape, which for steel averages about .0000065, and l is the measured length of the line. The true length is therefore

$$l_0 = l + lc(t - t_0)$$

If t is less than t_0, the correction is negative and should be subtracted from l.

EXAMPLE.—A line was measured with a tape that was standard at 62°. The temperature was 90°. The length, as measured, was 502.34 ft. If the coefficient of expansion of the tape was .0000065, what was the true length of the line?

SOLUTION.—Here, $c(t-t_0) = .0000065 \times (90-62) = .000182$. The correction $c(t-t_0)l$ is, practically, $.000182 \times 502$, the decimal .34 being dropped, as the product of it by .000182 is too small to be considered. Therefore, $l_0 = 502.34 + .000182 \times 502 = 502.43$ ft.

Correction for Pull.—If the length of the tape is denoted by L, the cross-section by A, and the modulus of elasticity by E, the true length L_0 of the tape stretched by a pull P is given by the formula

$$L_0 = L + \frac{P}{EA}L$$

If the length of a line as measured with the stretched tape is l, and the true length of the line is l_0, then

$$l_0 = l + \frac{P}{EA}l$$

For such steel as tapes are made of, E may be assumed without great error as 28,000,000 lb. per sq. in. A not unusual cross-section is about .002 sq. in. A tape 100 ft. long with such a cross-section would be lengthened about .036 ft. for a pull of 20 lb. above the normal. Hence, a line measured with such a tape under such a pull, and found to be 400 ft. long, would really be $400 + 4 \times .036 = 400.144$ ft. long.

Correction for Sag.—If a tape is held off the ground so that it is supported only at each end, it will sag and hang in a curve. The effect of sag is to shorten the distance between end graduations, the amount depending on the weight and length of the unsupported part of the tape, and on the pull exerted at the ends of the tape.

If L_0 denotes the unsupported length of the tape, w the weight of tape per unit of length, and P the pull, the shortening s due to the sag is given by the formula

$$s = \frac{w^2 L_0^3}{24P^2}$$

It should be observed that L_0 is the length of the *unsupported part*, which may not be the entire length of the tape.

Since, when the tape sags, the distance between its two supports, as indicated by the nominal length of the tape, is greater than the actual distance, or the length of the chord subtended by the arc, the correction for the sag is negative, and must be subtracted from the nominal length indicated by the tape. If the length of a line, as measured, contains n times the length L_0, and the sag is the same in all measurements, the correction for sag is

$$ns = \frac{nw^2 L_0^3}{24P^2}$$

Example.—A line as measured with a 100-ft. tape weighing .007 lb. per ft., with a pull of 14 lb., is found to be 400 ft. Determine the correction for sag.

Solution.—Here, $n=4$, $w=.007$, $L_0=100$, and $P=14$. Substituting these values in the formula,

$$ns = \frac{4 \times .007^2 \times 100^3}{24 \times 14^2} = .042 \text{ ft.}$$

If it is desired to pull the tape just enough to cause the stretch, which is a positive error, to balance the sag, which is a negative error, the proper pull P may be found by the following formula:

$$P = \sqrt[3]{\frac{w^2 L_0^2 A E}{24}}$$

Example.—The weight of a 100-ft. tape is .008 lb. per ft., and the sectional area is .002 sq. in. Taking E as 28,000,000 lb. per sq. in., determine the pull necessary to neutralize the sag.

Solution.—In this example, $w=.008$, $L_0=100$, $A=.002$, and $E=28,000,000$. Substituting these values in the formula,

$$P = \sqrt[3]{\frac{.008^2 \times 100^2 \times .002 \times 28,000,000}{24}} = 11.4 \text{ lb.}$$

ANGULAR MEASUREMENT

In city work, use is made of a transit having many features that contribute to greater accuracy. The least reading of the vernier is usually 30 or 20 sec., but sometimes angles are required to a smaller unit than the least reading of the vernier. These may be obtained by the *method of repetition* as follows: The transit is set up over the vertex of the angle with the verniers reading zero; the lower clamp being loosened and the upper set, the telescope is directed along the left-hand side of the angle. The lower clamp is then set, the upper loosened, and the telescope directed along the right-hand side of the angle. The upper clamp is now set, the vernier read, the lower clamp loosened, and the telescope directed along the left-hand side of the angle. The lower clamp is then set, the upper loosened, and the telescope directed along the right-hand side of the angle. The upper clamp is now set, the lower loosened, and the telescope directed again along the left-hand side of the angle; then the lower clamp is set, the upper loosened, and the telescope directed along the right-hand side of the angle. The process is repeated as often as necessary to obtain the required accuracy. The vernier is read after the final turning, when the telescope is set on the right-hand side of the angle, and the reading is divided by the number of turnings, including the first. The result will be the value of the angle, which, as a check, should closely approximate the first reading. This first reading is taken only for the purpose of checking the final result.

Theoretically, the number of measurements should be such that the sum will approximate a whole number of complete revolutions, so that all parts of the circle may be used in measuring; but, practically, three measurements are sufficient in all ordinary cases. In very precise work, the angle may be read as described, and then read again from right to left with the telescope inverted. This eliminates errors of pointing and adjustment of the line of collimation.

PRECISION

If a quantity, as a distance or an angle, is measured very accurately several times by the same method, it is usually found that the results vary slightly from one another. The true measure of the quantity is taken to be the mean of the different results obtained—that is, the sum of these results divided by their number. This mean is called the ***mean value***, or ***most probable value***.

By the law of probabilities it may be determined that the error made in using the mean value does not exceed a certain quantity, called its ***probable error***. This quantity may be positive or negative, that is, the exact value may be greater or smaller than the mean value. It serves as a measure of the accuracy obtained by the use of the mean value.

Let the probable error be denoted by p; the sum of the squares of the differences between the actual measurements and the mean value, the latter being called ***residuals***, by Σv^2; and the number of measurements made by m. Then,

$$p = \pm .6745\sqrt{\frac{\Sigma v^2}{m(m-1)}}$$

EXAMPLE.—A distance was measured four times, the results of the measurements being, respectively, 501.07, 501.06, 501.05, and 501.08 ft. Determine: (*a*) the mean value M of the distance; (*b*) the probable error p.

SOLUTION.—(*a*) Since 501 is common to all the measurements,

$$M = 501 + \frac{.07+.06+.05+.08}{4} = 501.065$$

(*b*) To apply the formula for p, $m=4$, $m-1=3$, and

$$v_1 = 501.065 - 501.07 = -.005$$
$$v_2 = 501.065 - 501.06 = .005$$
$$v_3 = 501.065 - 501.05 = .015$$
$$v_4 = 501.065 - 501.08 = -.015$$
$$\Sigma v^2 = (-.005)^2 + (.005)^2 + (.015)^2 + (-.015)^2 = .0005$$

Therefore, $p = \pm .6745\sqrt{\frac{.0005}{4\times 3}} = \pm .0044.$

Weighted Measurements.—If the measurements are not made under the same conditions, so that there are reasons to believe that some of them are more accurate than others, the results must be *weighted*. That measurement whose accuracy is supposed to be the least usually receives a weight of 1; a measurement whose accuracy appears to be twice as great receives a weight of 2; etc. After the measurements have been weighted, each measurement is multiplied by the number representing its weight, the products are added, and the sum is divided by the sum of the weight numbers. This result is the mean value, or most probable value, of the quantity. Thus, in the preceding example, if the first measurement is of the least weight, while the second is twice as great as the first, and the third and fourth are each two and one-half times as great as the first, the weights of the four measurements are respectively, 1, 2, 2.5, and 2.5, and the mean value M is

$$501.0+\frac{.07\times 1+.06\times 2+.05\times 2.5+.08\times 2.5}{1+2+2.5+2.5}=501.064.$$

If the weight of any measurement is denoted by h, then the probable error

$$p=\pm .6745\sqrt{\frac{\Sigma\,(hv^2)}{(\Sigma h-1)\,\Sigma h}},$$

in which $\Sigma\,(hv^2)$ is the sum of the products of the squares of the residuals by their corresponding weights, and Σh is the sum of all the weights.

EXAMPLE.—Determine the probable error p in the preceding example, the weights of the four measurements being, respectively, 1, 2, 2.5, and 2.5.

SOLUTION.—The mean value M has been found to be 501.064. The values of the residuals are as follows:

$$v_1=501.064-501.07=\ -.006$$
$$v_2=501.064-501.06=\ +.004$$
$$v_3=501.064-501.05=\ +.014$$
$$v_4=501.064-501.08=\ -.016$$

Then,

$$\Sigma\,(hv^2)=1\times(-.006)^2+2\times(+.004)^2+2.5\times(+.014)^2$$
$$+2.5\times(-.016)^2=.001198, \text{ and } (\Sigma h-1)\Sigma h=7\times 8$$

Substituting in the formula,

$$p=\pm .6745\sqrt{\frac{.001198}{8\times 7}}=\pm .0031.$$

Measure of Precision.—It is customary to express precision in terms of the probable error: when it is said that a line has been measured with a precision of $\frac{1}{5000}$, it is usually meant that the probable error derived from the series of measurements is not numerically greater than $\frac{1}{5000}$ of the determined length of the line. Thus, in the preceding example the precision was $.0031 \div 501.064 = \frac{1}{161634}$.

Precision Required.—In important cities, a precision of 1 in 50,000 should be obtained in land-surveying measurements; that is, the mean of two measurements of a given line should have a probable error of not more than $\frac{1}{50000}$ of the length of the line. This will generally be accomplished if the two measurements differ by not more than $\frac{3}{50000}$, or, say, $\frac{1}{15000}$, of the length of the line. This result is not very difficult to secure if the proper methods and instruments are used. In villages and small towns, a precision of $\frac{1}{5000}$ is ordinarily sufficient, but it is so easy to secure a better precision than this that no two measurements of the same line should differ by more than $\frac{1}{10000}$ of its length, giving a precision of the mean of the two measurements of about $\frac{1}{30000}$.

Precision in Angular Measurements.—In order that the direction of a line may be determined so that a distant end shall not depart from its true position by more than $\frac{1}{50000}$ of the length of the line, the angle on which the direction depends must be measured to about the nearest 4 sec. A transit reading to 30 sec. will permit an approximation to this result if the mean of three readings of the angle is used. An instrument reading to 20 sec. will ordinarily, by a triple measurement, permit a little closer result than the required one, and one reading to 10 sec. may give the requisite precision with a single measurement, though at least two measurements should be made for a check on the accuracy of the work.

Ordinarily, the position of a point can be more precisely determined by linear than by angular measurement, and, therefore, the former method of determination is in general to be preferred.

Adjustment of Measured Angles of a Triangle.—It is frequently necessary, in precise plane surveying, as in locating bridge piers, making topographical surveys of cities, etc., to measure triangles. When this need occurs, each angle of the triangle should be measured directly. If but two angles are measured and their sum is subtracted from 180° to get the third, all errors of measurement of the two angles are thrown into the third angle. When all the angles are measured to a high degree of precision, their sum will ordinarily be more or less than 180°, indicating an impossible triangle. To make the triangle possible, the angles are adjusted so that their sum shall be 180°. The adjustment is effected by dividing the total error equally among the three angles. It might seem that a distribution in some ratio to the size of the angles should be adopted; but the method applied considers that there is no more reason for making an error in measuring a large angle than in measuring a small angle, which is probably true.

PRACTICAL ASTRONOMY

DEFINITIONS AND TERMS

LATITUDE AND LONGITUDE.

If a meridian, that is, a circle passing through the axis of the earth, be passed at a given point of the earth's surface, the angular distance of the point from the equator, measured on the meridian, is the *latitude* of that point. A plane parallel to the equator cuts the earth's surface in a circle called a *parallel of latitude*. All the points on a parallel of latitude have the same latitude. The *longitude* of a place is the angle that the plane of the meridian of the place makes with the plane of a reference meridian (usually the meridian of Greenwich). This angle may be measured on the equatorial circle or on the parallel of latitude of the given place. Longitude is counted from the reference meridian toward the west.

THE CELESTIAL SPHERE

The *celestial sphere* is an imaginary sphere enclosing all the heavenly bodies. It is of such enormous dimensions that, in comparison with it, the earth may be considered as a mere dot.

The earth's axis produced indefinitely is called the *axis of the celestial sphere*. This axis intersects the celestial sphere in two points, called the *north pole* and the *south pole* of heavens. All the great circles of the celestial sphere passing through this axis are called *hour circles*. The circle in which the plane of the equator intersects the celestial sphere is called the *celestial equator*. The point on the equator that the sun in its apparent motion over the celestial sphere crosses on March 21, as it passes from the southern to the northern hemisphere, is called the *vernal equinox*.

REFERENCE CIRCLES

The accompanying illustration, which represents the celestial hemisphere, shows all the reference circles that are used for determining the position of a heavenly body. *O* is the position of the earth; *OP*, one-half of the axis of the celestial sphere,

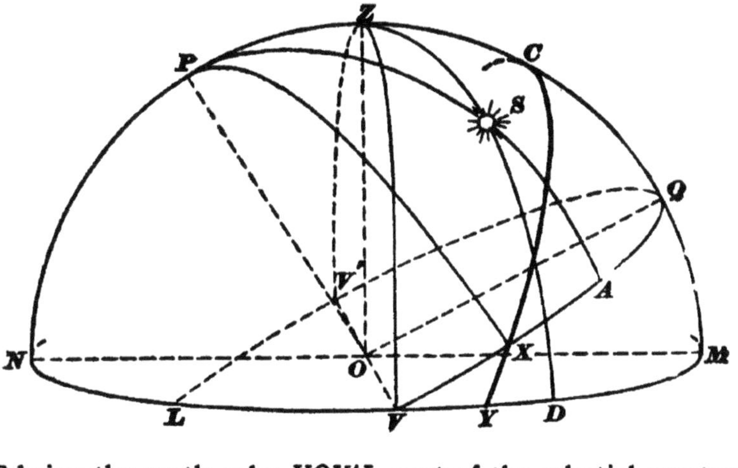

P being the north pole; *VQV'L*, part of the celestial equator; *X*, the vernal equinox; and *YXC*, part of the sun's path. *PX* is the hour circle passing through *X*, called the *equinoctial colure*. *S* is any star, and *PSA* is the hour circle passing through it. *XA* is the *right ascension* of the star, which is the arc on the equator measured eastwards from the vernal equinox to the hour circle passing through the star. AS is the *declination* of the star; that is, its angular distance from the equator. The declination is considered positive when the star is north and negative when south of the equator. The complement angle of the declination, SP, is called the *polar distance* of the star.

The *zenith* of a point on the earth's surface is the point in which the line passing through the center of the earth and the given point intersects the celestial sphere above the given point. The *horizon* is the plane passing through the given point and perpendicular to this line. In the illustration, Z is the zenith, and NVM is the celestial horizon.

The *celestial meridian* of a given point is a great circle passing through the zenith of the point and the poles. The celestial meridian cuts the horizon in two points N and M, called, respectively, the *north point* and the *south point*.

A *vertical circle* is one that passes through the zenith and is perpendicular to the horizon.

The *prime vertical* is the vertical circle at right angles to the meridian; it intersects the horizon in two points V and V', called the *west* and the *east point*, respectively.

The *altitude of a heavenly body* is its angular distance from the horizon, measured along the vertical circle passing through the body. The *zenith distance* is the angular distance of the star from the zenith, measured along the same circle. The zenith distance is the complement of the altitude. In the illustration, DS and SZ are, respectively, the altitude and zenith distance of S.

The *azimuth of a star* is the angle in the plane of the horizon intercepted by the planes of the meridian and the vertical circle passing through the star. It is measured from the north point toward the east or from the south point toward the west. NMD is the azimuth of S, measured from the north toward the east, and MD is the azimuth of S when measured from the south toward the west.

The *hour angle* of a star is the arc intercepted on the equator between the meridian and the foot of the hour circle passing through the star. It is measured from the meridian toward the west. In the illustration, QA is the hour angle of S.

TIME

The passing of a heavenly body across the meridian of a place is called its *culmination*, or *transit*. It is upper or lower culmination, according as it is then occupying the highest or the lowest position with regard to the horizon.

The interval of time that elapses between two successive upper or lower transits of a star over the same meridian is called a *sidereal day*. It begins, for any place, when the vernal equinox crosses the meridian above the pole. This instant is called *sidereal noon*. Sidereal hours, minutes, and seconds are reckoned from 0 to 24 hr., starting from sidereal noon. Time expressed in sidereal days and fractions (hours, minutes, seconds) is called *sidereal time*.

From this, it follows that sidereal time is the hour angle of the vernal equinox; also, that the right ascension of a star is equal to the sidereal time of its transit, or culmination. For any other position of the star, the sidereal time equals the algebraic sum of the right ascension and the hour angle of the star.

The interval between two successive upper transits of the sun is called a *true solar day*, or an *apparent day*. Owing to the fact that the motion of the sun is not uniform and that the solar days are not of equal duration, apparent time is not used for the ordinary affairs of life.

The *mean sun* is an imaginary body supposed to start from the vernal equinox at the same time as the true sun, and to move uniformly on the equator, returning to the vernal equinox with the true sun. The time between two successive upper transits of the mean sun is called a *mean solar day*, and time expressed in mean solar days is called *mean solar time*, or simply *mean time*. This is the time shown by ordinary clocks and watches.

A mean solar day is the mean of the duration of all the true solar days in a year (a year being the time in which either the true or the mean sun makes a complete circuit of the heavens). As there are 365.2422 true solar days and 366.2422 sidereal days in a year,

$$1 \text{ mean solar day} = \frac{366.2422}{365.2422} = 1.0027379 \text{ sidereal days}$$

$$= 24^{h}3^{m}56.55^{s}, \text{ sidereal time}$$

Likewise,

$$1 \text{ sidereal day} = \frac{365.2422}{366.2422} = .99726957 \text{ mean solar day}$$
$$= 23^h 56^m 4.09^s, \text{ mean solar time}$$

The *equation of time* is a certain quantity that must be added algebraically to the apparent solar time to obtain the corresponding mean time. The value of this quantity for each day of the year is given in the American Ephemeris, which is published yearly by the United States Government at Washington, D. C.

Civil Time and Astronomical Time.—By *civil time* is meant the time that is usually reckoned in ordinary life. For astronomical purposes, the day is considered to begin at noon, and hours are counted from 0 to 24. When time is reckoned in this manner it is called *astronomical time*. The civil day begins at 12 o'clock at night, and the astronomical day begins 12 hr. later. For instance, the date Oct. 17, $7^h\ 14^m\ 3^s$, astronomical time, means $7^h\ 14^m\ 3^s$ after noon of the civil date Oct. 17, and is in civil time, $7^h\ 14^m\ 3^s$ P. M. The astronomical date Feb. 20, $18^h\ 6^m\ 12^s$ means $18^h\ 6^m\ 12^s$ after noon of the civil date Feb. 20, or $6^h\ 6^m\ 12^s$ after midnight of Feb. 20; that is, Feb. 21, $6^h\ 6^m\ 12^s$ A. M.

Longitude and Time.—The mean sun describes a complete circle in 24 mean solar hours. In 1 hr. it moves over $\frac{360°}{24} = 15°$ of arc; in 1 min. of time, over 15′ of arc; and in 1 sec. of time, 15″ of arc.

Relation Between Time and Longitude.—Let A and B be two places on the earth's surface, B being west of A. Let their respective longitudes be g_a and g_b, and let the difference between g_b and g_a, expressed in measure of time, be d_g. Let, also, T_a be the time at A when the time at B is T_b. Then,

$$T_a = T_b + d_g \qquad (1)$$

and

$$T_b = T_a - d_g \qquad (2)$$

EXAMPLE 1.—The longitude of Washington, west of Greenwich, is $5^h\ 8^m\ 1^s$; that of San Francisco, $8^h\ 9^m\ 47^s$. What is the time at: (*a*) Washington when it is $9^h\ 3^m$ at San Francisco? (*b*) San Francisco when it is $19^h\ 54^m\ 30^s$ at Washington?

SOLUTION.—(*a*) Here *A*, the eastern locality, is Washington and *B* is San Francisco; also, $d_g = 8^h\ 9^m\ 47^s - 5^h\ 8^m\ 1^s = 3^h\ 1^m\ 46^s$. Therefore, applying formula 1,

$$T_a = 9^h\ 3^m + 3^h\ 1^m\ 46^s = 12^h\ 4^m\ 46^s.$$

(*b*) Applying formula 2,

$$T_b = 19^h\ 54^m\ 30^s - 3^h\ 1^m\ 46^s = 16^h\ 52^m\ 44^s.$$

Standard Time.—Time referred to the meridian of a given place is called *local time* of that place. To obviate complications in comparing local times of different localities, for use in ordinary affairs of life *standard times* have been adapted for regions between certain longitudes. The United States is divided into four zones, or sections of standard time. The time in each zone is referred to the meridian passing through its center. These central meridians are 15° or 1^h distant from each other and are, respectively, 75°, 90°, 105°, and 120° west of Greenwich; or, in hours, 5^h, 6^h, 7^h, and 8^h west of Greenwich. Each of these meridians controls the watch time of all places within 7½° on either side. This is shown as follows:

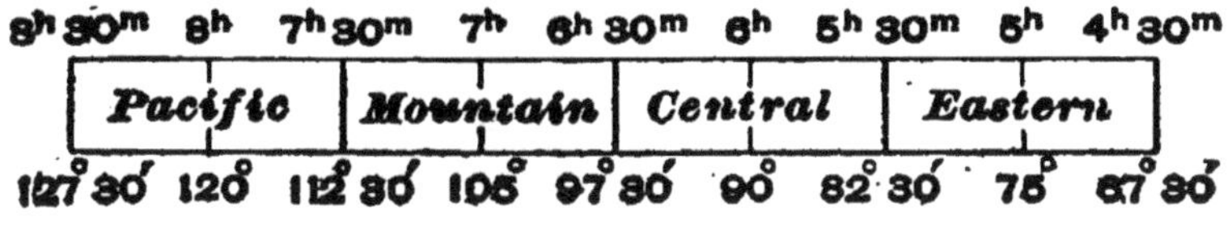

Time referred to the 75° meridian is called *eastern time;* to the 90° meridian, *central time;* to the 105° meridian, *mountain time;* and to the 120° meridian, *Pacific time.*

To Change Standard Time Into Local Time and Vice Versa. Standard time can be changed into local time or local time can be changed into standard time by applying formula 1 or formula 2, according as the given place is east or west of the reference meridian of the zone in which the place is located.

EXAMPLE.—The standard time, by a watch, at a place whose longitude is 81° 37′, is $9^h\ 37^m\ 45^s$ A. M.; what is the local time?

SOLUTION.—Since the longitude is 81° 37′, the place lies within the zone of the 75° meridian; and being west of the latter, formula 2 must be applied. In this case, $T_a = 9^h\ 37^m\ 45^s$ and $d_g = 81°\ 37' - 75° = 6°\ 37' = 26^m\ 28^s$. Therefore, $T_b = 9^h\ 37^m\ 45^s - 26^m\ 28^s = 9^h\ 11^m\ 17^s$ A. M.

DETERMINATION OF MERIDIAN

DETERMINATION BY OBSERVING POLARIS AT CULMINATION

The position of Polaris, or the north star, can easily be ascertained by means of the group of stars called the Dipper, or the Great Bear. As shown in the accompanying illustration, a straight line joining the stars, α and β, called the *pointers*, nearly intersects Polaris. There are two times during the day when the star crosses the meridian. It is then said to be at its *upper* or *lower culmination*, as the star is then occupying either the highest or the lowest position with reference to the horizon. When the star is in either one of these positions, the vertical plane passing through it and the observer's station is the meridian of the place, and its intersection with the horizon is therefore a true north-and-south line.

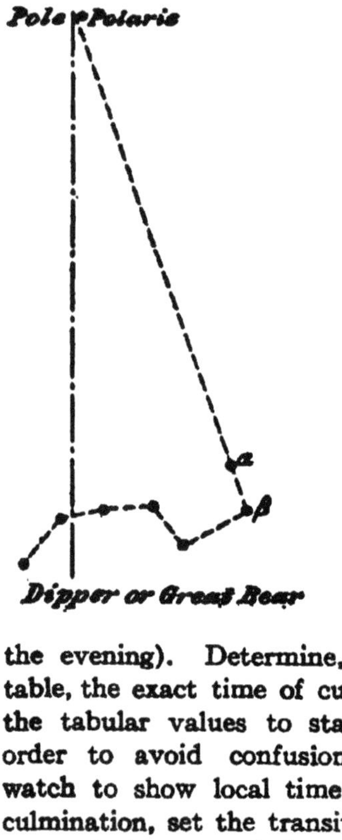

Field Work.—Select a date on which Polaris is at either lower or upper culmination during the night (preferably during the early part of the evening). Determine, by means of the accompanying table, the exact time of culmination, being careful to reduce the tabular values to standard civil time. It is safer, in order to avoid confusion, for the observer to set his watch to show local time. About 15m. before the time of culmination, set the transit in such a position that an unobstructed view toward the north may be obtained for a distance of between 300 and 500 ft. Drive a stake, and mark by a tack the exact point occupied by the instrument. About 5m. before the time of culmination, direct the telescope to the star, holding a lamp in front and a little toward one side of the

LOCAL MEAN ASTRONOMICAL TIME OF UPPER CULMINATION OF POLARIS

	1913		1914		1915		1916		1917		1918		1919		1920		1921		1922		Diff. for 1 Da.
	h.	m.	h.	m.	h.	m.	h.	m.	h.	m.	h.	m.	h.	m.	h.	m.	h.	m.	h.	m.	Minutes
Jan. 1..........	6	43.8	6	45.2	6	46.7	6	48.1	6	45.6	6	47.0	6	48.4	6	49.9	6	47.4	6	48.8	3.95
Jan. 15.........	5	48.5	5	49.9	5	51.4	5	52.8	5	50.3	5	51.7	5	53.1	5	54.6	5	52.1	5	53.5	3.95
Feb. 1..........	4	41.4	4	42.8	4	44.3	4	45.7	4	43.2	4	44.6	4	46.0	4	47.5	4	45.0	4	46.4	3.95
Feb. 15.........	3	46.1	3	47.5	3	49.0	3	50.4	3	47.9	3	49.3	3	50.7	3	52.2	3	49.7	3	51.1	3.94
Mar. 1..........	2	50.9	2	52.5	2	53.8	2	51.2	2	52.6	2	54.1	2	55.5	2	53.0	2	54.4	2	55.9	3.94
Mar. 15.........	1	55.7	1	57.1	1	58.6	1	55.9	1	57.4	1	58.9	2	0.3	1	57.7	1	59.2	2	0.7	3.93
Apr. 1..........	0	48.9	0	50.3	0	51.8	0	49.3	0	50.6	0	52.1	0	53.5	0	50.9	0	52.4	0	53.9	3.93
Apr. 15.........	23	49.9	23	51.3	23	52.8	23	50.3	23	51.6	23	53.1	23	54.6	23	52.0	23	53.4	23	54.9	3.92
May 1..........	22	47.2	22	48.6	22	50.0	22	47.5	22	48.9	22	50.3	22	51.8	22	49.3	22	50.7	22	52.2	3.92
May 15.........	21	52.3	21	53.7	21	55.1	21	52.6	21	54.0	21	55.4	21	56.9	21	54.4	21	55.8	21	57.3	3.92
Jun. 1..........	20	45.5	20	46.9	20	48.3	20	45.8	20	47.2	20	48.6	20	50.1	20	47.6	20	49.0	20	50.5	3.91
Jun. 15.........	19	50.7	19	52.1	19	53.5	19	51.0	19	52.4	19	53.8	19	55.3	19	52.8	19	54.2	19	55.7	3.91
Jul. 1..........	18	48.1	18	49.5	18	50.9	18	48.4	18	49.8	18	51.2	18	52.7	18	50.2	18	51.6	18	53.1	3.91
Jul. 15.........	17	53.4	17	54.8	17	56.2	17	53.7	17	55.1	17	56.5	17	58.0	17	55.5	17	56.9	17	58.4	3.92
Aug. 1..........	16	46.8	16	48.2	16	49.6	16	47.1	16	48.5	16	49.9	16	51.4	16	48.9	16	50.3	16	51.8	3.92
Aug. 15.........	15	52.0	15	53.4	15	54.8	15	52.3	15	53.7	15	55.1	15	56.6	15	54.1	15	55.5	15	57.0	3.92
Sept. 1.........	14	45.3	14	46.7	14	48.1	14	45.6	14	47.0	14	48.5	14	49.9	14	47.4	14	48.9	14	50.3	3.92
Sept. 15........	13	50.4	13	51.8	13	53.2	13	50.7	13	52.1	13	53.6	13	55.0	13	52.5	13	54.0	13	55.4	3.93
Oct. 1..........	12	47.6	12	49.0	12	50.4	12	47.9	12	49.3	12	50.8	12	52.2	12	49.7	12	51.2	12	52.6	3.93
Oct. 15.........	11	52.6	11	54.0	11	55.4	11	52.9	11	54.3	11	55.8	11	57.2	11	54.7	11	56.2	11	57.6	3.93
Nov. 1..........	10	45.8	10	47.2	10	48.6	10	46.1	10	47.5	10	49.0	10	50.4	10	47.9	10	49.4	10	50.8	3.94
Nov. 15.........	9	50.7	9	52.1	9	53.5	9	51.0	9	52.4	9	53.9	9	55.3	9	52.8	9	54.3	9	55.7	3.94
Dec. 1..........	8	47.5	8	48.9	8	50.3	8	47.8	8	49.2	8	50.7	8	52.1	8	49.6	8	51.1	8	52.5	3.94
Dec. 15.........	7	52.3	7	53.7	7	55.1	7	52.6	7	54.0	7	55.5	7	56.9	7	54.4	7	55.9	7	57.3	3.94

objective glass to illuminate the cross-hairs. Set both clamps, and with either tangent screw set the vertical cross-hair exactly on the star. The star will appear to be moving toward the left or toward the right according as it is approaching upper or lower culmination. Follow it in its motion by turning the tangent screw until the exact time of culmination (which, preferably, should be called out by an assistant). This completes the observation of the star. Now depress the telescope, direct it to a point on the ground about 400 or 500 ft. from the instrument, and have an assistant drive a tack in the top of a stake in line with the line of sight; this completes the operation. The line between the two stakes is a true north-and-south line, or true meridian.

Time of Culmination of Polaris.—The accompanying table contains the times of upper culmination of Polaris for the dates given. The lower culmination occurs nearly $11^h\ 58^m$ before and after the upper culmination, and can be determined from the latter. In the table the extreme right-hand column contains the difference between the times of culmination for any two succeeding days. Each difference applies to any day between the date horizontally opposite that difference in the left-hand column, and the following date. Thus, the difference 3.95^m, which is horizontally opposite Jan. 1, indicates that, between Jan. 1 and Jan. 15, the time of culmination decreases by 3.95 min. per day. For instance, the time of culmination on Jan. 8 is obtained by subtracting from the time of culmination for Jan. 1 the product $3.95^m \times 7$, or 27.65^m, the number of days elapsed from Jan. 1 to Jan. 8 being 7.

It should be borne in mind that the times given in the table are mean local times counted in the astronomical way; that is, from 0^h to 24^h, beginning at noon.

EXAMPLE.—Find the time of upper culmination of Polaris on Sept. 6, 1913.

SOLUTION.—Referring to the table,

Upper culmination, Sept. 1, 1913 .... $= 14^h 45.3^m$
Difference for 1 da............ $= 3.92^m$
Correction for 5 da......... $= 3.92^m \times 5 = 19.6^m$
Time of culmination on Sept. 6......... $= 14^h 25.7^m$

This means that upper culmination will occur when $14^h 25.7^m$ has elapsed since local noon Sept. 6; that is, at 2^h 25.7^m A.M., Sept. 7.

DETERMINATION BY OBSERVING POLARIS AT ELONGATION

When a star is at its extreme westerly or easterly position, it is said to be at *western* or *eastern elongation*. This position with reference to the meridian of the place is determined by the angle that a vertical plane passing through the star and the point of observation is making with the meridian. This angle is called the *azimuth of the star*, and its values for Polaris, for the years 1913 to 1922 and latitudes 5° to 74°, are given in the accompanying table.

Polaris is at eastern elongation about 5^h 55^m before it reaches its upper culmination; and at western elongation, 5^h 55^m after upper culmination. The times of elongation can, therefore, be readily determined from those of culmination taken from the table.

EXAMPLE.—Find the time of western elongation of Polaris on Mar. 1, 1914.

SOLUTION.—On referring to the table, it is found that the upper culmination is at 2^h 52.5^m, local astronomical time, or 2^h 52.5^m, P. M., local civil time. Polaris is at western elongation 5^h 55^m later or at 8^h 47.5^m P. M. local civil time.

Making the Observation and Marking the Meridian.—Determine the approximate time of elongation as just explained. About 20 min. before that time, set the transit over a point properly marked, and level it carefully. Set the vernier at zero. Direct the telescope to the star, and, with both clamps set, follow the star by means of the lower tangent screw. If the star is approaching eastern elongation, it will be moving to the right; if western, to the left. About the time of elongation, it will be noticed that the star ceases to move horizontally, and that its image appears to follow the vertical cross-hair of the instrument. The star has then reached its elongation and the observation is completed. Take the azimuth from the table. Depress the telescope, and turn it through an angle equal to the azimuth, to the west or to the east, according as the star was

AZIMUTHS OF POLARIS AT ELONGATION

Lat. Deg.	Year																			
	1913		1914		1915		1916		1917		1918		1919		1920		1921		1922	
	Deg.	Min.	Deg.	Min.	Deg.	Min.	Deg.	Min.	Deg.	Min.	Deg.	Min.	Deg.	Min.	Deg.	Min.	Deg.	Min.	Deg.	Min.
5	1	9.8	1	9.5	1	9.2	1	8.8	1	8.5	1	8.2	1	7.9	1	7.6	1	7.3	1	7.0
6		9.9		9.6		9.3		9.0		8.6		8.3		8.0		7.7		7.4		7.1
8		10.2		9.8		9.5		9.2		8.9		8.6		8.3		7.9		7.7		7.4
10	1	10.6	1	10.3	1	10.0	1	9.7	1	9.3	1	9.0	1	8.7	1	8.4	1	8.1	1	7.8
12		11.0		10.7		10.4		10.1		9.7		9.4		9.1		8.8		8.5		8.2
14		11.6		11.3		11.0		10.6		10.3		10.0		9.7		9.4		9.0		8.7
16		12.3		12.0		11.7		11.3		11.0		10.7		10.4		10.1		9.7		9.4
18		13.1		12.8		12.4		12.1		11.8		11.5		11.1		10.8		10.5		10.2
20	1	14.0	1	13.7	1	13.4	1	13.0	1	12.6	1	12.3	1	11.9	1	11.6	1	11.3	1	10.9
22		15.0		14.7		14.4		14.1		13.8		13.4		13.1		12.7		12.4		12.1
24		16.0		15.7		15.4		15.1		14.8		14.5		14.1		13.8		13.5		13.2
26		17.4		17.0		16.7		16.3		16.0		15.7		15.3		14.9		14.6		14.3
28		18.8		18.5		18.1		17.8		17.4		17.0		16.7		16.3		16.0		15.7

30	1	20.3	1	19.9	1	19.6	1	19.2	1	18.9	1	18.5	1	18.2	1	17.9	1	17.5	1	17.1
32		22.0		21.6		21.2		20.9		20.5		20.1		19.8		19.4		19.0		18.7
34		23.8		23.5		23.1		22.8		22.4		22.1		21.7		21.3		21.0		20.6
36		26.0		25.6		25.2		24.9		24.5		24.1		23.8		23.4		23.0		22.6
38		28.2		27.8		27.5		27.1		26.8		26.4		26.0		25.6		25.2		24.8
40	1	30.7	1	30.3	1	30.0	1	29.6	1	29.2	1	28.8	1	28.4	1	28.0	1	27.6	1	27.2
42		33.6		33.2		32.8		32.4		32.0		31.6		31.1		30.7		30.3		29.9
44		36.7		36.3		35.8		35.4		35.0		34.6		34.1		33.6		33.2		32.8
46		40.1		39.7		39.2		38.8		38.4		37.9		37.5		37.1		36.6		36.2
48		43.9		43.4		43.0		42.5		42.2		41.8		41.3		40.8		40.3		39.9
50	1	48.1	1	47.7	1	47.2	1	46.8	1	46.3	1	45.9	1	45.4	1	44.9	1	44.4	1	43.9
52		52.9		52.4		52.0		51.5		51.0		50.5		50.0		49.5		49.0		48.5
54		58.3		57.8		57.3		56.8		56.3		55.8		55.2		54.7		54.2		53.7
56	2	4.4	2	3.8	2	3.3	2	2.7	2	2.2	2	1.7	2	1.1	2	0.5	2	0.0		59.4
58		11.3		10.7		10.1		9.6		9.0		8.4		7.8		7.2		6.6	2	6.0
60	2	19.0	2	18.4	2	17.8	2	17.2	2	16.6	2	16.0	2	15.3	2	14.7	2	14.0	2	13.4
62		28.1		27.4		26.7		26.0		25.4		24.7		24.0		23.4		22.7		22.0
64		38.7		38.0		37.3		36.5		35.9		35.2		34.5		33.8		33.0		32.3
66		50.9		50.1		49.4		48.6		47.8		47.0		46.2		45.5		44.7		43.9
68	3	5.7	3	4.8	3	4.0	3	3.1	3	2.2	3	1.3	3	0.4		59.6		58.7		57.7
70	3	22.8	3	21.8	3	20.8	3	19.9	3	18.9	3	17.9	3	16.9	3	15.9	3	15.0	3	14.0
72		45.2		44.2		43.1		42.0		41.0		40.0		38.9		37.8		36.8		35.7
74	4	12.1	4	11.0	4	9.8	4	8.7	4	7.5	4	6.4	4	5.2	4	4.1	4	3.0	4	1.8

at eastern or western elongation. The line of sight will then be directed along the true meridian, and by marking another point 400 or 500 ft. from that occupied by the instrument, the direction of the true meridian will be established.

This is the most accurate method of determining the true meridian, and, where possible, should be used in preference to others.

DETERMINATION BY SOLAR OBSERVATION

One of the most convenient methods of determining the meridian is to measure the altitude of the sun at any hour angle with a transit. At the same time that the altitude is measured, determine also the horizontal angle between the sun and a fixed object, or reference mark. Then, the azimuth of the sun is calculated by the formula that follows. The azimuth of the reference mark is then equal to the algebraic sum of the azimuth of the sun and the measured angle between the sun and the mark. Finally, the true north-and-south line may be located from the azimuth of the reference mark.

Formula for Azimuth of the Sun.—Let a represent the required azimuth counted from north toward east; z, the zenith distance of the sun, which is equal to 90° minus the altitude; δ, declination of the sun; and ϕ, the latitude of the observer. Then,

$$\sin\frac{a}{2} = \sqrt{\frac{\cos\frac{1}{2}(z+\phi+\delta)\sin\frac{1}{2}(z+\phi-\delta)}{\sin z\cos\phi}}$$

Two values of $\frac{a}{2}$ will correspond to the computed $\sin\frac{a}{2}$; one angle will be acute and the other obtuse. The acute angle should be used for morning observations and the obtuse for afternoon observations.

Values of δ and ϕ.—The method just described requires that the declination of the sun at the time of observation, and the latitude of the place be known. The declination of the sun for every day of the year at the instant of Washington noon, together with the hourly change, is given in the Ephemeris, and has to be reduced to the time of observation as follows:

Rule.—*Change the local time to Washington time by adding algebraically to the former the longitude of the place counted from*

Washington. Take from the Ephemeris the declination corresponding to the preceding Washington noon and add algebraically the product of the hourly change by the time elapsed since Washington noon.

EXAMPLE.—Find the true declination of the sun for 9 A. M. Jan. 5, 1903, at Philadelphia.

SOLUTION.—Jan. 5, 9 A. M., civil time = Jan. 4, 21^h, astronomical time. The longitude of Philadelphia is $-7^m\ 37^s = -.127^h$. The Washington time corresponding to 9 A. M. is $21^h - .127^h = 20.873^h$. From the Ephemeris the declination at Washington at noon Jan. 4 is $-22°\ 47'\ 43''$, and the hourly change is $15.06''$. The algebraic increase is, therefore, $15.06 \times 20.873 = 5'\ 14''$; thus, the declination at 9 A. M. is $-22°\ 47'\ 43'' + 5'\ 14'' = -22°\ 42'\ 29''$.

DETERMINATION OF LATITUDE, AND CORRECTIONS FOR ALTITUDE

Approximate Determination of Latitude From Polaris.—In nearly all methods of determining the true meridian, the latitude of the place of observation must be known, at least approximately. In the majority of cases, the latitude can be taken from a map or book of reference. In case this cannot be done, a sufficiently close value may be obtained by measuring with a transit the altitude of Polaris, which is very nearly (within about 1°) equal to the latitude of the place.

This method of determining latitude is founded on the following very simple and useful principle:

Principle.—*The latitude of any place on the earth's surface is equal to the altitude of the pole with respect to the horizon of that place.*

For more accurate work, the tables given in the Ephemeris, entitled, For Finding the Latitude by Polaris, may be used. The simple directions for using them are there given in full.

Latitude by Solar Observation.—Latitude may be determined by measuring the sun's altitude, with the sextant or transit, at the instant of its passage across the meridian; that is, at apparent noon. The time of apparent noon may be determined by adding algebraically the equation of time to the noon of local mean time, as previously explained. Then begin the observations about 15 min. before apparent noon and repeat them every minute or two. At first the altitude will be increasing; then, it will be decreasing. The maximum altitude obtained will be the apparent meridian altitude. To this the corrections that follow must be applied, giving the true altitude. The true altitude is then subtracted from 90°, and the remainder is the zenith distance. The latitude is then equal to the algebraic sum of the zenith distance and the declination of the sun at the instant of apparent noon.

Corrections for Altitude.—The observed altitude of a heavenly body must be corrected: (1) for index error, (2) refraction, (3) parallax, and (4) semi-diameter.

1. The index error is a purely instrumental error and is explained under the heading Hydrographic Surveying.

2. *Refraction* is the change of direction of the rays of light when they pass from one medium into another of different density. Its amount for different altitudes is given in the accompanying table. It is subtractive. When the altitude is less than about 8° to 10°, the refraction becomes so uncertain that the measurement is of no value for accurate work.

3. *Parallax* is the difference in direction of a heavenly body as actually observed and the direction it would have if seen from the earth's center. This correction is necessary when

SUN'S PARALLAX IN ALTITUDE TO BE APPLIED TO ALL MEASURED ALTITUDES OF THE SUN

(Additive to observed altitude)

Altitude Degrees	Parallax Seconds	Altitude Degrees	Parallax Seconds	Altitude Degrees	Parallax Seconds
0	9	40	7	69	3
6	9	45	6	72	3
12	9	48	6	75	2
16	8	51	5	78	2
20	8	54	5	81	1
25	8	57	5	84	1
30	8	60	4	87	0
34	7	63	4	90	0
36	7	66	3		

MEAN REFRACTION TO BE APPLIED TO ALL MEASURED ALTITUDES

(*Subtractive from apparent altitude*)

App. Altitude	Refraction	App. Altitude	Refraction	App. Altitude	Refraction	App. Altitude	Refraction	App. Altitude	Refraction
° ′	′ ″	° ′	′ ″	° ′	′ ″	° ′	′ ″	° ′	′ ″
0 0	33 0			6 40	7 40	10 0	5 15	16 40	3 8
						10 10	5 10	16 50	3 6
		3 30	13 6			10 20	5 5	17 0	3 4
						10 30	5 0	17 10	3 3
				7 0	7 20	10 40	4 56	17 20	3 1
						10 50	4 51	17 30	2 59
						11 0	4 47	17 40	2 57
						11 10	4 43	17 50	2 55
		4 0	11 51	7 20	7 2	11 20	4 39	18 0	2 54
						11 30	4 34	18 10	2 52
						11 40	4 31	18 20	2 51
						11 50	4 27	18 30	2 49
1 0	24 29			7 40	6 45	12 0	4 23	18 40	2 47
						12 10	4 20	18 50	2 46
		4 30	10 48			12 20	4 16	19 0	2 44
						12 30	4 13	19 10	2 43
				8 0	6 29	12 40	4 9	19 20	2 41
						12 50	4 6	19 30	2 40
				8 10	6 22	13 0	4 3	19 40	2 38
						13 10	4 0	19 50	2 37
		5 0	9 54	8 20	6 15	13 20	3 57	20 0	2 35
						13 30	3 54	20 10	2 34
				8 30	6 8	13 40	3 51	20 20	2 32
						13 50	3 48	20 30	2 31
2 0	18 35	5 20	9 23	8 40	6 1	14 0	3 45	20 40	2 29
						14 10	3 43	20 50	2 28
				8 50	5 55	14 20	3 40	21 0	2 27
						14 30	3 38	21 10	2 26
		5 40	8 54	9 0	5 48	14 40	3 35	21 20	2 25
						14 50	3 33	21 30	2 24
				9 10	5 42	15 0	3 30	21 40	2 23
						15 10	3 28	21 50	2 21
		6 0	8 28	9 20	5 36	15 20	3 26	22 0	2 20
						15 30	3 24	22 10	2 19
				9 30	5 31	15 40	3 21	22 20	2 18
						15 50	3 19	22 30	2 17
3 0	14 36	6 20	8 3	9 40	5 25	16 0	3 17	22 40	2 16
						16 10	3 15	22 50	2 15
				9 50	5 20	16 20	3 12	23 0	2 14
						16 30	3 10	23 10	2 13

TABLE—(*Continued*)

App. Altitude	Refraction	App. Altitude	Refraction	App. Altitude	Refraction	App. Altitude	Refraction	App. Altitude	Refraction
° '	' "	° '	' "	° '	' "	° '	' "	° '	' "
23 20	2 12	26 40	1 53	34 0	1 24	48 0	0 51	68 0	0 23
23 30	2 11	26 50	1 52	34 30	1 23	49 0	0 49	69 0	0 22
23 40	2 10	27 0	1 51	35 0	1 21	50 0	0 48	70 0	0 21
23 50	2 9	27 15	1 50	35 30	1 20	51 0	0 46	71 0	0 19
24 0	2 8	27 30	1 49	36 0	1 18	52 0	0 44	72 0	0 18
24 10	2 7	27 45	1 48	36 30	1 17	53 0	0 43	73 0	0 17
24 20	2 6	28 0	1 47	37 0	1 16	54 0	0 41	74 0	0 16
24 30	2 5	28 15	1 46	37 30	1 14	55 0	0 40	75 0	0 15
24 40	2 4	28 30	1 45	38 0	1 13	56 0	0 38	76 0	0 14
24 50	2 3	28 45	1 44	38 30	1 11	57 0	0 37	77 0	0 13
25 0	2 2	29 0	1 42	39 0	1 10	58 0	0 35	78 0	0 12
25 10	2 1	29 30	1 40	39 30	1 9	59 0	0 34	79 0	0 11
25 20	2 0	30 0	1 38	40 0	1 8	60 0	0 33	80 0	0 10
25 30	1 59	30 30	1 37	41 0	1 5	61 0	0 32	81 0	0 9
25 40	1 58	31 0	1 35	42 0	1 3	62 0	0 30	82 0	0 8
25 50	1 57	31 30	1 33	43 0	1 1	63 0	0 29	83 0	0 7
26 0	1 56	32 0	1 31	44 0	0 59	64 0	0 28	84 0	0 6
26 10	1 55	32 30	1 30	45 0	0 57	65 0	0 26	86 0	0 4
26 20	1 55	33 0	1 29	46 0	0 55	66 0	0 25	88 0	0 2
26 30	1 54	33 30	1 26	47 0	0 53	67 0	0 24	90 0	0 0

the sun is observed; its values for different altitudes are given in the accompanying table. It is additive.

4. The correction for *semi-diameter* is also necessary when the sun is observed, owing to the fact that either the upper or the lower edge of the disk, instead of the center, is observed. This correction may be taken from the Ephemeris in the same manner as the sun's declination. For the purpose of ordinary calculations, however, this may be taken from the following table:

Time of year (approx.)	Jan. 1,	Apr. 1,	July 1,	Oct. 1
Sun's semi-diameter	16' 18"	16' 2"	15' 45"	16' 2"

It is additive when the lower limb is observed, and subtractive when the upper one is observed.

Corrections for Observation of the Sun for Azimuth.—When the sun is observed for azimuth, a correction for semi-diameter must also be applied to the reading of the horizontal circle; this may be found by dividing the correction for altitude by the cosine of the sun's altitude. This correction is to be added to the reading of the horizontal circle if the hair is placed tangent to the left edge of the sun, and subtracted from the reading of the horizontal circle if the hair is placed tangent to the right edge of the sun.

In making observations of the sun for azimuth, the errors of adjustment, the index error, and the correction for semidiameter may be eliminated by the following method, which assumes that the vertical circle of the transit is complete.

The instrument is set up with the horizontal plate reading 0° when sighting at the azimuth mark. For forenoon work, the sun should be so sighted that it occupies position *1*, Fig. 1,

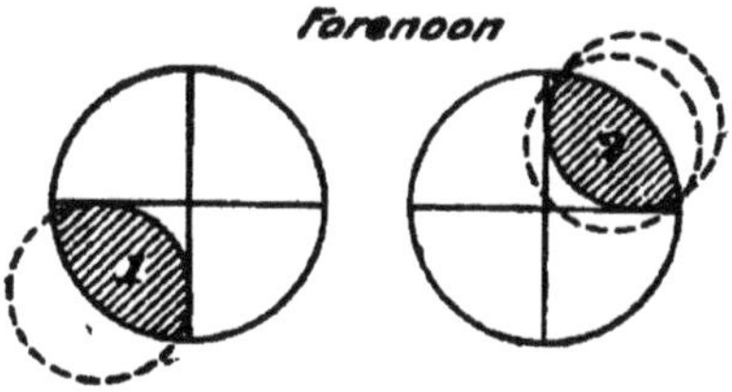

Fig. 1

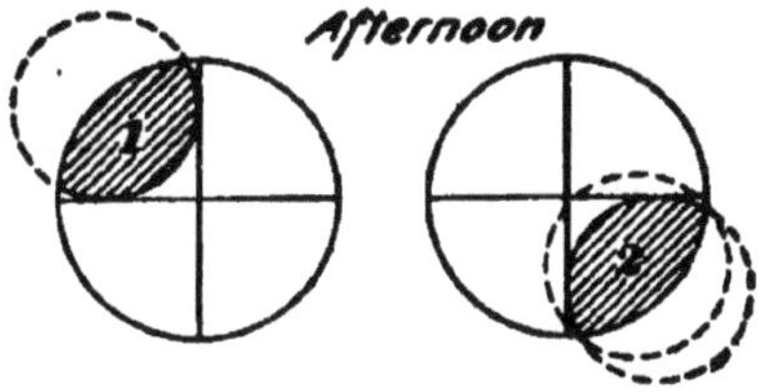

Fig. 2

with reference to the cross-hairs. The time, vertical angle, and horizontal angle are noted. Then the upper plate is loosened, the instrument turned 180° in azimuth, the telescope inverted, and the sun sighted again, as in position *2*, Fig. 1. In position *1*, the sun is moving toward both hairs; in position *2*, the telescope should be set approximately as shown by the dotted circle, so that the sun will clear both hairs at the same instant. For afternoon work, the positions shown in Fig. 2 should be used. The observations are taken in *pairs;* if the second observation of a pair cannot be obtained promptly after the first one (owing to a passing cloud, or some other cause), the first must be ignored and considered as useless.

It should be noted that the reversal of the transit between the observations eliminates the index error of the vertical circle, the error of level in the horizontal axis of the telescope, and the error of collimation of the telescope. By sighting in diagonal corners of the field of view and taking the mean of the observations, the corrections (both horizontal and vertical) due to the semi-diameter of the sun are eliminated. To simplify the notes, 180° should be added to (or subtracted from) the horizontal plate reading when the instrument is inverted.

EXAMPLE.—The following measurements were taken in the manner just described. The four means of the circle readings were formed in the field. The declination of the sun was −9° 30′ 5″, and the approximate latitude +39° 57′. Find the azimuth of the reference mark.

Telescope	Time P. M.	Vertical Circle	Horizontal Circle
Direct..........	3:27	19° 39′ 00″	99° 52′ 00″
Inverted........	3:29	19 52 00	99 49 00
Mean...........	3:28	19 45 30	99 50 30
Direct..........	3:32	18 46 00	100 55 30
Inverted........	3:34	19 3 00	100 49 00
Mean...........	3:33	18 54 30	100 52 15
Direct..........	3:36	18 4 30	101 46 00
Inverted........	3:38	18 23 30	101 35 00
Mean...........	3:37	18 14 00	101 40 30
Direct..........	3:40	17 26 30	102 29 30
Inverted........	3:42	17 43 00	102 21 00
Mean...........	3:41	17 34 45	102 25 15

SOLUTION.—

Mean of the four vertical circle readings..	18° 37′ 11″
Refraction.........................	−2 48
Parallax..........................	+8
True altitude of center...............	18° 34′ 31″
Zenith distance = 90° − true altitude.....	71° 25′ 29″

To find the azimuth of the sun: $z = 71° 25' 29''$; $\phi = 39° 57' 0''$; $\delta = -9° 30' 5''$; $\frac{1}{2}(z+\phi+\delta) = 50° 56' 12''$; $\frac{1}{2}(z+\phi-\delta) = 60° 26' 17''$. Substituting these values in the formula for the azimuth of the sun,

$$\sin \tfrac{1}{2} a = \sqrt{\frac{\cos 50° 56' 12'' \sin 60° 26' 17''}{\sin 71° 25' 29'' \cos 39° 57'}}$$

The two values of $\tfrac{1}{2}\, a$ are 60° 17′ 15″ and 119° 42′ 45″ (=180°−60° 17′ 15″). As the observations were made in the afternoon, the obtuse angle should be used. This gives $a = 2 \times 119° 42' 45'' = 239° 25' 30''$. The mean of the four horizontal readings is 101° 12′ 8″. Subtracting this from the azimuth of the sun, the azimuth of the reference mark is found to be 239° 25′ 30″ − 101° 12′ 8″ = 138° 13′ 22″.

RAILROAD CURVES

CIRCULAR CURVES

DEFINITIONS

The line of a railroad consists of a series of straight lines connected by curves. Each two adjacent lines are united by a curve having the radius best adapted to the conditions of the surface. The straight lines are called *tangents*, because they are tangent to the curves that unite them.

Railroad curves are usually circular and are divided into three general classes, namely, simple, compound, and reverse curves.

A *simple curve* is a curve having but one radius, as the curve *AB*, Fig. 1, whose radius is *AC*.

A *compound curve* is a continuous curve composed of two or more arcs of different radii, as the curve *CDEF*, Fig. 2, which

FIG. 1

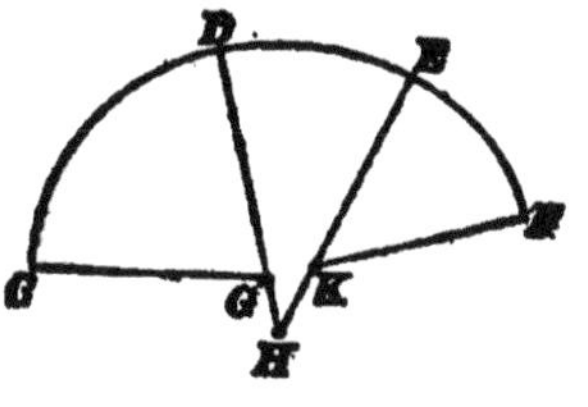

FIG. 2

FIG. 3

is composed of the arcs *CD*, *DE*, and *EF*, whose respective radii are *GC*, *HD*, and *KE*. In the general class of compound curves may be included what are known as ***easement curves***, ***transition curves***, and ***spiral curves***, now used very generally on the more important railroads.

A ***reverse curve*** is a continuous curve composed of the arcs of two circles of the same or different radii, the centers of which lie on opposite sides of the curve, as in Fig. 3. The two arcs composing the curve meet at a common point or point of reversal *M*, at which point they are tangent to a common line perpendicular to the line joining their centers. Reverse curves are becoming less common on railroads of standard gauge.

GEOMETRY OF CIRCULAR CURVES

The following principles of geometry are of special importance as relating to curves:

1. A tangent to a circle is perpendicular to the radius at its tangent point. Thus, in Fig. 4, *AF* is perpendicular to *BO* at its tangent point *B*, and *ED* is perpendicular to *CO* at *C*.

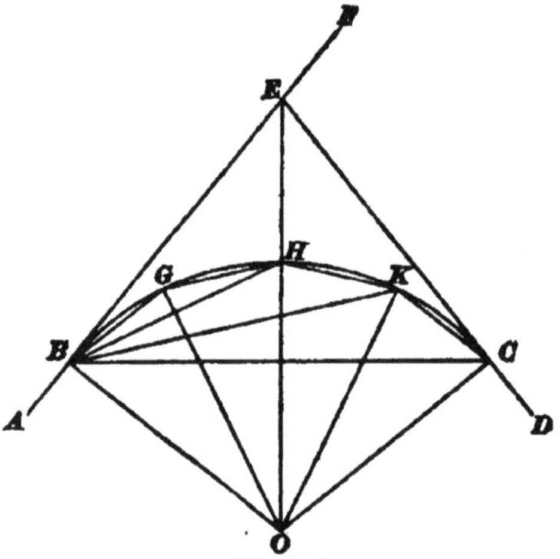

FIG. 4

2. Two tangents to a circle from any point without the circle are equal in length, and make equal angles with the chord joining their points of tangency. Thus *BE* and *CE* are equal, and the angles *EBC* and *ECB* are equal.

3. An angle not exceeding 90° formed by a chord and the tangent at one of its extremities is equal to one-half the central angle subtended by the chord. Thus, the angle *EBC* = *ECB* = one-half *BOC*.

4. An angle not exceeding 90° having its vertex in the circumference of a circle and subtended by a chord of the circle, is equal to one-half the central angle subtended by the chord. Thus, the angle *GBH*, whose vertex *B* is in the

circumference, is subtended by the chord *GH* and is equal to one-half the central angle *GOH*, subtended by the same chord *GH*.

5. Equal chords of a circle subtend equal angles at its center and also in its circumference, if the angles lie in corresponding segments of the circle. Thus, if *BG*, *GH*, *HK*, and *KC* are equal, $BOG = GOH$, $GBH = HBK$, etc.

6. The angle *FEC*, called the ***angle of intersection***, of two tangents of a circle is equal to the central angle subtended by the chord joining the two points of tangency. Thus, the angle $CEF = BOC$.

7. A radius that bisects any chord of a circle is perpendicular to the chord.

8. A chord subtending an arc of 1° in a circle having a radius = 100 ft. is very closely equal to 1.745 ft.

ELEMENTS AND METHODS OF LAYING OUT A CIRCULAR CURVE

The ***degree of curvature*** of a curve is the central angle subtending a chord of 100′. Thus, if, in Fig. 4, the chord *BG* is 100 ft. long and the angle *BOG* is 1°, the curve is called a ***one-degree curve;*** but if, with the same length of chord, the angle *BOG* is 4°, the curve is called a ***four-degree curve.***

The ***deflection angle*** of a chord is the angle formed between any chord of a curve and a tangent to the curve at one extremity of the chord. It is equal to one-half the central angle subtended by the chord. The deflection angle for a chord of 100 ft. is called the ***regular deflection angle***, and is equal to one-half the degree of curvature. The deflection angle for a ***subchord***—that is, for a chord less than 100 ft.—is equal to one-half the degree of curvature multiplied by the length of the subchord expressed in chords of 100 ft. The length c of a subchord or of any chord is given by the equation

$$c = 2R \sin D,$$

in which R is the radius and D the deflection angle of that chord.

Relation Between Radius and Deflection Angle.—From the equation just given,

$$R = \frac{c}{2 \sin D}$$

TABLE OF RADII AND DEFLECTIONS

Degree °	Degree ′	Radii	Chord Deflection	Tangent Deflection	Degree °	Degree ′	Radii	Chord Deflection	Tangent Deflection
					3	0	1,910.08	5.235	2.618
0	5	68,754.94	.145	.073		5	1,858.47	5.381	2.690
	10	34,377.48	.291	.145		10	1,809.57	5.526	2.763
	15	22,918.33	.436	.218		15	1,763.18	5.672	2.836
	20	17,188.76	.582	.291		20	1,719.12	5.817	2.908
	25	13,751.02	.727	.364		25	1,677.20	5.962	2.981
	30	11,459.19	.873	.436		30	1,637.28	6.108	3.054
	35	9,822.18	1.018	.509		35	1,599.21	6.253	3.127
	40	8,594.41	1.164	.582		40	1,562.88	6.398	3.199
	45	7,639.49	1.309	.654		45	1,528.16	6.544	3.272
	50	6,875.55	1.454	.727		50	1,494.95	6.689	3.345
	55	6,250.51	1.600	.800		55	1,463.16	6.835	3.417
1	0	5,729.65	1.745	.873	4	0	1,432.69	6.980	3.490
	5	5,288.92	1.891	.945		5	1,403.46	7.125	3.563
	10	4,911.15	2.036	1.018		10	1,375.40	7.271	3.635
	15	4,583.75	2.182	1.091		15	1,348.45	7.416	3.708
	20	4,297.28	2.327	1.164		20	1,322.53	7.561	3.781
	25	4,044.51	2.472	1.236		25	1,297.58	7.707	3.853
	30	3,819.83	2.618	1.309		30	1,273.57	7.852	3.926
	35	3,618.80	2.763	1.382		35	1,250.42	7.997	3.999
	40	3,437.87	2.909	1.454		40	1,228.11	8.143	4.071
	45	3,274.17	3.054	1.527		45	1,206.57	8.288	4.144
	50	3,125.36	3.200	1.600		50	1,185.78	8.433	4.217
	55	2,989.48	3.345	1.673		55	1,165.70	8.579	4.289
2	0	2,864.93	3.490	1.745	5	0	1,146.28	8.724	4.362
	5	2,750.35	3.636	1.818		5	1,127.50	8.869	4.435
	10	2,644.58	3.781	1.891		10	1,109.33	9.014	4.507
	15	2,546.64	3.927	1.963		15	1,091.73	9.160	4.580
	20	2,455.70	4.072	2.036		20	1,074.68	9.305	4.653
	25	2,371.04	4.218	2.109		25	1,058.16	9.450	4.725
	30	2,292.01	4.363	2.181		30	1,042.14	9.596	4.798
	35	2,218.09	4.508	2.254		35	1,026.60	9.741	4.870
	40	2,148.79	4.654	2.327		40	1,011.51	9.886	4.943
	45	2,083.68	4.799	2.400		45	996.87	10.031	5.016
	50	2,022.41	4.945	2.472		50	982.64	10.177	5.088
	55	1,964.64	5.090	2.545		55	968.81	10.322	5.161

TABLE—(*Continued*)

Degree ° ′	Radii	Chord Deflection	Tangent Deflection	Degree ° ′	Radii	Chord Deflection	Tangent Deflection
6 0	955.37	10.467	5.234	9 0	637.27	15.692	7.846
5	942.29	10.612	5.306	5	631.44	15.837	7.918
10	929.57	10.758	5.379	10	625.71	15.982	7.991
15	917.19	10.903	5.451	15	620.09	16.127	8.063
20	905.13	11.048	5.524	20	614.56	16.272	8.136
25	893.39	11.193	5.597	25	609.14	16.417	8.208
30	881.95	11.339	5.669	30	603.80	16.562	8.281
35	870.79	11.484	5.742	35	598.57	16.707	8.353
40	859.92	11.629	5.814	40	593.42	16.852	8.426
45	849.32	11.774	5.887	45	588.36	16.996	8.498
50	838.97	11.919	5.960	50	583.38	17.141	8.571
55	828.88	12.065	6.032	55	578.49	17.286	8.643
7 0	819.02	12.210	6.105	10 0	573.69	17.431	8.716
5	809.40	12.355	6.177	10	564.31	17.721	8.860
10	800.00	12.500	6.250	20	555.23	18.011	9.005
15	790.81	12.645	6.323	30	546.44	18.300	9.150
20	781.84	12.790	6.395	40	537.92	18.590	9.295
25	773.07	12.936	6.468	50	529.67	18.880	9.440
30	764.49	13.081	6.540	11 0	521.67	19.169	9.585
35	756.10	13.226	6.613	10	513.91	19.459	9.729
40	747.89	13.371	6.685	20	506.38	19.748	9.874
45	739.86	13.516	6.758	30	499.06	20.038	10.019
50	732.01	13.661	6.831	40	491.96	20.327	10.164
55	724.31	13.806	6.903	50	485.05	20.616	10.308
8 0	716.78	13.951	6.976	12 0	478.34	20.906	10.453
5	709.40	14.096	7.048	10	471.81	21.195	10.597
10	702.18	14.241	7.121	20	465.46	21.484	10.742
15	695.09	14.387	7.193	30	459.28	21.773	10.887
20	688.16	14.532	7.266	40	453.26	22.063	11.031
25	681.35	14.677	7.338	50	447.40	22.352	11.176
30	674.69	14.822	7.411	13 0	441.68	22.641	11.320
35	668.15	14.967	7.483	10	436.12	22.930	11.465
40	661.74	15.112	7.556	20	430.69	23.219	11.609
45	655.45	15.257	7.628	30	425.40	23.507	11.754
50	649.27	15.402	7.701	40	420.23	23.796	11.898
55	643.22	15.547	7.773	50	415.19	24.085	12.043

TABLE—(*Continued*)

Degree	Radii	Chord Deflection	Tangent Deflection	Degree	Radii	Chord Deflection	Tangent Deflection
° ′				° ′			
14 0	410.28	24.374	12.187	17 0	338.27	29.562	14.781
10	405.47	24.663	12.331	10	335.01	29.850	14.925
20	400.78	24.951	12.476	20	331.82	30.137	15.069
30	396.20	25.240	12.620	30	328.68	30.425	15.212
40	391.72	25.528	12.764	40	325.60	30.712	15.356
50	387.34	25.817	12.908	50	322.59	31.000	15.500
15 0	383.06	26.105	13.053	18 0	319.62	31.287	15.643
10	378.88	26.394	13.197	10	316.71	31.574	15.787
20	374.79	26.682	13.341	20	313.86	31.861	15.931
30	370.78	26.970	13.485	30	311.06	32.149	16.074
40	366.86	27.258	13.629	40	308.30	32.436	16.218
50	363.02	27.547	13.773	50	305.60	32.723	16.361
16 0	359.26	27.835	13.917	19 0	302.94	33.010	16.505
10	355.59	28.123	14.061	10	300.33	33.296	16.648
20	351.98	28.411	14.205	20	297.77	33.583	16.792
30	348.45	28.699	14.349	30	295.25	33.870	16.935
40	344.99	28.986	14.493	40	292.77	34.157	17.078
50	341.60	29.274	14.637	50	290.33	34.443	17.222

If D_{100} is the deflection angle for a chord of 100 ft., then

$$R = \frac{50}{\sin D_{100}}$$

For a 1° curve, $D_{100} = 30'$ and $R = 5{,}730$, nearly. For curves less than 10°, the radius may be taken as $\frac{5{,}730}{D_c}$, in which D_c is the degree of curvature. The accompanying table gives the length of the radius, in feet, for degrees of curvature ranging by intervals of 5′ and 10′ from 0′ to 20°.

Tangent Distance.—The point where a curve begins is called the *point of curve*, and is designated by the letters P. C.; and the point where the curve terminates is called the *point of tangency*, and is designated by the letters P. T. The point of intersection of the tangents is called the *point of intersection;* it is designated by the letters P. I.

The distance of the P. C. or P. T. from the P. I. is called the *tangent distance*, and the chord connecting the P. C. and P. T. of a curve is commonly called its *long chord*. This term is also applied to chords more than one station long.

If I denotes the angle of intersection and R the radius of the curve, then the tangent distance

$$T = R \tan \tfrac{1}{2} I$$

Laying Out a Curve With a Transit.—When the angle of intersection I has been measured and the degree of curve decided upon, the radius of the curve can be taken from the table of radii and deflections or it can be figured by the formula

$$R = \frac{5{,}730}{D_c}$$

The tangent distance is then computed and measured back on each tangent from the P. I., thus determining the P. C. and P. T. Subtracting the tangent distance from the station number of the P. I. will give the station number of the P. C. Ordinarily, this will not be an even or full station. The length of the curve is then computed by dividing the angle I by the degree of curve, the quotient giving the length of the curve in stations of 100 ft. and decimals thereof. After having found the length of the curve, compute the deflection angles for the chords joining the P. C. with all the station points; set the transit at the P. C.; set the vernier at zero, sight to the intersection point, and turn off successively the deflection angles, at the same time measuring the chords and marking the stations. The station of the P. T. is found by adding the length of curve in chords of 100 ft. to the station of the P. C.

If the entire curve cannot be run from the P. C. on account of obstructions to the view, run the curve as far as the stations are visible from the P. C. and run the remainder of the curve from the last station that can be seen. Suppose that in the 10° curve shown in Fig. 5 the station at H, 200 ft. from the P. C., which is at B, is the last point on the curve that can be set from the P. C. A plug is driven at H and centered carefully by a tack driven at the point. The transit is now moved forwards and set up at H. Since the deflection angle EBH is 10° to the right, an angle of 10° is turned to the left from zero and the vernier clamped. The instrument is then sighted to a flag at B, the lower clamp set, and by means of the lower tangent screw the cross-hairs are made to bisect the flag exactly. The vernier clamp is then loosened, the vernier set at zero, and the telescope plunged. The line of sight will then be on the tangent IP, and the deflection angles to K and C can be turned off from this tangent, and the stations at K and C located in the same manner that the stations at G and H were

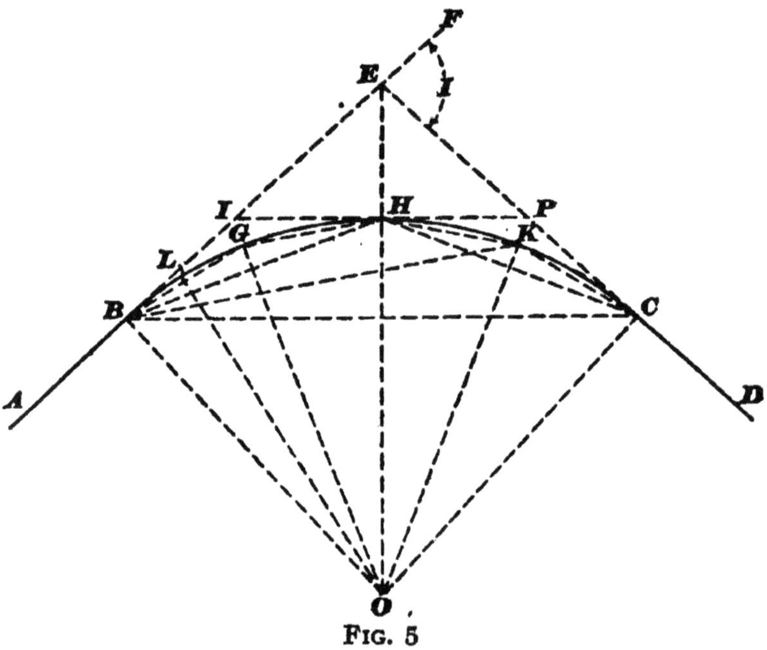

FIG. 5

located from B, because the angle at IHB between the tangent IH and the chord BH is equal to the angle EBH between the tangent EB and the same chord.

This method of setting the vernier for the backsight when the instrument is moved forwards to a new instrument point on the curve is sometimes called the *method by zero tangent.* The essential principle of the method is that the vernier always reads zero when the instrument is sighted on the tangent to the curve at the point where the instrument is set, and the deflection angles are made to read from the tangent to the curve at this point in the same manner as if this point were the P. C. of the curve.

Tangent and Chord Deflections.—Let *AB*, Fig. 6, be a tangent joining the curve *BCEH* at *B*. If the tangent *AB* is prolonged to *D*, the perpendicular distance *DC* from the tangent to the curve is called a *tangent deflection*. If the chord *BC* is prolonged to the point *G*, so that $CG = CE$, the distance

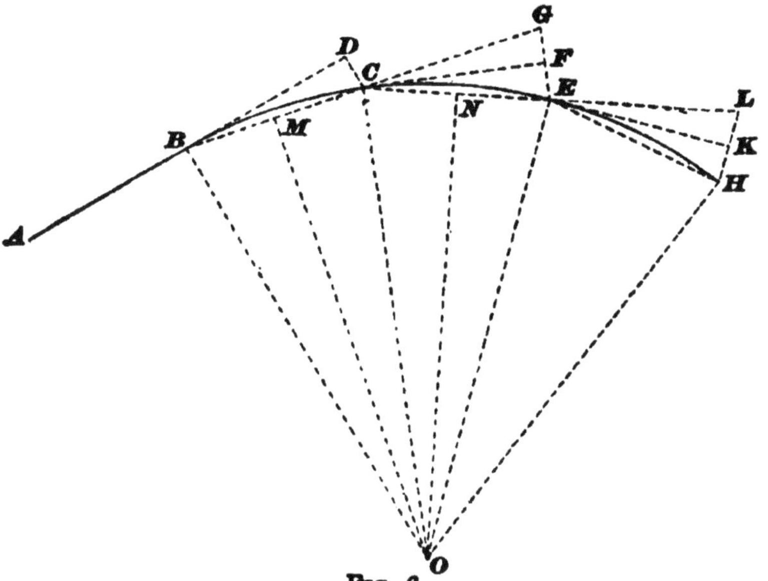

FIG. 6

GE is called a *chord deflection*. If the radius *R* of the curve and the length of the chord *c* are known, the tangent deflection *f* can be determined by the formula

$$f = \frac{c^2}{2\ R}$$

This formula can be used for any length of chord or radius. If $CE = BC$, the chord deflection $= 2f = \frac{c^2}{R}$. For this condition, the table of radii and deflections gives the chord deflection and tangent deflection for 100-ft. chords and for degrees of curvature varying by intervals of 5′ and 10′ from 5′ to 20°.

When the two chords preceding the station considered are of unequal lengths, the chord deflection $= \frac{c_1(c_1+c_2)}{2R}$, where c_1 is the length of the first chord and c_2 the length of the second chord preceding the station considered. When the tangent deflection f is known, the chord deflection

$$d_0 = f\left(1+\frac{c_2}{c_1}\right)$$

Special Values of Chord and Tangent Deflection.—For a chord of 100 ft. preceded by one of the same length the chord deflection for a 1° curve is 1.745; for a 2° curve, it is twice that amount, or 3.49; and so on. The tangent deflection, being half the chord deflection, will be .873 ft. for a 1° curve, 1.745 for a 2° curve, etc. The tangent deflection for a chord of any length equals the tangent deflection for a chord of 100 ft. multiplied by the square of the given chord expressed as the decimal part of a chord of 100 ft.

Application of Chord and Tangent Deflection.—Let it be required to restore center stakes on the 4° curve, Fig. 7, at each full station. The points A and B determine the direc-

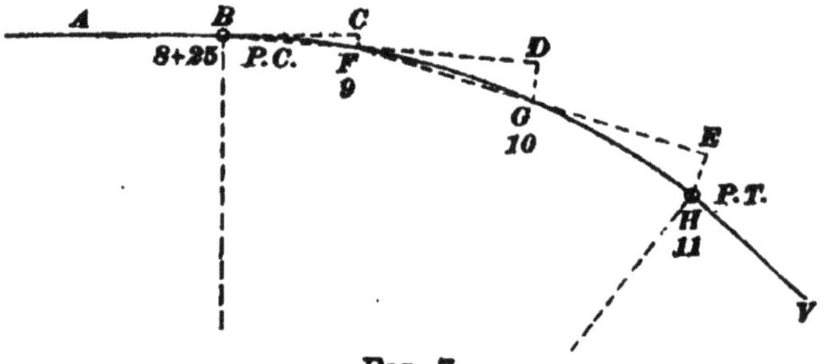

FIG. 7

tion of the tangent, the point B being the P. C., which is at Station 8+25. For a 4° curve the regular chord deflection for 100 ft. is $4\times1.745=6.98$ ft., and the tangent deflection is 3.49 ft. The distance from P. C. to the next full station is 75 ft.; hence, the tangent deflection $CF = .75^2\times3.49 = 1.96$ ft. The point F is found by first measuring 75 ft. from B, thus locating the point C in the line AB prolonged, then from C measuring $CF = 1.96$ ft., at right angles to BC; the point F

thus determined will be Station 9. Next the chord *BF* is prolonged 100 ft. to *D*; *BF* is only 75 ft., *DG* is computed from the preceding formula; thus, $d_0 = 3.49\ (1+\frac{75}{100}) = 6.11$. This distance is measured at right angles to *BD*; the point *G* thus determined will be Station 10. The point *H*, which is Station 11, and the P. T. of the curve, is determined in the same manner, except that, as the chords *FG* and *GH* are each 100 ft. long, the regular chord deflection of 6.98 ft. is used for *EH*. A stake is driven at each station thus located. Although a chord deflection is not at right angles to the chord theoretically, yet the deflection is so small, as compared with the length of the chord, that for curves of ordinary degree it is usually measured at right angles.

Middle Ordinate.—The *middle ordinate* of a chord is the ordinate to the curve at the middle point of the chord. The following formulas give the relation between the length of the chord *c*, the radius of the curve *R*, and the middle ordinate *m*.

$$m = R - \sqrt{R^2 - \frac{c^2}{4}}$$

$$c = 2\sqrt{2Rm - m^2}$$

$$R = \frac{c^2}{8m} + \frac{m}{2}$$

To Determine Degree of Curve From Middle Ordinate.—It is sometimes necessary to determine the radius or the degree of a curve in an existing track when no transit is available. By measuring the middle ordinate of any convenient chord, the degree of the curve can be calculated from the relative values of the ordinate and chord. As the track is likely not to be in perfect alinement, it is well to measure the middle ordinate of different chords in different parts of the curve; as, also, the middle ordinate of a chord measured to the inner rail will somewhat exceed the middle ordinate of the same chord measured to the outer rail, the ordinate of each chord should be measured to both rails and the average of the two taken as the value of the ordinate. Having measured the middle ordinate of one or more chords, the degree of curve D_0 can be found by the formula

$$D_0 = \frac{45{,}840\ m}{c^2}$$

The following rule is sometimes applied in determining the degree of curve:

Rule.—*Measure the middle ordinate to a chord of 67.71 ft.; express it in feet and decimals of a foot, and multiply by 10; the result will be the degree of the curve.*

Other Ordinates.—Any ordinate y to the curve at a distance a from the middle point of a chord may be determined by means of the formula:

$$y = \sqrt{R^2 - a^2} - R + m$$

By using long chords, a curve may be laid out or obstacles passed by means of ordinates.

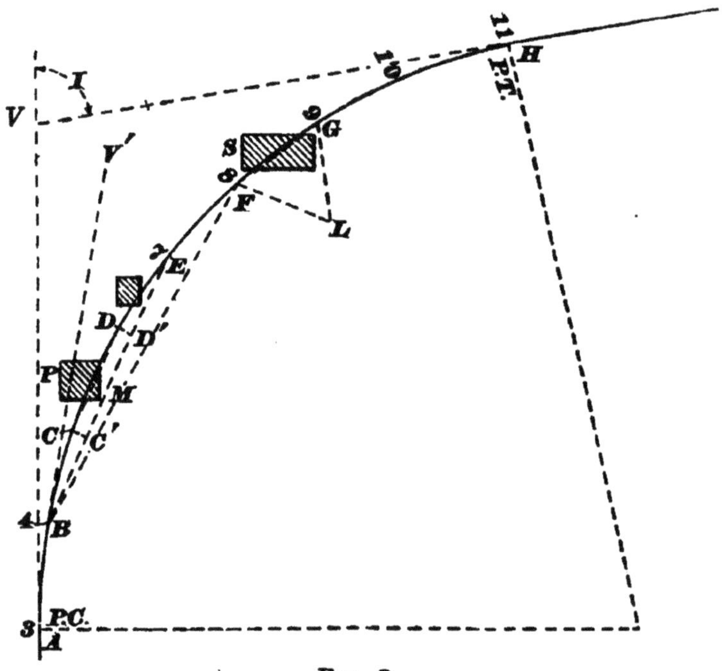

FIG. 8

Suppose that it is required to run out the curve AEH, Fig. 8, with several obstacles in the direct line of the curve, as shown, Station 3 being the P. C., and the regular stations on the curve being in the positions indicated by the numbers 4, 7, 8, etc. The positions of Stations 5 and 6 are indicated by the letters C and D. The stations are to be located in their proper positions on the curve, between the obstructions, wherever it is possible to do so. In addition to this, it is customary to mark with a tack or otherwise the point where the line of the curve intersects each obstruction.

Beginning at the point of curve A, which is at Station 3, the curve can be run in as far as the first obstruction, which is the building P, setting the stakes on the curve at Stations 4 and 5, and a tack in the side of the building P at the point where the line of curve intersects it, according to the deflection angle as determined by its distance from Station 5. It is not possible to proceed further in the regular manner, however, because Station 6 cannot be seen from the P. C. Therefore, it is necessary to locate Station 7 by deflection angle $V'BE$, from B or Station 4, to determine the chord *4–7*, which, in this case, is a long chord of 3 stations, and to calculate the ordinates $D'D$ and $C'C$ by substituting for a in the preceding formula the value of $MC' = MD' =$ half a station or 50′.

Fig. 8 shows also another method of passing a building, as S, namely, by running an equilateral triangle FLG. In this method, the instrument is set up at Station 8 and sighted back to the P. C. Then, the telescope is reversed and the deflection angle for Station 9 is turned off the same as if no obstruction existed. The telescope will then be sighted on the line FG, although the point G will not be visible. The angle GFL, equal to 60°, is then turned, and the point L is located so that $FL = FG = 100'$. The instrument is next moved to L, and the line LG is run, making 60° with FL. On this line the distance $LG = 100'$ is measured, giving the point G, which is Station 9. The transit is then set up at this point and sighted to L, and an angle of 60° is turned off to the right, giving the direction of the line *9–8*, the intersection of which with S is marked. The remainder of the curve may be run in the following manner: Set the vernier at an angle equal to the deflection angle of the chord *9–8* to the left from the zero; clamp the upper plate, sight at the point set in the line *9–8*; then clamp the lower plate and set vernier at zero. The line of sight will then be in the tangent at point *9*, and by plunging the telescope the remainder of the curve can be run as if the point *9* were the P. C.

FIELD NOTES FOR CURVES

Various styles of field notebooks are published, in which the pages are ruled to suit the different kinds and methods of field work. The following, which are the field notes of a portion of a line containing a curve, represent a good form for recording the field notes of a curve that is run in by the method of zero tangent.

In the first column are recorded the station numbers; in the second column, the deflections with the abbreviations P.C. and P. T., together with the degree of curve and the abbreviation *R* or *L*, according as the line curves to the right or left. At each transit point on the curve, the total or central angle from the P. C. to that point is calculated and recorded in the third column. This total angle is double the deflection angle between the P. C. and the transit point. In the accompanying notes, there is but one intermediate transit point between the

Station	Deflection	Tot. Angle	Mag. Bearing	Ded. Bearing	Remarks	June 20, 1912
9						
8						
7						
6+95	4°54' P.T.	15°00'	N.35°20'E.	N.35°15'E.		
6+50	4°00'					70°00'
6	3°00'					
5+50	2°00'				5+80	Centerline of Highway
5	1°00'				5+60	
4+50	2°36'	5°12'				
4	1°36'				Int. Angle = 15°00'	4° Curve R
3+50	0°36'				T = 188.61 ft.	Def. Angle for 50 ft. = 1°00'
3+20	P.C. 4° R.				P.C. = 3+20	Def. Angle for 1 ft. = 1.2'
3					Length of Curve = 375 ft.	
2					P.T. = 6+95	
1						
0			N20°15'E.	N20°15'E.		

P. C. and the P. T. The deflection from the P. C. at Station 3+20 to the intermediate transit point at Station 4+50 is 2° 36′. The total angle is double this deflection, or 5° 12′, which is recorded on the same line in the third column. The record of total angles at once indicates the stations at which transit points are placed. The total angle at the P. T. will be the same as the angle of intersection, provided the work is

SUPERELEVATION (FEET) OF OUTER RAIL ON CURVES

Degree of Curve	Velocity, in Miles per Hour											
Degrees	10	15	20	25	30	35	40	45	50	60	70	75
1	.006	.013	.023	.036	.051	.070	.091	.116	.143	.206	.283	.325
2	.011	.026	.046	.071	.103	.140	.183	.231	.285	.410	.567	.650
3	.017	.039	.069	.107	.154	.210	.274	.346	.427	.612	.842	
4	.023	.051	.091	.143	.206	.280	.365	.461	.568	.811	1.125	
5	.029	.064	.114	.179	.257	.349	.455	.574	.707	1.006		
6	.034	.077	.137	.214	.308	.418	.545	.687	.844	1.196		
7	.040	.090	.160	.250	.359	.487	.634	.798	.979			
8	.046	.103	.183	.285	.410	.556	.723	.908	1.112			
9	.051	.116	.206	.320	.460	.624	.811	1.017				
10	.057	.129	.228	.356	.511	.692	.898	1.124				
11	.063	.142	.251	.391	.561	.760	.984					
12	.069	.154	.274	.427	.611	.826	1.069					
14	.080	.180	.319	.497	.711	.959						
16	.091	.206	.365	.567	.809	1.088						
18	.102	.231	.410	.637	.906							
20	.114	.256	.455	.707	1.002							
25	.141	.318	.563	.775								
30	.168	.380	.672	.844								
35	.195	.441	.778									
40	.222	.501	.881									
50	.276	.618										

correct. When the curve is finished, the transit is set up at the P. T., and the bearing of the forward tangent taken, which affords an additional check upon the previous calculations. The magnetic bearing is recorded in the fourth column, and the deduced, or calculated, bearing is recorded in the fifth column.

SUPERELEVATION OF OUTER RAIL

The difference between the elevation of the outer rail and that of the inner rail of a circular track is called the *superelevation of the outer rail.* If the degree of curve is denoted by D and the velocity, in miles per hour, by V, then the superelevation e, in feet, is $e = .000058\ DV^2$

The accompanying table gives the values of e, corresponding to all values of D and V, that are likely to be required in practice. This table is computed from a more accurate formula than the one just given. The formula given is, however, sufficiently exact and may be used if no tables are at hand.

TRANSITION SPIRAL

DEFINITIONS, PRINCIPLES, AND FORMULAS

Transition curves are introduced for the purpose of connecting a tangent with a circular curve in such a manner that the change of direction and elevation from one to the other takes place gradually. A *transition spiral* is a transition curve in which the degree of curve at any point increases directly as the distance of this point, measured along the curve, from the tangent. The degree of curve is zero at the tangent, and, at the point at which the spiral meets the circular curve, it is equal to the degree of the circular curve.

The point at which the transition spiral joins the tangent is called the *point of spiral*, and it is denoted by P. S_1. The point at which the transition spiral joins the circular curve is called the *second point of spiral;* this point is denoted by P. S_2.

The *unit degree of curve of spiral* is the degree of curve of the spiral at a point 100 ft., or one station, from the point of spiral; it is equal to the degree of curve of the simple circular curve divided by the total length of the spiral, measured in stations of 100 ft. At any other point of the spiral, the degree of curve is equal to the unit degree of curve multiplied by the distance of the point from the P. S_1, also measured in stations of 100 ft.

Let D_c denote the degree of circular curve; D, the degree of

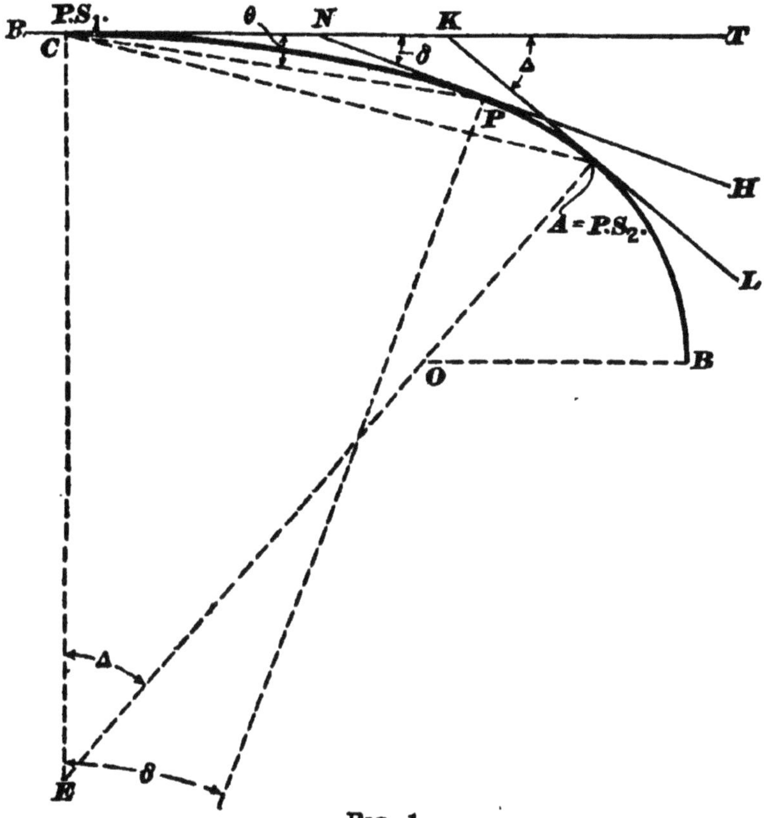

FIG. 1

curve at any point of the spiral, distant l stations from the P. S_1; L, the total length of the spiral in stations; and a the unit degree of spiral. Then,

$$a=\frac{D_c}{L}$$

and

$$D=al$$

The superelevation of the outer rail on the spiral is proportional to l; it is zero at P. S_1, and attains the value e, the superelevation of the circular curve, at the P. S_2. At any intermediate point distant l stations from the P. S_1, it is therefore equal to

$$e_1 = e \times \frac{l}{L}$$

Angle of Deviation and Angle of Deflection.—Let CA, Fig. 1, be a spiral connecting the tangent RT with the circular curve AB. Let P be any point on the spiral and HN a tangent to the spiral at the point P.

The angle that a tangent drawn to the spiral at any point P forms with the original tangent RT is called the *deviation angle* for the point P. It is represented by the Greek small letter δ (called *delta*).

When the point P coincides with the P. S_2, the deviation angle becomes LKT, which is represented by the Greek capital letter Δ (called *delta*).

Since $LKT = AEC$, it follows that Δ is the whole *central angle* of the spiral, which measures the whole change in direction of the track between the original tangent and the P. S_2.

The angle between the original tangent and a chord drawn from the P. S_1 to any point of the spiral is called the *deflection angle* to this point. It is represented by the Greek letter θ (called theta). In Fig. 1, TCP is the deflection angle for the point P. It is the angle that must be deflected at the P. S_1 from the original tangent in order to locate the point P of the spiral.

$\frac{1}{3}\delta$ Degrees	N Minutes	$\frac{1}{3}\delta$ Degrees	N Minutes
3	.0	8	.7
4	.1	9	1.0
5	.2	10	1.4
6	.3	11	1.9
7	.5	12	2.4

By using the preceding notation, the following formulas are derived:

$$\delta = \frac{1}{2} a l^2$$

$$\Delta = \frac{1}{2} a L^2$$

and $\theta = \frac{1}{3}\delta - N$, in which the value of N can be taken from the accompanying table. Intermediate values of N may be found by interpolation. Angle $NPC = \delta - \theta$. This is the angle that must be deflected from the direction of PC to bring the line of sight tangent to the spiral at P.

EXAMPLE.—A spiral 600 ft. long connects a tangent with a 12° curve. Find the angle of deviation and deflection angle, and angle *NPC* for a point 580 ft. from the P. S_1.

SOLUTION.—The unit degree of curve,

$$a = \frac{D_c}{L} = \frac{12°}{6} = 2°$$

and $l = 5.8$ stations; hence, $\delta = \frac{1}{2} \times 2° \times 5.8^2 = 33°\ 38.4'$.

In determining the deflection angle θ, it is known that $\frac{1}{3}\delta = 11°\ 12.8'$. Interpolating from the table,

$$N = 1.9' + \frac{12.8}{60} \times (2.4' - 1.9') = 2.0'$$

Therefore, $\theta = \frac{1}{3}\delta - N = 11°\ 12.8' - 2.0' = 11°\ 10.8'$.

Coordinates of the Spiral.—Let *P*, Fig. 2, be any point of a spiral, and *PR* the perpendicular distance from this point to the

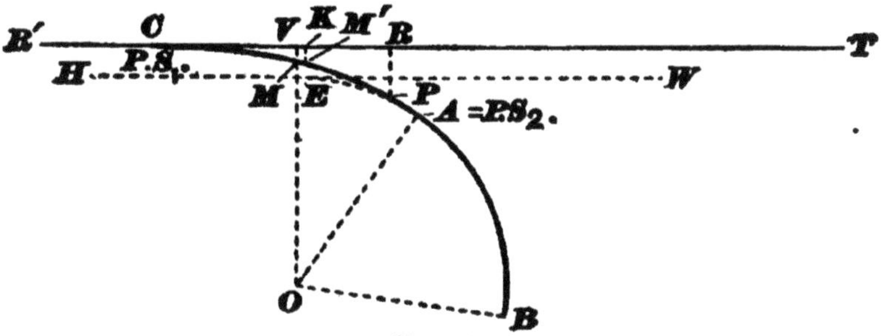

FIG. 2

original tangent. This perpendicular is represented by y, and its value is given by the formula

$$y = .291\, a\, l^3 - a^3 M$$

in which a and l have the same meanings as before.

l	*M*	*l*	*M*	*l*	*M*
3.0	.003	5.5	.241	8.0	3.314
3.5	.010	6.0	.442	8.5	5.065
4.0	.026	6.5	.775	9.0	7.557
4.5	.059	7.0	1.301	9.5	11.033
5.0	.124	7.5	2.109	10.0	15.800

The value of *M* corresponding to any value of *l* may be taken from the accompanying table.

The distance *CR*, measured along the original tangent from the P. S_1 to the foot of the perpendicular *PR*, is represented by x. This distance is somewhat shorter than the distance *CP* measured along the curve: the difference in length between *CR* and *CP* is called the *x correction*, and is given by the formula

$$x \text{ correction} = .000762\ a^2l^5$$

This formula gives the quantity to be subtracted from *CP*, expressed in feet, to obtain the length *CR*, in feet.

EXAMPLE.—Find the values of *PR* and *CR* to a point of the spiral 310 ft. from the P. S_1 in the preceding example.

SOLUTION.—In this example, $a = 2°$, and $l = 3.1$, and from the table, using interpolation, $M = .003 + \frac{1}{5}(.010 - .003) = .004$.

Substituting these values,

$$y = .291 \times 2 \times 3.1^3 - 2^3 \times .004 = 17.31 \text{ ft.}$$

Substituting known values in the formula for the x cor.,

$$x \text{ cor.} = .000762 \times 2^2 \times 3.1^5 = .9 \text{ ft.}$$

The distance $l = 310$ ft.; therefore, the distance $CR = 310 - .9 = 309.1$ ft.

The Spiral Offset and *t* Correction.—Let the circular curve *BA*, Fig. 2, be produced backwards until at a point *E* it becomes parallel to the original tangent—that is, until the tangent *HW* to the circular curve becomes parallel to *R'T*.

The point *E* at which a spiraled circular curve, if produced backwards, becomes parallel to the original tangent is called the *point of curve*, and is denoted by P. C.

The offset *EV* from the point of curve to the original tangent is called the *spiral offset*. It is represented by *F*, and its value, in feet, is given by the formula

$$F = .072709\ aL^3$$

If *M'*, Fig. 2, is the middle point of the spiral—that is, a point half way between the P. S_1 and the P. S_2—it will always be found that the spiral offset cuts the spiral at a point *M* that is a very short distance to the left of *M'*. The distance *CV* will therefore always be slightly less than the distance *CM'*. The difference between the half length of spiral, *CM'*, and the distance *CV* from the P. S_1 to the foot of the spiral offset is called the *t correction;* it is denoted by t, and its value, in feet, is given by the formula

$$t = .000127\ a^2L^5$$

This correction must be subtracted from the half length of spiral, expressed in feet, to obtain the distance *CV*, in feet.

The values of *F* and *t* are given in the fifth and eighth columns of the tables for transition spirals, which follow. The value of *l* in the first column, corresponding to which is found *F* and the *t* correction, is to be taken as the whole length of the spiral.

EXAMPLE.—Find the distances *EV* and *CV* for a spiral 400 ft. long that connects with a 2° curve.

SOLUTION.—Here, $a = \frac{D}{L} = \frac{2^\circ}{4} = \frac{1}{2}^\circ$.

The whole length of spiral is 4 sta. Therefore, substituting in the formula, $F = .072709 \times \frac{1}{2} \times 4^3 = .072709 \times \frac{1}{2} \times 64 = 2.33$ ft.

By the formula for the *t* correction, $t = .000127 \times \frac{1}{2}^3 \times 4^5 = .033$ ft. Therefore, $CV = \frac{1}{2} \times 400$ ft. $- .033$ ft. $= 199.97$ ft.

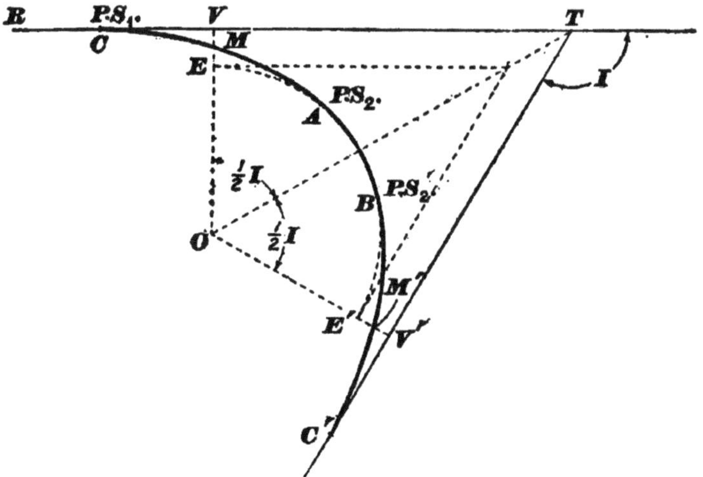

FIG. 3

The Middle Point of the Spiral Offset.—If M', Fig. 2, is the middle point of the spiral, and *M'K* is the offset from the original tangent, *M'K* is almost exactly equal to one-half the spiral offset *VE*. The distance *CK* from the P. S_1 to the foot of *M'K* is almost exactly equal to the distance *CV* from the

P. S_1 to the foot of the spiral offset. Consequently, the spiral offset and the spiral very nearly bisect each other; the point *M* at which the spiral cuts the offset is almost exactly half way between the P. C. and the original tangent.

Tangent Distance.—The *tangent distance* of a transition spiral is the distance of the P. S_1 from the point of intersection of the tangents at the points of spirals. When the lengths of the two spirals are equal (Fig. 3),

$$TC = \frac{1}{2}\text{ length of spiral} - t\text{ cor.} + (R+F)\tan\frac{1}{2}I$$

in which *R* is the radius of the circular curve and *F* the spiral offset *EV*.

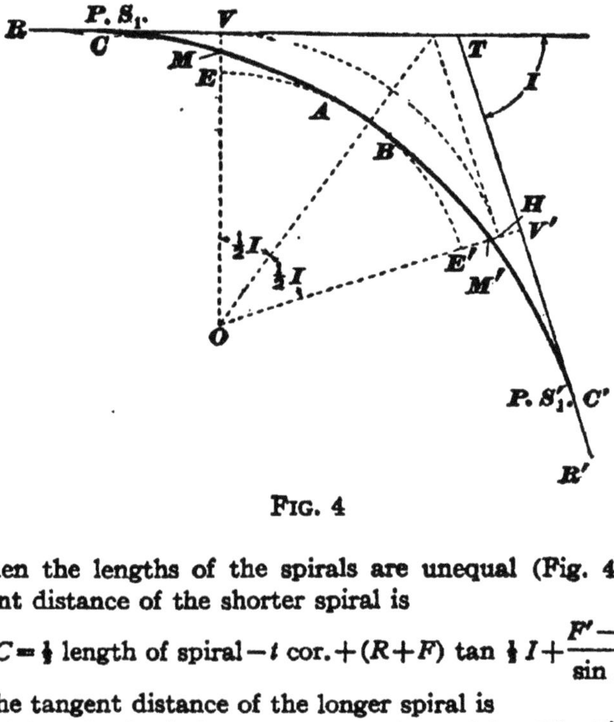

FIG. 4

When the lengths of the spirals are unequal (Fig. 4), the tangent distance of the shorter spiral is

$$TC = \frac{1}{2}\text{ length of spiral} - t\text{ cor.} + (R+F)\tan\frac{1}{2}I + \frac{F'-F}{\sin I}$$

and the tangent distance of the longer spiral is

$$TC' = \frac{1}{2}\text{ length of spiral} - t\text{ cor.} + (R+F)\tan\frac{1}{2}I - (F'-F)\cot I$$

F' and *F* denote, respectively, the spiral offsets of the longer and shorter spirals.

TABLES FOR TRANSITION SPIRALS

The following tables contain the data required for the laying out of eleven different spirals. The unit degree of spiral is marked at the top of each table. The column headed l contains the length, in feet, between the P. S_1 and the points on the spiral, and the one headed d gives the degrees of curve of spiral at these points. The third column gives the corresponding deviation angles; the fourth the deflection angle; and the remaining columns give the values of the spiral offset F, the coordinate y, and the corrections x and t, all in feet. As an illustration of the use of these tables, let the preceding example be solved by means of them. Since $a = \frac{1}{2}°$, reference is made to the table for $a = 0° 30'$, where it is found that for $l = 400$ ft., the corresponding value of $F = 2.33$, and that of t cor. $= .03$. Then, as before, $EV = 2.33$ ft. and $CV = 199.97$ ft.

LAYING OUT A SPIRAL IN THE FIELD

Let RT and $R'T$, in the accompanying illustration, be the two tangents that are to be connected with the circular curve

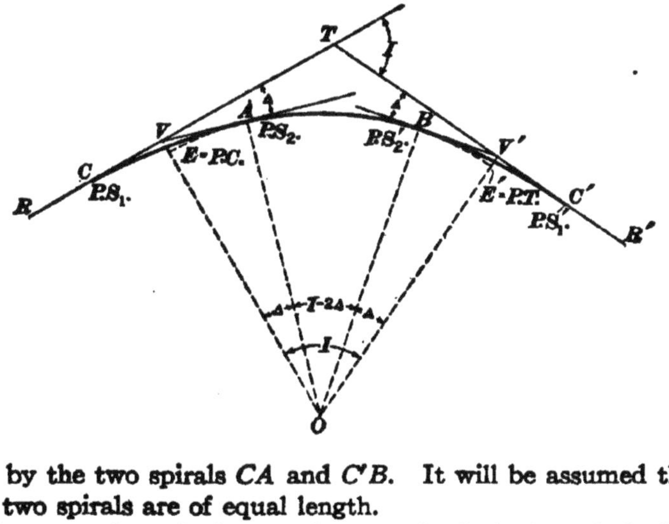

AB by the two spirals CA and $C'B$. It will be assumed that the two spirals are of equal length.

Compute the unit degree of curve of spiral, the spiral offset $VE = V'E'$, and the distance $CV = C'V'$, or obtain these quantities with the help of the tables and compute the distance

TABLE FOR TRANSITION SPIRALS

a = 0° 30′. 1° in 200 ft.

l	*d*	*δ*	*θ*	*F*	*y*	*x* cor.	*t* cor.
	° ′	° ′	° ′	Ft.	Ft.	Ft.	Ft.
25	0 7.5	0 0.9	0 0.3	.00	.00	.00	.00
50	15	3.8	1.3	.00	.02	.00	.00
75	22.5	8.4	2.8	.02	.06	.00	.00
100	30	15	5	.04	.15	.00	.00
125	0 37.5	0 23.4	0 7.8	.07	.29	.00	.00
150	45	33.8	11.3	.12	.49	.00	.00
175	52.5	45.9	15.3	.20	.78	.00	.00
200	1 00	1 00	20	.29	1.16	.01	.00
225	1 7.5	1 15.9	0 25.3	.41	1.66	.01	.00
250	15	33.8	31.3	.57	2.27	.02	.00
275	22.5	53.4	37.8	.76	3.03	.03	.01
300	30	2 15	45	.98	3.93	.05	.01
325	1 37.5	2 38.4	0 52.8	1.25	5.00	.07	.01
350	45	3 3.8	1 1.3	1.56	6.23	.10	.02
375	52.5	30.9	10.3	1.92	7.67	.14	.02
400	2 00	4 00	20	2.33	9.31	.19	.03
425	2 7.5	4 30.9	1 30.3	2.79	11.16	.26	.04
450	15	5 3.8	41.3	3.31	13.25	.35	.06
475	22.5	38.4	52.8	3.89	15.58	.46	.08
500	30	6 15	2 5	4.54	18.16	.59	.10
525	2 37.5	6 53.4	2 17.8	5.26	21.03	.75	.13
550	45	7 33.8	31.3	6.04	24.17	.95	.16
575	52.5	8 15.9	45.3	6.91	27.62	1.20	.20
600	3 00	9 00	3 00	7.84	31.36	1.48	.24
625	3 7.5	9 45.9	3 15.3	8.87	35.45	1.81	.30
650	15	10 33.8	31.3	9.97	39.85	2.21	.37
675	22.5	11 23.4	47.8	11.16	44.63	2.66	.44
700	30	12 15	4 4.9	12.45	49.73	3.20	.53
725	3 37.5	13 8.4	4 22.7	13.83	55.22	3.81	.64
750	45	14 3.8	41.2	15.30	61.09	4.51	.75
775	52.5	15 0.9	5 00.1	16.88	67.37	5.31	.89
800	4 00	16 0	19.8	18.56	74.05	6.22	1.04

TABLE—(*Continued*)

$a=0° 40'$. *1° in 150 ft.*

l	d	δ	θ	F	y	x cor.	t cor.
	° ′	° ′	° ′	Ft.	Ft.	Ft.	Ft.
25	0 10	0 1.3	0 0.4	.00	.00	.00	.00
50	20	5	1.7	.01	.02	.00	.00
75	30	11.3	3.8	.02	.08	.00	.00
100	40	20	6.7	.05	.19	.00	.00
125	0 50	0 31.3	0 10.4	.10	.38	.00	.00
150	1 00	45	15	.16	.65	.00	.00
175	10	1 1.3	20.4	.26	1.04	.01	.00
200	20	20	26.7	.39	1.55	.01	.00
225	1 30	1 41.3	0 33.8	.55	2.21	.02	.00
250	40	2 5	41.7	.76	3.03	.03	.01
275	50	31.3	50.4	1.01	4.04	.05	.01
300	2 00	3 00	1 00	1.31	5.23	.08	.01
325	2 10	3 31.3	1 10.4	1.66	6.66	.12	.02
350	20	4 5	21.7	2.08	8.31	.18	.03
375	30	41.3	33.8	2.56	10.23	.25	.04
400	40	5 20	46.7	3.10	12.40	.35	.06
425	2 50	6 1.3	2 .4	3.72	14.88	.47	.08
450	3 00	45	15	4.41	17.66	.62	.10
475	10	7 31.3	30.4	5.19	20.76	.82	.14
500	20	8 20	46.7	6.05	24.20	1.06	.18
525	3 30	9 11.3	3 3.8	7.01	28.02	1.35	.22
550	40	10 5	21.7	8.05	32.19	1.70	.28
575	50	11 1.3	40.4	9.20	36.78	2.12	.36
600	4 00	12 0	59.9	10.45	41.76	2.63	.44
625	4 10	13 1.3	4 20.3	11.83	47.20	3.22	.54
650	20	14 5	41.6	13.29	53.05	3.93	.66
675	30	15 11.3	5 3.6	14.88	59.41	4.73	.78
700	40	16 20	26.4	16.60	66.20	5.69	.94

TABLE—(*Continued*)

$a = 0° 48'$. *1° in 125 ft.*

l	d	δ	θ	F	y	x cor.	t cor.
	° ′	° ′	° ′	Ft.	Ft.	Ft.	Ft.
25	0 12	0 1.5	0 0.5	.00	.00	.00	.00
50	24	6	2	.01	.03	.00	.00
75	36	13.5	4.5	.02	.10	.00	.00
100	48	24	8	.06	.23	.00	.00
125	1 00	0 37.5	0 12.5	.11	.46	.00	.00
150	12	54	18	.20	.79	.00	.00
175	24	1 13.5	24.5	.31	1.25	.01	.00
200	36	36	32	.47	1.86	.02	.00
225	1 48	2 1.5	0 40.5	.66	2.65	.03	.00
250	2 00	30	50	.91	3.64	.05	.01
275	12	3 1.5	1 0.5	1.21	4.84	.08	.01
300	24	36	12	1.57	6.28	.12	.02
325	2 36	4 13.5	1 24.5	2.00	7.99	.18	.03
350	48	54	38	2.49	9.97	.26	.04
375	3 00	5 37.5	52.5	3.07	12.27	.36	.06
400	12	6 24	2 8	3.72	14.88	.50	.08
425	3 24	7 13.5	2 24.5	4.47	17.85	.68	.11
450	36	8 6	42	5.31	21.18	.90	.15
475	48	9 1.5	3 0.5	6.23	24.90	1.18	.20
500	4 00	10 0	20	7.26	29.02	1.52	.25
525	4 12	11 1.5	3 40.5	8.41	33.60	1.92	.33
550	24	12 6	4 2	9.66	38.62	2.44	.41
575	36	13 13.5	4 24.5	11.02	44.08	3.07	.51
600	48	14 24	4 48	12.50	50.06	3.78	.62
625	5 00	15 37.5	5 12.5	14.15	56.55	4.63	.77
650	12	16 54	5 38	15.90	63.55	5.63	.95
675	24	18 13.5	6 4	17.80	71.09	6.81	1.13
700	36	19 36	6 32	19.84	79.20	8.13	1.36

TABLE—(*Continued*)

$a = 1° 0'$. *1° in 100 ft.*

l	d	δ	θ	F	y	x cor.	t cor.
	°	° ′	° ′	Ft.	Ft.	Ft.	Ft.
20	.2	1.2	0.4	.001	.002	.000	.000
40	.4	4.8	1.6	.005	.019	.000	.000
60	.6	0 10.8	0 3.6	.016	.063	.000	.000
80	.8	19.2	6.4	.037	.149	.000	.000
100	1.0	30	10	.073	.291	.001	.000
120	.2	43.2	14.4	.126	.503	.002	.000
140	.4	58.8	19.6	.199	.798	.004	.000
160	1.6	1 16.8	0 25.6	.298	1.191	.008	.001
180	.8	37.2	32.4	.424	1.696	.014	.002
200	2.0	2 00	40	.582	2.327	.024	.004
220	.2	25.2	48.4	.774	3.097	.039	.006
240	.4	52.8	57.6	1.005	4.020	.061	.010
260	2.6	3 22.8	1 7.6	1.278	5.111	.090	.015
280	.8	55.2	18.4	1.596	6.383	.131	.022
300	3.0	4 30	30	1.963	7.850	.185	.031
320	.2	5 7.2	42.4	2.382	9.53	.255	.043
340	.4	46.8	55.6	2.857	11.42	.346	.058
360	3.6	6 28.8	2 9	3.391	13.56	.460	.077
380	.8	7 13.2	24.4	3.988	15.94	.603	.100
400	4.0	8 00	40	4.651	18.59	.779	.130
420	.2	49.2	56.4	5.38	21.51	.994	.166
440	.4	9 40.8	3 13.6	6.19	24.73	1.254	.209
460	4.6	10 34.8	31.6	7.07	28.24	1.57	.26
480	.8	11 31.2	50.4	8.03	32.07	1.94	.32
500	5.0	12 30	4 10	9.07	36.23	2.37	.40
520	.2	13 31.2	30.4	10.20	40.73	2.89	.48
540	.4	14 34.8	51.4	11.42	45.59	3.49	.58
560	5.6	15 40.8	5 13.4	12.74	50.83	4.18	.70
580	.8	16 49.2	36.2	14.14	56.40	4.98	.83
600	6.0	18 00	59.7	15.65	62.39	5.89	.98

TABLE—(*Continued*)

a = 1° 15′. 1° in 80 ft.

l	*d*	*δ*	*θ*	*F*	*y*	*x* cor.	*t* cor.
	° ′	° ′	° ′	Ft.	Ft.	Ft.	Ft.
20	15	1.5	0.5	.00	.00	.0	.0
40	30	6	2	.00	.02	.0	.0
60	0 45	0 13.5	0 4.5	.02	.08	.0	.0
80	1 00	24	8	.05	.19	.0	.0
100	15	37.5	12.5	.09	.36	.0	.0
120	30	54	18	.16	.63	.0	.0
140	45	1 13.5	24.5	.25	1.00	.0	.0
160	2 00	36	0 32	.37	1.49	.0	.0
180	15	2 1.5	40.5	.53	2.12	.0	.0
200	30	30	50	.73	2.90	.0	.0
220	45	3 1.5	1 00.5	.97	3.87	.0	.0
240	3 00	36	12	1.25	5.02	.0	.0
260	15	4 13.5	24.5	1.59	6.38	.1	.0
280	30	54	38	1.99	7.98	.2	.0
300	45	5 37.5	52.5	2.45	9.81	.3	.0
320	4 00	6 24	2 8	2.98	11.91	.4	.0
340	15	7 13.5	24.5	3.57	14.28	.5	.0
360	30	8 6	42	4.23	16.95	.7	.1
380	45	9 1.5	3 00.5	4.97	19.92	.9	.2
400	5 00	10 00	20	5.80	23.23	1.2	.2
420	15	11 1.5	40.5	6.72	26.86	1.6	.3
440	30	12 6	4 2	7.74	30.87	2.0	.3
460	45	13 13.5	24.5	8.84	35.25	2.4	.4
480	6 00	14 24	48	10.03	40.02	3.0	.5
500	15	15 37.5	5 12.5	11.33	45.20	3.7	.6
520	30	16 54	38	12.74	50.79	4.5	.8
540	45	18 13.5	6 4	14.26	56.84	5.4	.9
560	7 00	19 36	32	15.90	63.34	6.5	1.1
580	15	21 1.5	7 00	17.65	70.26	7.8	1.3
600	30	22 30	29	19.52	77.68	9.2	1.5

TABLE—(*Continued*)

$a = 1° 40'$. 1° in 60 ft.

l	d	δ	θ	F	y	x cor.	t cor.
	° ′	° ′	° ′	Ft.	Ft.	Ft.	Ft.
20	20	2	0.5	.00	.00	.0	.0
40	40	8	3	.00	.03	.0	.0
60	1 00	0 18	0 6	.03	.10	.0	.0
80	20	32	10.5	.06	.25	.0	.0
100	40	50	16.5	.12	.48	.0	.0
120	2 00	1 12	24	.21	.84	.0	.0
140	20	38	32.5	.33	1.33	.0	.0
160	40	2 8	0 42.5	.50	1.98	.0	.0
180	3 00	42	54	.70	2.82	.0	.0
200	20	3 20	1 6.5	.97	3.88	.0	.0
220	40	4 2	20.5	1.29	5.15	.1	.0
240	4 00	48	36	1.67	6.69	.2	.0
260	20	5 38	52.5	2.13	8.52	.2	.0
280	40	6 32	2 10.5	2.65	10.64	.4	.0
300	5 00	7 30	30	3.26	13.07	.5	.0
320	20	8 32	50.5	3.96	15.87	.7	.1
340	40	9 38	3 12.5	4.75	19.02	.9	.2
360	6 00	10 48	36	5.64	22.56	1.3	.2
380	20	12 2	4 00.5	6.63	26.53	1.7	.3
400	40	13 20	26.5	7.73	30.92	2.2	.4
420	7 00	14 42	54	8.96	35.73	2.8	.5
440	20	16 8	5 22.5	10.30	41.07	3.5	.6
460	40	17 38	52	11.75	46.86	4.3	.7
480	8 00	19 12	6 24	13.35	53.16	5.4	.9
500	20	20 50	56	15.07	60.01	6.6	1.1
520	40	22 32	7 30	16.94	67.36	8.0	1.3
540	9 00	24 18	8 5	18.95	75.31	9.6	1.6
560	20	26 8	42	21.13	83.88	11.5	1.9
580	40	28 2	9 19.5	23.42	92.92	13.7	2.3
600	10 00	30 00	59	25.91	102.66	16.2	2.7

TABLE—(*Continued*)

$a = 2° 0'$. 1° in 50 ft.

l	d	δ	θ	F	y	x cor.	t cor.
	° ′	° ′	° ′	Ft.	Ft.	Ft.	Ft.
20	24	2.5	1	.00	.00	.0	.0
40	48	9.5	3	.01	.04	.0	.0
60	1 12	0 21.5	0 7	.03	.13	.0	.0
80	36	38.5	13	.07	.30	.0	.0
100	2 00	1 00	20	.15	.58	.0	.0
120	24	26.5	29	.25	1.00	.0	.0
140	48	57.5	39	.40	1.60	.0	.0
160	3 12	2 33.5	0 51	.59	2.38	.0	.0
180	36	3 14.5	1 5	.85	3.39	.1	.0
200	4 00	4 00	20	1.16	4.65	.1	.0
220	24	50.5	37	1.54	6.19	.2	.0
240	48	5 45.5	55	2.00	8.04	.2	.0
260	5 12	6 45.5	2 15	2.55	10.22	.4	.0
280	36	7 50.5	37	3.18	12.75	.5	.0
300	6 00	9 00	3 00	3.91	15.68	.7	.1
320	24	10 14.5	25	4.75	19.03	1.0	.2
340	48	11 33.5	51	5.70	22.81	1.4	.2
360	7 12	12 57.5	4 19	6.77	27.05	1.8	.3
380	36	14 26.5	49	7.95	31.79	2.4	.4
400	8 00	16 00	5 20	9.28	37.04	3.1	.5
420	24	17 38.5	53	10.73	42.79	4.0	.7
440	48	19 21.5	6 27	12.34	49.14	5.0	.8
460	9 12	21 9.5	7 3	14.09	56.05	6.3	1.0
480	36	23 2.5	40	15.99	63.55	7.7	1.3
500	10 00	25 00	8 19	18.05	71.72	9.4	1.6
520	24	27 2.5	9 00	20.27	80.04	11.4	1.9
540	48	29 9.5	42	22.68	89.88	13.8	2.3
560	11 12	31 21.5	10 26	25.27	99.97	16.5	2.8
580	36	33 38.5	11 10.5	28.01	110.62	19.6	3.3
600	12 00	36 00	58	30.97	122.13	23.2	3.9

TABLE—(*Continued*)

$a = 2° 30'$. 1° in 40 ft.

l	d	δ	θ	F	y	x cor.	t cor.
	° ′	° ′	° ′	Ft.	Ft.	Ft.	Ft.
20	30	3	1	.00	.00	.0	.0
40	1 00	12	4	.01	.05	.0	.0
60	30	0 27	0 9	.04	.16	.0	.0
80	2 00	48	16	.09	.37	.0	.0
100	30	1 15	25	.18	.73	.0	.0
120	3 00	48	36	.31	1.25	.0	.0
140	30	2 27	49	.50	2.00	.0	.0
160	4 00	3 12	1 4	.74	2.97	.0	.0
180	30	4 3	21	1.06	4.24	.1	.0
200	5 00	5 00	40	1.45	5.81	.2	.0
220	30	6 3	2 1	1.93	7.74	.2	.0
240	6 00	7 12	24	2.51	10.05	.4	.0
260	30	8 27	49	3.19	12.77	.6	.1
280	7 00	9 48	3 16	3.98	15.94	.8	.1
300	30	11 15	45	4.89	19.59	1.2	.2
320	8 00	12 48	4 16	5.94	23.76	1.6	.3
340	30	14 27	49	7.12	28.46	2.2	.4
360	9 00	16 12	5 24	8.46	33.74	2.9	.5
380	30	18 3	6 1	9.95	39.64	3.7	.6
400	10 00	20 00	40	11.60	46.16	4.9	.8
420	30	22 3	7 21	13.39	53.28	6.2	1.0
440	11 00	24 12	8 4	15.39	61.12	7.8	1.3

$a = 3° 20'$. 1° in 30 ft.

l	d	δ	θ	F	y	x cor.	t cor.
	° ′	° ′	° ′	Ft.	Ft.	Ft.	Ft.
20	40	4	1	.00	.01	.0	.0
40	1 20	16	5	.02	.06	.0	.0
60	2 00	0 36	0 12	.05	.21	.0	.0
80	40	1 4	21	.12	.50	.0	.0
100	3 20	40	33	.24	.97	.0	.0
120	4 00	2 24	48	.42	1.68	.0	.0
140	40	3 16	1 5	.67	2.66	.0	.0
160	5 20	4 16	25	.99	3.97	.1	.0
180	6 00	5 24	48	1.41	5.65	.2	.0
200	40	6 40	2 13	1.94	7.75	.3	.0

TABLE—(*Continued*)

l	d	δ	θ	F	y	x cor.	t cor.
	° ′	° ′	° ′	Ft.	Ft.	Ft.	Ft.
220	7 20	8 4	41	2.58	10.31	.4	.1
240	8 00	9 36	3 12	3.35	13.38	.7	.1
260	40	11 16	45	4.25	17.00	1.0	.2
280	9 20	13 4	4 21	5.31	21.20	1.4	.2
300	10 00	15 00	5 00	6.53	26.05	2.0	.3
320	40	17 4	41	7.92	31.57	2.8	.5
340	11 20	19 16	6 25	9.49	37.80	3.8	.6
360	12 00	21 36	7 11	11.25	44.78	5.1	.8
380	40	24 4	8 00	13.22	52.53	6.6	1.1
400	13 20	26 40	52	15.39	61.10	8.6	1.4
420	14 00	29 24	9 47	17.79	70.49	10.9	1.8
440	40	32 16	10 43	20.41	80.74	13.7	2.3

a = 5° 0′. 1° in 20 ft.

l	d	δ	θ	F	y	x cor.	t cor.
	° ′	° ′	° ′	Ft.	Ft.	Ft.	Ft.
20	1 00	6	2	.00	.01	.0	.0
40	2 00	24	8	.02	.09	.0	.0
60	3 00	0 54	0 18	.08	.31	.0	.0
80	4 00	1 36	32	.19	.74	.0	.0
100	5 00	2 30	50	.36	1.45	.0	.0
120	6 00	3 36	1 12	.62	2.51	.0	.0
140	7 00	4 54	38	.99	3.99	.1	.0
160	8 00	6 24	2 8	1.48	5.96	.2	.0
180	9 00	8 6	42	2.11	8.49	.4	.0
200	10 00	10 0	3 20	2.90	11.62	.6	.1
220	11 00	12 6	4 2	3.86	15.44	1.0	.2
240	12 00	14 24	48	5.01	20.01	1.5	.3
260	13 00	16 54	5 38	6.37	25.38	2.2	.4
280	14 00	19 36	6 32	7.94	31.62	3.3	.6
300	15 00	22 30	7 29	9.76	38.83	4.6	.8
320	16 00	25 36	8 31	11.82	46.92	6.3	1.1
340	17 00	28 54	9 37	14.15	56.05	8.6	1.4
360	18 00	32 24	10 46	16.75	66.31	11.3	1.9
380	19 00	36 6	12 00	19.65	77.35	14.8	2.5
400	20 00	40 0	13 17	22.87	89.83	19.0	3.2

TABLE—(*Concluded*)

$a = 10° 0'$. 1° in 10 ft.

l	d	δ	θ	F	y	x cor.	t cor.
	° ′	° ′	° ′	Ft.	Ft.	Ft.	Ft.
20	2	12	4	.01	.02	.0	.0
40	4	48	16	.05	.19	.0	.0
60	6 00	1 48	0 36	.16	.63	.0	.0
80	8	3 12	1 4	.37	1.49	.0	.0
100	10	5 0	40	.73	2.91	.1	.0
120	12	7 12	2 24	1.26	5.02	.2	.0
140	14	9 48	3 16	1.99	7.97	.4	.1
160	16 00	12 48	4 16	2.97	11.87	.8	.1
180	18	16 12	5 24	4.23	16.87	1.4	.2
200	20	20 0	6 39	5.79	23.07	2.4	.4
220	22	24 12	8 3	7.69	30.58	3.9	.6
240	24	28 48	9 35	9.96	39.49	6.0	1.0
260	26 00	33 48	11 14	12.61	49.67	8.9	1.5
280	28	39 12	13 1	15.67	61.40	12.9	2.1
300	30	45 0	14 55	19.23	75.07	18.1	3.1

$CT = C'T'$. Run the two tangents to their point of intersection T, measure back from T the distances TC and TC', and at C and C' set stakes marked P. S_1.

Set up the transit at P. S_1, sight on T, and then set stakes on the spiral exactly as on a simple circular curve, except that the deflection angle for each stake is computed by the formula or taken from the tables. When the stake at A (marked P. S_2) has been set, move the transit to A, backsight on P. S_1, and deflect from this direction the angle necessary to bring the telescope tangent to the simple circular curve at A. This angle is equal to the angle of deviation Δ minus the angle of deflection VCA. Run in the circular curve as usual.

When the stake at B (marked P. S_2') has been set, move the transit to C', backsight on T, and stake out the second spiral in exactly the same manner as the first, using the deflection angles computed for the first spiral. When the last stake along $C'B$ has been set, backsight on T, and continue the survey along the tangent $C'R'$.

EXAMPLE.—Two tangents that intersect at an angle of 80° 20′ are to be connected with a 6° circular curve by two equal spirals, each 300 ft. long. The tangents intersect at Sta. 36. Lay out the two spirals and the circular curve.

SOLUTION.—The unit degree of spiral $a=\frac{D_c}{L}=\frac{6°}{3}=2°$; the spiral offset $F=.072709\ aL^3=.072709\times2\times3^3=3.93$ ft.; $CV=\frac{1}{2}$ length $-t$ cor. $=150-.000127\ a^2L^5=150-0.1=149.9$. $R=\frac{5{,}730}{D_c}$ $=\frac{5{,}730}{6}=955$ ft. $CT=\frac{1}{2}$ length $-t$ *cor.* $+(R+F)$ $\tan\frac{80°\ 20'}{2}$ $=149.9+(955+3.93)\tan 40°\ 10'=959.3$ ft.

Since T is at Station 36, the station number of the P. S_1 is $36-(9+59.3)=26+40.7$.

It will be assumed that stakes are set 50 ft. apart on the spirals and at the even stations on the circular curve. The spiral deflections are then figured as shown in example under the heading Angle of Deviation and Angle of Deflection. They are:

to first stake,	0° 5′	(*A*) Angles to be deflected from the tangent. Vernier set at 0° 0′.
to second stake,	0° 20′	
to third stake,	0° 45′	
to fourth stake,	1° 20′	
to fifth stake,	2° 5′	
to P. S_2 at 29+40.7,	3° 0′	

The deviation angle $\Delta=\frac{1}{2}\ aL^2=\frac{1}{2}\times2\times3^2=9°$. Therefore, the central angle of circular curve $=I-2\Delta=80°\ 20'-2\times9°$ $=62°\ 20'$. The length of AB is therefore $62°\ 20'\div6=10.389$ Sta. and the station number of B is $29+40.7+(10+38.9)$ $=39+79.6$.

The angle between the chord CA and the tangent to the circular curve at A is $\Delta-\theta=9°-3°=6°$.

Transit at P. S_2.—The deflection angles to the stakes on the circular curve are as follows:

to Sta. 30, $.593\times3°=$	1° 47′;	to Sta. 35,	16° 47′	(*B*) Angles to be deflected from tangent to circular curve. Vernier set at 6° 0′.
to Sta. 31,	4° 47′;	to Sta. 36,	19° 47′	
to Sta. 32,	7° 47′;	to Sta. 37,	22° 47′	
to Sta. 33,	10° 47′;	to Sta. 38,	25° 47′	
to Sta. 34,	13° 47′;	to Sta. 39,	28° 47′	
		to *B*,	31° 10′	

Transit at P. S_1'.—The angles to be deflected are the same as at P. S_1. The station number of P. S_1' is $(39+79.6)+3 = 42+79.6$.

The Field Work.—Run the two tangents to their intersection. Measure back from T the distances $TC = TC' = 959.3$ ft., and set stakes marked P. S_1 at C and C'. Set the transit at C with the vernier at 0° 0′; sight on T and deflect the angles (A) to locate the first spiral. When the stake at A (marked P. S_2) has been set, move to this point, set the vernier at 6° 0′, backsight on C, turn the telescope until the vernier reads 0° 0′, and from this direction deflect the angles (B) to locate the circular curve. When the stake B (marked P. S_2) has been set, move the transit to C', set the vernier at 0° 0′, backsight on T, and deflect the angles (A) to locate the second spiral.

SELECTION OF SPIRALS

For a given velocity of train, in miles per hour, V, and the degree of curve of the circular curve D_c, the best length of spiral, in stations is found by the following formula:

$$L = \frac{V^2 D_c}{108{,}000}.$$

EXAMPLE.—Find the theoretically best length of spiral to connect with a 6° curve, the maximum train velocity being 40 mi. per hr.

SOLUTION.—Substituting the value of 40 for V and 6 for D_c,

$$L = \frac{40^2 \times 6}{108{,}000} = 3.556 \text{ Sta.}$$
$$= 355.6 \text{ ft.}$$

Maximum Train Speed Miles per Hour	Unit Degree of Curve of Spiral
75	30′ or less
60	30′ or less
50	1° or less
40	2° or less
30	3° 20′ or less
25	5° or less
20	10° or less

Table of Minimum Spiral Lengths.—The accompanying table, from Talbot's "Transition Spiral," gives the values of a corresponding to the least length of spiral that the engineer should endeavor to insert. The spiral may be longer than the length obtained from this table, but it should not be shorter, unless topographical conditions make it necessary to use a shorter spiral than the minimum given in the table.

The least length corresponding to any value of a is found from the formula

$$L = \frac{D_c}{a}$$

EXAMPLE.—Find the least length for the spiral in the preceding example.

SOLUTION.—The velocity is 40 mi. per hr.; therefore, from the table, $a = 2°$, and $L = 6° \div 2 = 3$ sta. = 300 ft.

EARTHWORK

FIELD WORK

Cuts and Fills.—In building a railroad, cuts and fills are introduced to equalize the irregularities of the natural soil. Figs. 1 and 2 show a typical fill and cut in ordinary firm earth or gravel.

Slope Ratio.—In cuts in the hardest rock, the average slope is usually made ¼:1; that is, ¼ horizontal to 1 vertical. As

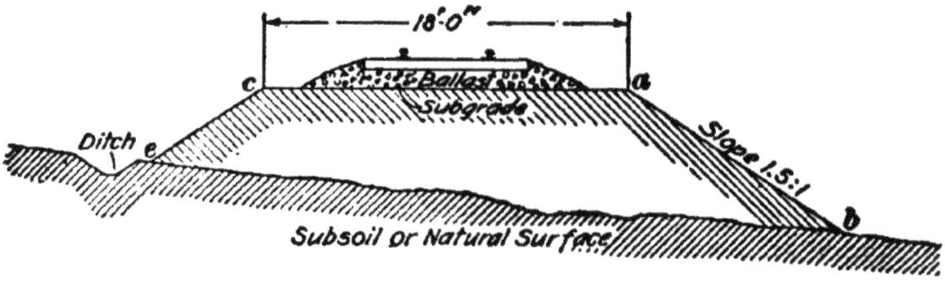

FIG. 1

the soil becomes less firm the slope must be flattened until, for a soil of firm earth or gravel, a slope of 1 to 1 may be permissible, although a slope of 1½:1 is commonly adopted. In very soft soil, the slope ratio is sometimes cut down even as far as 4 horizontal to 1 vertical. The standard practice in a fill is 1½ horizontal to 1 vertical. When a fill is made of the material from a rock cut, it is possible to make a stable embankment with a slope ratio of 1:1. On side-hill work, where a slope ratio of 1½:1 or even 1:1 might require a very

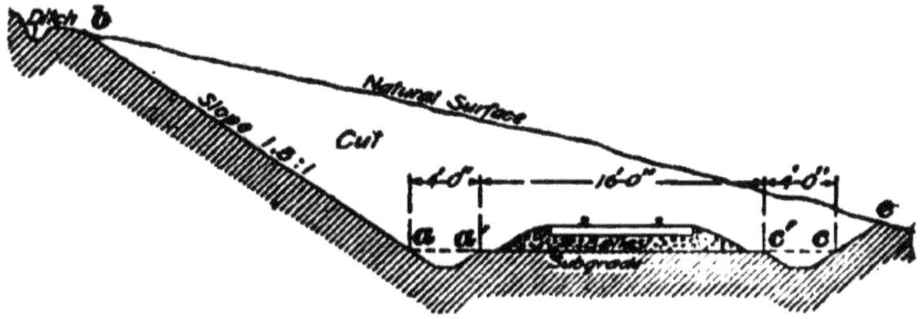

FIG. 2

long slope, it is often advisable to make a rough dry wall of the stones from a rock cut that will have a slope ratio of ½:1, or it may even be steeper.

Width of Excavations and Embankments.—The width required for a standard-gauge single-track roadbed may be estimated as follows (see Figs. 1 and 2): The tie will be between 8 and 9 ft. long, usually 8 ft. 6 in. At the ends of the ties, the ballast will slope down to subgrade. The extra width required for this will be about 1 or 2 ft. at each end of the tie. Usually, the embankment is widened for about 2 ft. beyond the ballast on each side. The absolute minimum for the width of subgrade for a fill is, therefore, 8½ ft.$+2\times$(1$+$2) ft., or about 14½ ft. This width would be used only for light-traffic, cheaply constructed roads; 16 to 18 ft. is far more common, while 20 ft. and even more is frequently used, as the danger of accident due to a washing out of the embankment is materially reduced by widening the roadbed.

In cuts, the proper width for two ditches should be added. Unless the soil is especially firm, the ditches should have a side slope of 1.5:1. If the ditch is 12 in. wide at the base and 12 in. deep, with side slopes of 1.5:1, each ditch will require a total width of 4 ft. This will add 8 ft. to the width of the cut at the elevation of subgrade. The usual distance between track centers for double track is 13 ft. Therefore, whatever rate of side slopes and width of ditches is required for single-track work, the width for double-track work must be 13 ft. greater. When excavation is made through rock, the side slopes of the ditches may properly be made much steeper; the danger of scouring during heavy rain storms being eliminated, the total required width may be very materially reduced from the figures just given. The heavy expense of excavating through solid rock requires that such economy shall be used if possible.

Grade Profile.—For the purpose of constructing a road as well as for calculating the earthwork, a grade profile is prepared by setting stakes on the center line at every full station and also at all intermediate points at which the inclination of the natural surface of the ground changes abruptly; then, by leveling, the elevation of the natural surface at each stake is determined and plotted, as explained under Leveling. The established grade is then drawn in. It consists of a series of straight lines, the elevations of the ends of which are clearly indicated. These elevations are those of the subgrade *ac*, Figs. 1 and 2.

A short portion of a profile is shown in Fig. 3. The horizontal line *XX'* represents a reference plane, and the broken line

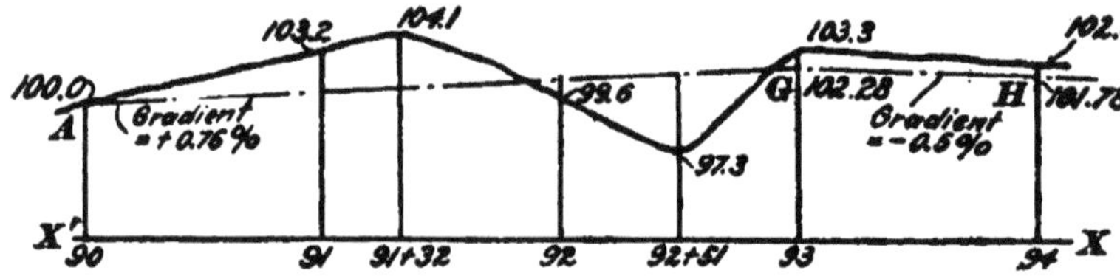

FIG. 3

AGH shows the position of the established grade. The station numbers are written along the line *XX'*, and the elevations of the corresponding points of the established grade are written along the grade line. Thus, in Fig. 3, the elevation of subgrade at Sta. 90, or *A*, is 100 ft.; at Sta. 93, or *G*, it is 102.28 ft.; and at Sta. 94, or *H*, it is 101.78 ft.

The *gradient* of the established grade is the per cent. of rise or fall of grade; that is, the number of feet by which the elevation increases or decreases in 100 ft. It is usually marked on the grade line in the manner shown in Fig. 3. The ***depth of center stake*** is the difference between the elevation of the natural surface at any stake and the elevation of the subgrade. The elevation of the natural surface is found in the level notes, while the elevations of the subgrade are computed from the gradients and also entered in the level notes. The difference for each stake is then figured and entered in a column headed Depth of Center Stake, being preceded by the letter *C* or *F* to indicate cut or fill.

EXAMPLE.—Stakes are set at the stations indicated in the first column of the accompanying field notes. The gradient is +.76% from Sta. 90 to Sta. 93, and −.50% beyond Sta. 93. The elevation of the established grade at Sta. 90 is 100.00 ft.; the elevation of the natural surface at each stake is given in the third column. Find the center depth at each stake. (See Fig. 3.)

Station	Subgrade	Elevation	Depth of Center Stake
94	101.8	102.6	C .8
93	102.3	103.3	C 1.0
92+51	101.9	97.3	F 4.6
92	101.5	99.6	F 1.9
91+32	101.0	104.1	C 3.1
91	100.8	103.2	C 2.4
90	100.0	100.0	0

SOLUTION.—The elevations of the subgrade at the station stakes are determined as follows:

Station	*Elevation*
91	100.00+1.00× .76=100.8
91+32	100.00+1.32× .76=101.0
92	100.00+2.00× .76=101.5
92+51	100.00+2.51× .76=101.9
93	100.00+3.00× .76=102.3
94	102.28+1.00×−.50=101.8

The center depth is the difference between the corresponding numbers in the second and third columns. This is a fill if the subgrade is higher than the natural surface; otherwise, it is a cut.

Slope Stakes.—In addition to center stakes, slope stakes are used to mark the points where the side slopes of a cut or a fill intersect the natural surface of the ground. In Fig. 4, *c* is the center stake and *m* and *m'* are the slope stakes.

The method of locating slope stakes is as follows, all letters referring to Fig. 4:

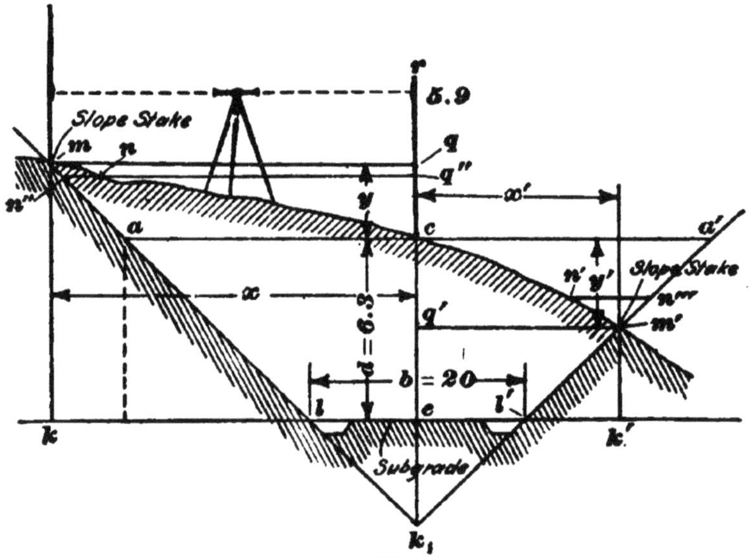

FIG. 4

Let *b* be the width *l l'* of the roadbed; *d*, the depth *ce* of the center stake; and *s* the slope ratio $= lk \div mk = l'k' \div m'k'$. For the upper stake at *m*, let *x* be the distance *mq* from the slope stake to the center line; $y+d$, the elevation of *m* above the subgrade $= qc+ce = mk$. Similarly for the lower stake at *m'*, let *x'* be the horizontal distance *m'q'* from *m'* to the center line, and let $d-y' = m'k'$, the elevation of *m'* above the subgrade.

Then, $$x = \frac{b}{2} + s \times d + s \times y \quad (1)$$

and $$x' = \frac{b}{2} + s \times d - s \times y' \qquad (2)$$

If the natural surface *mcm'* is a level line, so that *q*, *c*, and *q'* are at the same elevation, then $y = o$, $y' = o$, and

$$x = x' = ca = ca' = \frac{b}{2} + s \times d \qquad (3)$$

Formulas 1 and 2 are called *slope-stake equations* and formula 3 is called the *level-section equation*. The latter formula is available when the ground is nearly level. When the ground is sloping or irregular, formula 1 is employed, but not directly, as the value of y is not known until after the stake has been located. The distance x or x' is determined by successive trials. Suppose, for example, that, in Fig. 4, $d = 6.3$, and let the rod reading on the point c be 5.9. Suppose, also, that $s = 1.5:1$ and $b = 20$. Then, if the ground were level, by formula 3,

$$ac = \frac{20}{2} + 1.5 \times 6.3 = 19.5 \text{ ft.}$$

To find the location of *m*, the rodman will hold the rod at some point more than 19.5 from *c r*. Suppose that he holds it at *n*, 20 ft. from *c r*, and that the reading on the rod in this position is 2.8. Then, the height of this point above c equals the reading on c minus the reading on *n*, or $5.9 - 2.8 = 3.1$ ft. The computed distance from the rod to *c r* is by formula 1, $\frac{20}{2} + 1.5 \times 6.3 + 1.5 \times 3.1 = 24.1$ ft. Since the measured distance (20 ft.) is much smaller than this, the rod must be moved much farther out.

Suppose that the rod is carried out 7 ft. so that the measured distance to *c r* is 27 ft., and suppose that the reading on the rod in this position is .8 ft. The elevation of this trial point above c will be $5.9 - .8 = 5.1$ ft., and by formula 1, the computed distance x is $\frac{20}{2} + 1.5 \times 6.3 + 1.5 \times 5.1 = 27.2$ ft. This agrees so closely with the measured distance that the slope stake may be driven at this point.

The lower slope stake at *m'* is set in the same manner as the upper, except that the distance of each trial point below c is measured, and formula 2 is used in computing the corresponding value of x'. The distance of the trial point from *c r* will in this case be taken less than the distance *ca'* computed by formula 3. As in the preceding case, if the measured distance from *cr* to the trial point is less than the computed distance, the point should be moved out; if greater, it should be moved in.

Form of Notes in Cross-Section Work.—When each slope stake has been set as just explained, its distance from the center line and the elevation of the stake above or below subgrade are entered in the field book in the form of a fraction. The numerator of this fraction is the distance of the stake above or below subgrade, and the denominator is the distance of the stake from the center line. Thus, if the slope stakes in the preceding example are set at Sta. 131, the complete entry in the notebook will be as follows:

Station	Subgrade	Elevation	Center Depth	Left	Right
132	149.80	159.7	C 9.9		
131	148.80	155.1	C 6.3	$\frac{C\ 11.4}{27.2}$	$\frac{C\ 2.3}{13.5}$
130	147.80	147.2	F .6		

The fraction $\frac{C11.4}{27.2}$ indicates that the left slope stake at *m*, Fig. 4, is 27.2 ft. from the center line of the roadbed and 11.4 ft. above subgrade. Similarly, the fraction $\frac{C2.3}{13.5}$ indicates that the right slope stake *m'* is 13.5 ft. to the right of the center line and 2.3 ft. above subgrade. These expressions are called *slope-stake fractions.*

When the ground between the slope stakes and the center stake is irregular, the elevations and distances from the center of the intermediate points where the ground changes abruptly are determined and also entered in the notebook in the form of fractions.

COMPUTATION OF VOLUME

In calculating the cubical contents of earthwork, the volumes between two consecutive cross-sections are considered as prismoids whose bases are such sections as *mcm'l'l*, Fig. 4, and whose lengths are the distances between the cross-sections. These are usually 100 ft., unless the surface of the ground is rough and irregular, when sections at intervals of less than 100 ft. are taken. If A_1 and A_2 are the areas of the bases of a prismoid, A_m the area of a section midway between the bases, and l the perpendicular distance between them, the approximate volume V_a of the prismoid, as figured by the end-area method, is

$$V_a = \frac{l}{2}(A_1 + A_2) \qquad (1)$$

and the true area, as figured by the prismoidal formula, is

$$V = \frac{l}{6}(A_1 + 4\,A_m + A_2) \qquad (2)$$

Prismoidal Correction.—Formula 1 will usually give fairly good results; for accurate work, however, formula 2 is used. This formula requires that the dimensions of the middle section whose area is A_m shall be determined. This may be done by averaging the dimensions of the bases from which A_m might be computed. It is much simpler, however, to figure the approximate volume V_a by formula 1, and then, if desired, apply a correction equal to the algebraic difference between the volume V and V_a; the result obtained will be the same as if formula 2 were used. This difference is called the prismoidal correction.

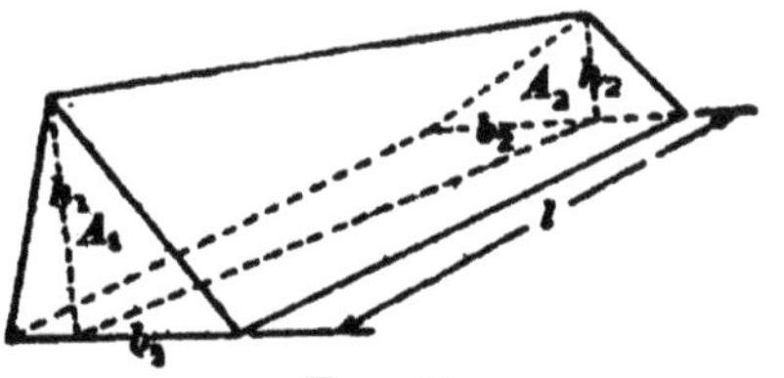

FIG. 5

Correction for a Triangular Prismoid.—Fig. 5 shows a triangular prismoid, the dimensions of which are marked. Its approximate volume as computed by formula 1 is

$$V_a = \frac{l}{2}\left(\frac{b_1h_1}{2} + \frac{b_2h_2}{2}\right)$$

and the prismoidal correction is

$$C=\frac{l}{12}(b_1-b_2)\ (h_2-h_1)$$

The true volume of the triangular prismoid is, therefore,

$$V=V_a+C$$

A study of the correction will show that, if either the bases, or the altitudes of the two end sections are equal, one of the factors (b_1-b_2) or (h_2-h_1) will become zero, and therefore the correction becomes zero. It shows also that, when one or both of these factors are small, the correction is a correspondingly small quantity; and that, when (as is usually the case) the breadth and height at one section are both smaller or both larger than the breadth and height at the other section, the correction is *negative*. Thus, if b_2 is less than b_1 and h_2 is less than h_1, then b_1-b_2 is positive, h_2-h_1 is negative, and, therefore, C is negative. But when C is negative, V_a is greater than the true volume V; that is, the method of averaging end areas usually gives a result that is too large. When the difference of the breadths and heights is very large, the correction is very large, and V_a is very greatly in error. Thus, for a pyramid, in which both b_2 and h_2 are zero, the correction is

$$\frac{l}{12}(b_1-0)\ (0-h_1)=-\frac{b_1h_1l}{12}$$

The true volume is $\frac{1}{6}b_1h_1l$, and therefore, the error in the value of V_a is one-half or 50%, of the true volume. This extreme case shows the importance of computing the prismoidal correction when the areas of the bases are very unequal.

EXAMPLE.—The dimensions of the bases of a triangular prismoid are: $b_1=18$ ft., $h_1=8$ ft., $b_2=12$ ft., and $h_2=9$ ft. Find the volume of this prismoid, in cubic yards, if the length of the prismoid is 100 ft.

SOLUTION.—The areas of the bases are: $A_1=\frac{1}{2}\times18\times8=72$ sq. ft., and $A_2=\frac{1}{2}\times12\times9=54$ sq. ft. Substituting these values in the preceding formula for V_a, and dividing by 27 to reduce to cubic yards,

$$V_a=\frac{100}{2}\times(72+54)\div27=233.33 \text{ cu. yd., nearly}$$

Substituting the given values in the formula for C, and dividing by 27 to reduce to cubic yards,

$$C=\frac{100}{12}\times(18-12)\times(9-8)\div27=1.85 \text{ cu. yd.}$$

Therefore,

$$V = 233.33 + 1.85 = 235.18, \text{ say } 235, \text{ cu. yd.}$$

Correction for Curvature.—Besides the prismoidal correction, a correction for curvature is sometimes required in calculations of earthwork on a curve.

In Fig. 6, let rr_1 be the curved center line of the roadbed, O the center of this circular curve, and R its radius. Let A_1 be the area of the cross-section $mnpq$, G its center of gravity, and e_1 the horizontal distance from G to the center of the roadbed, which distance is called the *eccentricity* of the section. Similarly, let A_2 be the area of the section $m_1n_1p_1q_1$, G_1 its

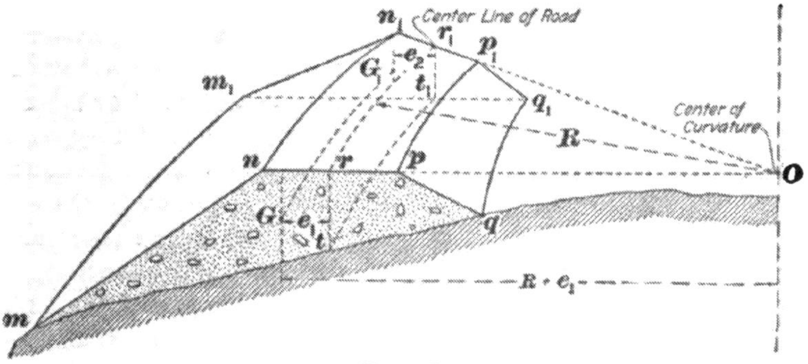

FIG. 6

center of gravity, and e_2 the eccentricity of that section. The general formula for curvature correction is, then,

$$C_c = \frac{l}{2R}(A_1e_1 + A_2e_2)$$

If G and G_1 lie on the outside of the curved center line of the roadbed, C_c is to be added to the volume calculated as for a straight track. If G and G_1 are on the inside of this curved center line, the correction C_c is to be subtracted.

The expression for C_c shows that the larger the eccentricities of the end sections, the larger C_c will be, and that, if the radius of the curve is very large, C_c will be very small. For curves of very large radii, the correction is usually so small that it may be neglected. When the area of that part $rpqt$ of the end section lying on the inside of the center of the track is approximately equal to the portion of the area $rtmn$ lying outside of the center, the eccentricity is small, and the correction may usually be neglected, even with curves of short radii. But when the eccentricity is large (as is usually the case in side-hill work), the curvature correction may be a very considerable percentage of the volume, and should not be neglected, especially if the radius of the curve is small.

To apply the general formula for curvature correction, the eccentricities e_1 and e_2 are required. These can be determined by using the methods employed in finding the center of gravity of plane figures. The section is divided into triangles and their areas are referred to the vertical axis through the center of the track; then the coordinate of the center of gravity of the total area with regard to this axis is found, which coordinate is the eccentricity of the section.

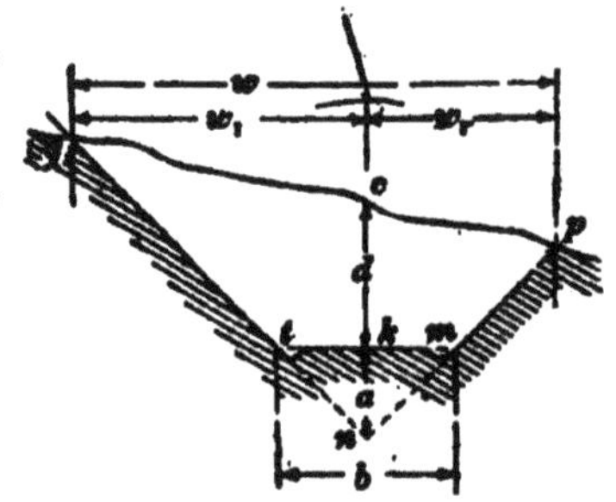

FIG. 7

Three-Level Sections. Where the surface of the ground is fairly regular, it is sufficiently accurate to determine the elevation of the center point and the distances and elevations of the two slope stakes. The method assumes that the straight lines cq and cp, Fig. 7, that join the center with the slope stakes are on the surface of the ground. When this method is used, the sections are called *three-level sections.*

To calculate the volume of a prismoid whose bases are three-level sections distant l from each other, let, in Fig. 7, the area $qcpn = A_t$ and the area of $tmn = T$. Then, using the notation of the figure and the sign (′) to denote corresponding values at the other base, the approximate volume is

$$V_a = \frac{l}{2}(A_t + A_t' - 2T)$$

or

$$V_a = \frac{l}{4}\left[(a+d)w + (a+d')w' - 2ab\right]$$

and the prismoidal correction is

$$C = \frac{l}{12}(w-w')\,(d'-d)$$

In calculating the correction for curvature in three-level section work, it is sufficiently accurate to use in the general formula for curvature correction the values e_1, e_2 and A_1, A_2 for the full sections $qcpn = A_t$ instead of the actual area $qcpmt$. The values of e_1 and e_2 are then too small, and the resulting error nearly neutralizes the one due to the inclusion in the area of the triangle tmn. The eccentricity of the area $qcpn$ is $e_1 = \frac{1}{3}(w_l - w_r)$, and, using the same notation as before, the curvature correction becomes

$$C_c = \frac{l}{6R}\left[A_t(w_l - w_r) + A_t'(w_l' - w'_r)\right]$$

The form in which the computation of volume should be arranged when the cross-sections are three-level sections is shown in the table on page 205. The figures in the first four columns are written while the survey is being made; those in columns 5, 6, and 7 are used for computing the average-end area volume V_a; those in columns 8, 9, and 10 are employed in computing the prismoidal correction; and the figures in the last two columns are used for computing the correction for curvature.

The values of V_a for the prismoids included between the successive cross-sections are found as follows: Since the results always are expressed in cubic yards, the preceding formula for V_a becomes, for the volume between two full stations ($l = 100$),

$$V_a = \frac{100}{4\times 27}(a+d)w + \frac{100}{4\times 27}(a+d')w' - \frac{2\times 100}{4\times 27}\times a \times b$$

If the slope $s = 1\frac{1}{2}$:1 and the width of the roadbed $b = 22$ ft., then a for all stations is

$$a = \frac{1}{2}b \div s = \frac{\frac{1}{2}\times 22}{\frac{3}{2}} = 7.3 \text{ ft.}$$

The sums of the constant depth a and the variable depths d in the second column are written in the fifth column. Thus, at Sta. 22, $a+d = 7.3+6.2 = 13.5$ ft.; at Sta. 23, $a+d = 7.3 + 9.4 = 16.7$ ft. The total width at each station is written in the sixth column. Since, in Fig. 7, $w = w_l + w_r$, and since the measured distances w_l and w_r are the denominators of the fractions in columns 3 and 4 respectively, it is only necessary to add the two denominators at each station to obtain the numbers in column 6. Thus, at Sta. 22, $w = 16.1 + 30.2 = 46.3$; at Sta. 23, $w = 18.2 + 31.4 = 49.6$ ft.

To compute the value of V_a between Sta. 22 and Sta. 23, the proper values must be substituted in the formula for V_a. This gives

$$V_a = \frac{100}{4 \times 27} \times 13.5 \times 46.3 + \frac{100}{4 \times 27} \times 16.7 \times 49.6 - \frac{2 \times 100}{4 \times 27} \times 7.3 \times 22 = 579 + 767 - 297 = 1{,}049 \text{ cu. yd.}$$

The number 579 is written in column 7 (*a*) opposite Sta. 22, and 767 in the same column opposite Sta. 23. The result, 1,049 cu. yd., is written opposite Sta. 23, in column 7 (*b*).

In a similar manner, for the volume of the prismoid between Sta. 23 and Sta. 24,

$$V_a = \frac{100}{4 \times 27} \times 16.7 \times 49.6 + \frac{100}{4 \times 27} \times 19.1 \times 64 - \frac{2 \times 100}{4 \times 27} \times 7.3 \times 22$$

The first term of this expression has already been computed, and its value, 767 cu. yd. has been written in column 7 (*a*) opposite Sta. 23. The last term is the constant volume 297 cu. yd. It is therefore necessary to compute the second term only. Its value is found to be 1,132 cu. yd., and this is written in column 7 (*a*) opposite Sta. 24. Then, $V_a = 767 + 1{,}132 - 297 = 1{,}602$ cu. yd., and this result is written in column 7 (*b*).

It is thus seen that, at each station, it is necessary to compute but one term of the formula for V_a; this term is the value of $\frac{100}{4 \times 27}(a+d)w$ for that station. The value of this term for each station is written in column 7 (*a*). If the stations are 100 ft. apart, any number in column 7 (*b*) is obtained by adding the number opposite and the one preceding it in column 7 (*a*) and subtracting 297 cu. yd. from the resulting sum. The result so obtained is the value of V_a for a prismoid 100 ft. long. But if the two stations are less than 100 ft. apart, the result must be multiplied by the ratio of their distance to 100 ft. to obtain the volume of the prismoid. This volume is then written in column 7 (*b*). For example, for the prismoid

FORM OF NOTES FOR THREE-LEVEL GROUND

1 Station	2 Center Depth	3 Left	4 Right	5 $(a+d)$	6 w	7 Volumes (a)	7 Volumes (b)	8 $w-w'$	9 $d'-d$	10 Prismoidal Correction C	11 w_l-w_r	12 Curvature Correction
25	C 2.4	C .6 / 11.9	C 4.7 / 18.1	9.7	30.0	269	426	+18.0	−5.7	−21	− 6.1	− 2
24+35	C 8.1	C 5.9 / 19.9	C 11.4 / 28.1	15.4	48.0	684	532	+16.0	−3.7	− 6	− 8.2	− 3
24	C 11.8	C 8.8 / 24.2	C 19.2 / 39.8	19.1	64.0	1,132	1,602	−14.4	+2.4	−11	−15.6	−11
23	C 9.4	C 4.8 / 18.2	C 13.6 / 31.4	16.7	49.6	767	1,049	− 3.3	+3.2	− 3	−13.2	− 7
22	C 6.2	C 3.4 / 16.1	C 12.8 / 30.2	13.5	46.3	579					−14.1	
										−41		−23

Volume by average end **areas** ... 3,609
Prismoidal correction........... −41
Curvature correction........... −23

Volume by prismoidal formula.. 3,545

Roadbed 22 ft. wide. Slope ratio = 1.5 to 1. 7° curve to the right

between Sta. 24 and Sta. 24+35, there should be obtained, provided the prismoid is 100 ft. long, 1,132+684−297 = 1,519 cu. yd. As the length is but 35 ft., the actual value of V_a is $\frac{35}{100} \times 1{,}519 = 532$ cu. yd., which is written in column 7 (*b*).

It is usually more convenient to compute all the numbers in each column before passing on to the next column. When column 7 (*b*) has been filled up, the number in this column opposite each station is the approximate number of cubic yards, computed by average end areas, contained between that station and the preceding station. Thus, 1,048 is the approximate number of cubic yards between Sta. 23 and Sta. 22; 531 is the approximate number between Sta. 24+35 and Sta. 24; etc. The total approximate number of cubic yards, between Sta. 22 and Sta. 25, as computed by average end areas, is, therefore, 1,049+1,602+532+426 = 3,609 cu. yd.

The prismoidal correction must now be computed.

Since the result is to be expressed in cubic yards, the preceding formula for *C* becomes

$$C = \frac{l}{12 \times 27}(w - w')\ (d' - d)$$

The successive values of $w - w'$ in column 8 are obtained by subtracting each number in column 6 from the number just below it in this column. Thus, for the prismoid between Sta. 22 and Sta. 23, $w = 46.3$, $w' = 49.6$; and $w - w' = -3.3$ ft. Similarly, the values of $d' - d$ in column 9 are obtained by subtracting each number in column 2 from the number just above it in this column. Thus, for the first prismoid, $d = 6.2$, $d' = 9.4$, and $d' - d = +3.2$ ft.

The numbers in column 10 are the computed values of the prismoidal correction *C*. Thus, for the first prismoid, since $l = 100$

$$C = \frac{100}{12 \times 27} \times -3.3 \times 3.2 = -3 \text{ cu. yd.}$$

for the second prismoid,

$$C = \frac{100}{12 \times 27} \times -14.4 \times 2.4 = -11 \text{ cu. yd.},$$

and similarly for the remaining prismoids.

The volume of the first prismoid, as obtained by the

prismoidal formula, is, therefore, 1,049 − 3 = 1,046 cu. yd.; that of the second, 1,602 − 11 = 1,591 cu. yd., etc.

Now assume that the portion of the track just calculated is on a 7° curve to the right. Applying the formula for C_c for stations of 100 ft. and in cubic yards,

$$C_c = \frac{100}{3R}\left[\frac{A_t(w_l - w_r)}{2\times 27} + \frac{A'_t(w'_l - w'_r)}{2\times 27}\right]$$

At Sta. 22, $w_l = 16.1$, $w_r = 30.2$, and, hence, $w_l - w_r = 16.1 - 30.2 = -14.1$. At Sta. 23, $w_l' = 18.2$, $w_r' = 31.4$, and, hence, $w_l' - w_r' = 18.2 - 31.4 = -13.2$. The values of $\frac{100A_t}{2\times 27}$ and $\frac{100A_t'}{2\times 27}$ are those already tabulated in column 7 (*a*); thus, $\frac{100A_t}{2\times 27} = 579$ and $\frac{100A'_t}{2\times 27} = 767$. Substituting all of these values, and the value of $R = 819$ for a 7° curve,

$$C_c = \frac{1}{3\times 819}\times(579\times -14.1 + 767\times -13.2) = -7 \text{ cu. yd.}$$

Since w_l and w_l' are smaller, respectively, than w_r and w_r', the centers of gravity of the sections lie on the right of the center line of the roadbed; and, as the curve turns to the right, the centers of gravity lie inside of the center line, and the correction is to be subtracted. The volume for this section computed by the prismoidal formula is 1,049 − 3 = 1,046 cu. yd., and, corrected for curvature, the final result is 1,046 − 7 = 1,039 cu. yd. The curvature corrections for other sections are figured in a similar manner, except for sections less than 100 ft. long, when the result must be multiplied by the ratio of the length of the section to 100 ft. To find, for instance, the curvature correction for the section between Sta. 24 and Sta. 24 + 35, determine, as before, the correction just as if the station were 100 ft. long and multiply the result by $\frac{35}{100}$. Thus,

$$C_c = \frac{1}{3\times 819}(1{,}132\times -15.6 + 684\times -8.2)\times \tfrac{35}{100} = -3 \text{ cu. yd.}$$

As in the previous case, the actual volume is less than the one computed for a straight track; therefore, the actual volume $V = 532 - 6 - 3 = 523$ cu. yd.

Irregular Sections.—When the cross-sections are irregular, the process of determining the volume of earthwork is essentially the same as for three-level sections, except that a more accurate method for computing the areas of the cross-sections is applied. After the area has been determined, the volume by the end-area method, and the prismoidal and curvature corrections are obtained just as if the figures were three-level sections.

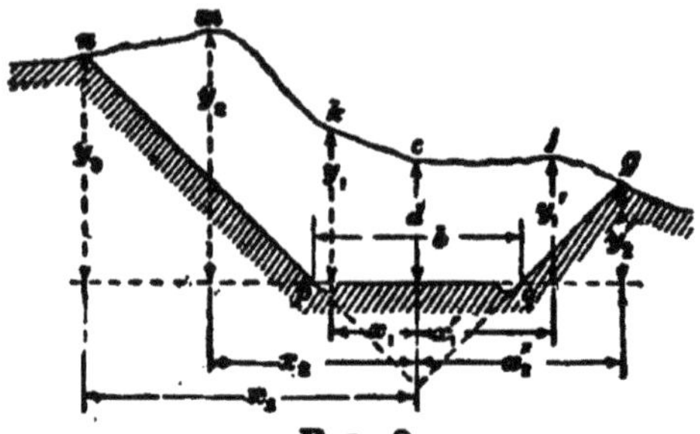

FIG. 8

Areas of Irregular Sections.—Using the notation of Fig. 8, the area *npqgfckm* is determined by the formula

$$A=\tfrac{1}{2}(\tfrac{1}{2}by_3+x_3y_2+x_2y_1+x_1d+dx_1'+y_1'x_2'$$
$$+\tfrac{1}{2}by_2'-y_3x_2-y_2x_1-x_1'y_2')$$

This long expression for the area may be very easily formed as follows: Write the successive slope-stake fractions in order, in a horizontal row, beginning with the extreme left slope stake; and for the center stake write the fraction $\frac{d}{0}$. At the beginning and end of the row, write the fraction $\frac{0}{\frac{1}{2}b}$. Thus, the fraction for the stake at *n* is $\frac{y_3}{x_3}$; for the point *m*, it is $\frac{y_2}{x_2}$, etc.; so that the row of fractions for Fig. 8 will be as follows:

$$\frac{0}{\frac{1}{2}b}\times\frac{y_3}{x_3}\times\frac{y_2}{x_2}\times\frac{y_1}{x_1}\times\frac{d}{0}\times\frac{y_1'}{x_1'}\times\frac{y_2'}{x_2'}\times\frac{0}{\frac{1}{2}b}$$

Next, multiply each denominator by the numerator that follows it and each numerator by the denominator that follows it, giving to those products connected with full lines the plus sign, and to those connected with dotted lines the minus sign. One-half of the algebraic sum of these products will be the desired area. This is evident, since, proceeding according to the directions, the positive products are

$$\tfrac{1}{2}by_3,\ x_3y_2,\ x_2y_1,\ x_1d,\ dx_1',\ y_1'x_2',\ \text{and}\ y_2'\tfrac{1}{2}b$$

and the negative products are

$$-y_3x_2, -y_2x_1,\ \text{and}\ -x_1'y_2'$$

One-half of the algebraic sum of these is identical with the second member of the preceding formula.

NOTE.—The method just described for determining areas of irregular sections is general and may also be used for three-level sections.

Following is an illustrative example showing the application of the preceding method of determining the areas of irregular sections. The field notes are given in the accompanying table. The station numbers in column 1 run from the

FIELD NOTES

1 Station	2 Center Cut or Fill	3 Left	4 Right
129	C 8.3	C 12.7 / 31.0 C 16.0 / 15.0 C 12.2 / 10.5	C 4.1 / 8.2 C 6.0 / 21.0
+ 40	C 13.2	C 22.8 / 46.2 C 20.4 / 31.0 C 18.2 / 19.5	C 12.8 / 13.7 C 10.4 / 27.6
128	C 10.9	C 18.6 / 39.9	C 8.0 / 4.2 C 8.5 / 24.8
127	C 8.6	C 14.6 / 33.9	C 12.4 / 30.6
126	C 4.2	C 9.6 / 26.4	C 2.1 / 15.1

Roadbed 24 feet wide in cut. Slope 1.5 : 1.

bottom of the page upwards, so that when one stands on the line of the road looking forwards, the slope-stake fractions, which give for each point the height and distance from the center, will have on the notebook the same relative position as they have on the ground. These figures for the left-hand side are always given at the extreme left of the space in column

COMPUTATION

1 Station	2 Double Plus Areas	3 Double Minus Areas	4 Cubic Yards (a)	 (b)	5 w	6 $w-w'$	7 $d'-d$	8 Prismoidal Correction
129	152.4 496.0 183.0 87.2 68.1 86.1 72.0	190.5 168.0 49.2	683	1,215	52.0	+21.8	−4.9	−20
128+40	273.6 942.5 564.2 257.4 180.8 353.3 124.8	706.8 397.8 142.5	1,342	896	73.8	−9.1	+2.3	− 3

128	223.2 434.9 45.8 198.4 102.0	35.7	897	1,711	64.7	−.2	+2.3	0
127	175.2 291.5 263.2 148.8		814	1,105	64.5	−23.0	+4.4	−31
126	115.2 110.9 63.4 25.2		291		41.5			
				$V_a = 4{,}927$ $C = -54$ $V = 4{,}873$				−54

3. The line between columns 3 and 4 may then represent the center line; the intermediate points between the left-hand slope stake and the center are given in their order in column 3. Similarly, the points on the right side are placed in column 4. The figures for the right-hand slope stake are always placed at the extreme right-hand side of that column.

The preceding table shows how the computations are arranged. Take, for example, the section between Sta. 128 +40 and Sta. 129. To find the end area at Sta. 128+40, the following fractions are written:

$$\frac{0}{12.0} \times \frac{22.8}{46.2} \times \frac{20.4}{31.0} \times \frac{18.2}{19.5} \times \frac{18.2}{0} \times \frac{12.8}{13.7} \times \frac{10.4}{27.6} \times \frac{0}{12.0}$$

The products of the numbers connected by full lines, 12.0 ×22.8, 46.2×20.4, etc., are written in column 2, and the products of those connected by dotted lines, 22.8×31.0, 20.4×19.5, etc., are written in column 3. The sum of the double plus areas is 2,696.6, and the sum of the double minus areas is 1,247.1. The area of the section is, therefore, $\frac{1}{2}\times(2{,}696.6-1{,}247.1)=724.8$ sq. ft.

The area at Sta. 129 is obtained in a similar manner; thus, $\frac{1}{2}(1{,}144.8-407.7)=368.6$ sq. ft.

The volume for a 100-ft. section as figured by the average end-area method is

$$V_a=\frac{l}{2}(A_1+A_2)=\frac{100}{2\times 27}A_1+\frac{100}{2\times 27}A_2$$

For Sta. 128+40,

$$\frac{100}{2\times 27}A_1=\frac{100}{2\times 27}\times 724.8=1{,}342 \text{ cu. yd.}$$

And for Sta. 129,

$$\frac{100}{2\times 27}A_2=\frac{100}{2\times 27}\times 368.6=683 \text{ cu. yd.}$$

These figures are entered in column 4 (*a*) of the table of computations.

If the prismoid were 100 ft. long, the volume V_a would be 683+1,342=2,025 cu. yd. As the prismoid is but 60 ft.

long, the volume is $\frac{60}{100}\times 2{,}025 = 1{,}215$ cu. yd., and this number is written in column 4 (*b*) opposite Sta. 129.

The computation for the other stations is made in a similar way. It will be observed that the sections at Sta. 126 and Sta. 127 are three-level sections, and that in this case there are no minus areas.

The sum of the numbers in column 4 (*b*) is 4,927 cu. yd.; the prismoidal correction, which is figured according to the formula for *C* under the heading Three-Level Section, is -54 cu. yd.; the final volume between Sta. 126 and Sta. 129, is, therefore, $4{,}927 - 54 = 4{,}873$ cu. yd.

Side-Hill Work.—When both the cut and the fill occur in the same section, as in Fig. 9, the areas, volumes, and their corrections are determined for the fill and the cut separately. For

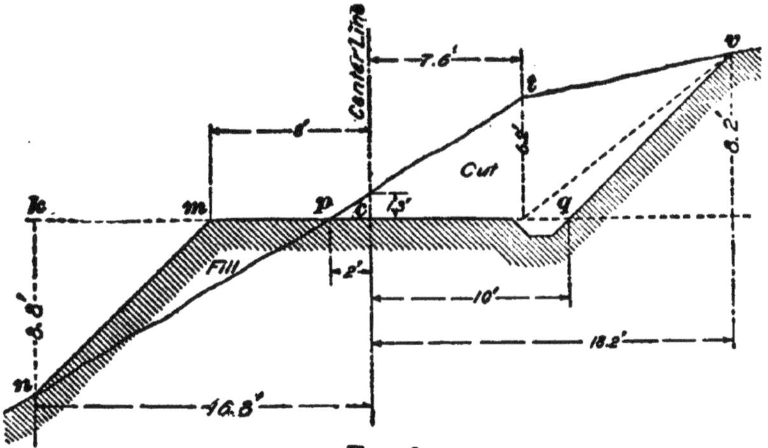

FIG. 9

the purpose of calculating the prismoidal and curvature corrections, each part of the section, cut or fill, is considered as a triangle and the formulas previously given are used.

For calculating the areas, it is also frequently sufficient to consider that the section in either fill or cut is triangular. This is, however, not exact enough when the ground is very irregular. In Fig. 9, the area of the fill would be taken as that of the triangle *mnp*, while for determining the area of the cut the method of irregular sections would be used.

Suppose that the shoulder m, Fig. 9, of the slope is 8 ft. from the center; that the fill begins at 2 ft. from the center, and is a rock fill with a slope of 1:1, and that the slope stake n is 16.8 ft. from the center. Then, $mk = ck - cm = 16.8 - 8.0 = 8.8$ ft.; and, since the slope $km + nk$ is 1:1, the vertical distance nk of n below subgrade will also be 8.8 ft. The area of the fill, is, then, $\frac{8.8 \times 6}{2} = 26.4$ sq. ft.

In determining the area of the cut, it will be observed that the fraction for the point p is $\frac{0}{2}$; that for t is $\frac{C6.2}{7.6}$; and that for v is $\frac{C8.2}{18.2}$. The center depth is 1.3 ft., and the distance $cq = \frac{1}{2}b$ is 10 ft. The notes for the entire section shown in Fig. 9, will therefore be as given in the following table:

Station	Center Depth	Left		Right	
33	C 1.3	$\frac{F\ 8.8}{16.8}$	$\frac{0}{2.0}$	$\frac{C\ 6.2}{7.6}$	$\frac{C\ 8.2}{18.2}$

The series of fractions will therefore be, considering only the section of cut,

$$\frac{0}{10} \quad \frac{0}{2} \quad \frac{1.3}{0} \quad \frac{6.2}{7.6} \quad \frac{8.2}{18.2} \quad \frac{0}{10}$$

The double areas are as follows:

Plus Areas	*Minus Areas*
2.6	62.3
9.9	
112.8	
82.0	
Sum = 207.3	

The desired area for cut is, therefore, $\frac{1}{2} \times (207.3 - 62.3) = 72.5$ sq. ft.

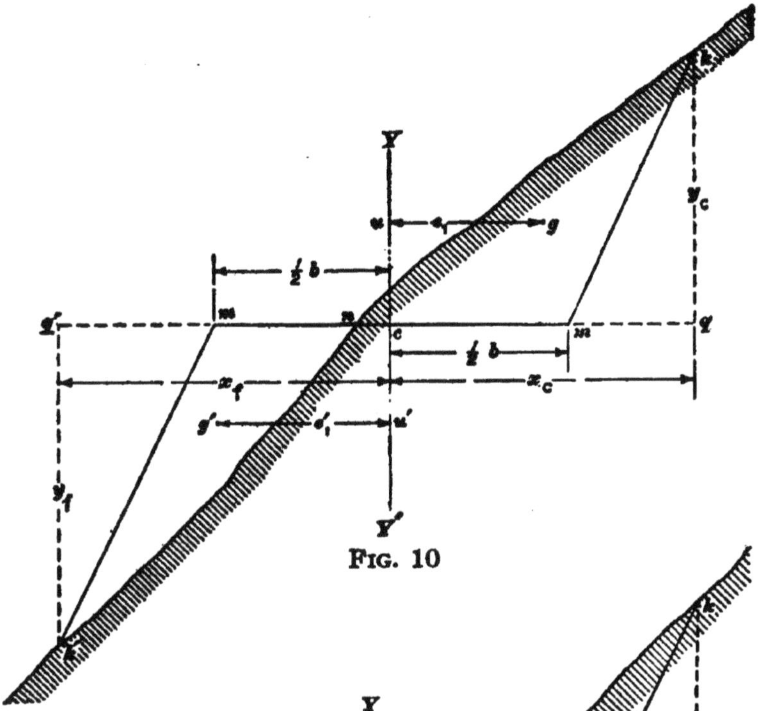
Y
u
e_1
g
$\frac{1}{2}b$
y_c
q'
m
n
c
m
q
$\frac{1}{2}b$
x_f
x_c
g'
e'_1
u'
y_f
k
k'
Y'

FIG. 10

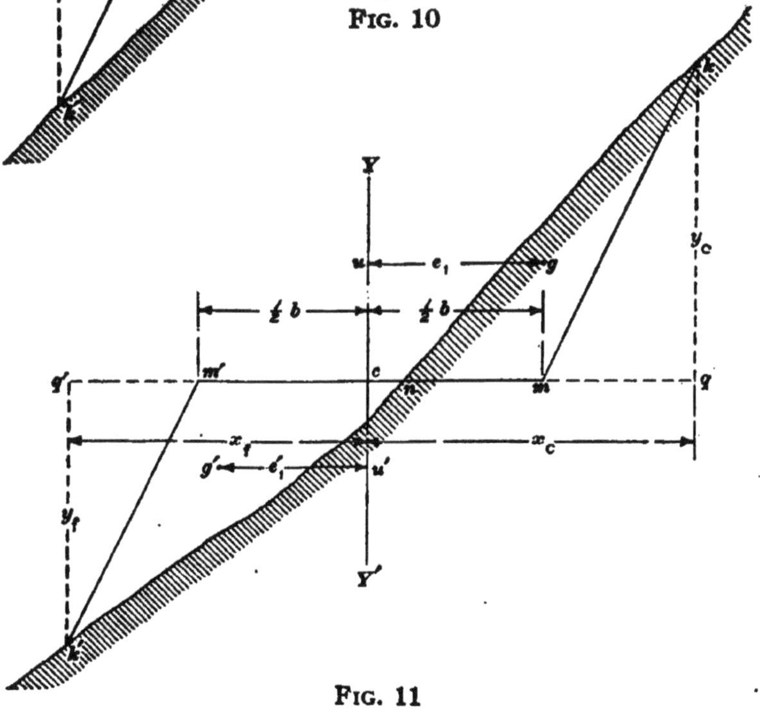
Y
u
e_1
g
$\frac{1}{2}b$
$\frac{1}{2}b$
y_c
q'
m'
c
n
m
q
x_f
x_c
g'
e'_1
u'
y_f
k
k'
Y'

FIG. 11

Eccentricity in Side-Hill Work.—As stated before, in making the correction for curvature in side-hill work the sections of fill or cut are considered as triangles and the following formula is used:

$$C_c = \frac{l}{2R}(A_1e_1 + A_2e_2)$$

The values of A_1 and A_2 are readily obtained as areas of triangles. For finding the eccentricities, two cases are to be distinguished in either cut or fill. Using the notation of Figs. 10 and 11, in which g and g' are the centers of gravity at the cuts and fills considered as triangles, the formulas for e_1 and e_1' Fig. 10, where the central stake lies in the cut, are

$$e_1 = gu = \tfrac{1}{3}(x_c + \tfrac{1}{2}b - nc)$$

and

$$e_1' = g'u' = \tfrac{1}{3}(x_f + \tfrac{1}{2}b + nc)$$

When the central stake lies in the fill, as in Fig. 11,

$$e_1 = gu = \tfrac{1}{3}(x_c + \tfrac{1}{2}b + nc)$$

and

$$e_1' = g'u' = \tfrac{1}{3}(x_f + \tfrac{1}{2}b - nc)$$

As will be noted, the value of $\tfrac{1}{2}b$ to be substituted in the formulas is not the same for cut as for fill.

CHANGE IN VOLUME OF EARTHWORK

Shrinkage of Earthwork.—When earth is excavated and formed into an embankment the volume of earth is at first larger than the original excavation, but, after some time, it shrinks to a volume less than that of the original excavation. The accompanying table contains for various kinds of soils, in the second column, the approximate number of cubic yards of embankment that can be formed from 1,000 cu. yd. of excavation. In the third column is given the number of cubic yards of excavation required for each 1,000 cu. yd. of embankment, and in the fourth column is shown the per cent. of shrinkage.

Growth of Rock.—The material from a rock excavation has a larger volume than the original volume in the cut, and there is practically no subsequent shrinkage. The following table shows the approximate number of cubic yards of embankment that can be formed from 1,000 cu. yd. of excavation, the

SHRINKAGE OF EARTHWORK

Character of Material	Embankment Obtained From 1,000 Cu. Yd. of Excavation Cubic Yards	Excavation Required for 1,000 Cu. Yd. of Embankment Cubic Yards	Shrinkage Per Cent.
Sand and gravel.	920	1,087	8
Clay............	900	1,111	10
Loam...........	880	1,136	12
Wet soil.........	850	1,200	15

number of cubic yards of excavation required for 1,000 cu. yd. of embankment, and the per cent. of growth for the various sizes of hard rock.

GROWTH OF ROCK

Character of Material	Embankment Obtained From 1,000 Cu. Yd. of Excavation Cubic Yards	Excavation Required for 1,000 Cu. Yd. of Embankment Cubic Yards	Growth Per Cent.
Hard rock, large fragments....	1,600	625	60
Hard rock, medium fragments.	1,700	587	70
Hard rock, small fragments....	1,800	556	80

HAULAGE

Limit of Free Haul.—Specifications for earthwork usually allow the contractor extra compensation for transporting material beyond a certain distance, say 800, or, perhaps, 1,000 ft., which distance is called the *limit of free-haul.* No deduction is made for hauls that are less than the specified limit; but in cases of long hauls, he receives compensation for *overhaul* only; that is, only for the distance exceeding the free-haul limit. The allowance is made per cubic yard for each station of 100 ft.

Computation of Haulage.—If, in the profile shown in the accompanying illustration, the material of the cut is deposited in the position ODN, the *total haulage*—that is, the sum of all products obtained by multiplying each volume by the distance through which it is hauled—will be

$$\text{volume } CMO \times ZZ' = \text{volume } OND \times ZZ',$$

G and G' being, respectively, the centers of gravity of the cut CMO and the embankment ODN.

But, as the short hauls are not averaged against those which are beyond the limit of free haul, the contractor is entitled to

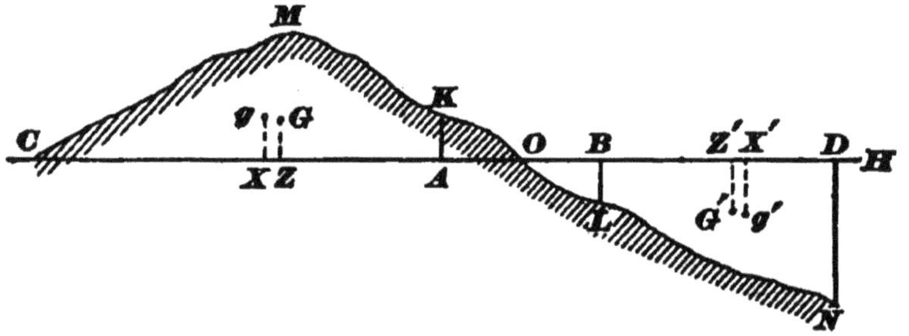

extra compensation when the distance CD exceeds the limit of free haul. To calculate the overhaul, two points A and B must be found whose distance apart equals the limits of free haul and which are situated so that the volume AKO equals that of OBL. The remaining part $CMKA$, which is to be placed in the position $BLND$ is to be considered as overhaul.

If g and g' are, respectively, the centers of gravity of these volumes, and V the cubical contents of each, then the haulage of this volume is $V \times XX'$. Of this, the distance AB is to be hauled free of charge, and the overhaul is therefore

$$O = V \times XX' - V \times AB = V(XX' - AB) = V \times XA + V \times X'B$$

Since V = volume of $CMKA$ = volume of $BLDN$, the simple rule for figuring overhaul is to compute the total haulage of the cut $CMKA$ to the point A and the total haulage of the fill $BDNL$ to the point B, and then add the results. These values of $V \times XA$ and $V \times X'B$ are found as follows:

Let v = volume of any prismoid in cut;
a = area of its end section nearest to A;
a' = area of its end section most remote from A;
m = distance from A to middle section of prismoid;
l = length of prismoid, in feet;
x = distance from center of gravity of this prismoid to point A.

Then, $$x = m + \frac{l}{6} \times \frac{a' - a}{a' + a}$$

The overhaul of this prismoid from its position in the cut to the point A will therefore be, since overhaul is reckoned in stations,

$$\frac{vx}{100} = \frac{v}{100}\left(m + \frac{l}{6} \times \frac{a' - a}{a' + a}\right)$$

By this formula, the overhaul for each prismoid of the cut is computed for the transportation of this material to the point A. In a similar manner, the overhaul for the transportation of each prismoid to its position in the fill $BLND$ from the point B is found. The sum of the overhauls for all the prismoids of the cut and fill is the desired total overhaul.

If a part of the cut, for example MZO, is hauled in one direction, and the remainder MZC in the other, the overhaul for each part of the cut must be computed separately.

EXAMPLE.—Let $CMKA$ in the preceding illustration represent the cut for which the computations on pages 210 and 211 are shown, C being Sta. 126 and A Sta. 129. Let, also, the length of free haul be 600 ft., B being Sta. 135, and let the volumes and end areas of the prismoids beyond Sta. 135 be as follows:

Station	End Area	Volume	
		(*a*)	(*b*)
137	769	1,424	2,105
136	368	681	2,262
135	854	1,581	

Sum = 4,367

If 1c. is paid for each cubic yard hauled one station in the overhaul, find the total allowance for overhaul if the shrinkage of the material in the embankment is 10%.

SOLUTION.—The foregoing formula must be applied to each of the prismoids.

1. *For the Cut.*—Following is the tabulation of the end areas and volumes; the end areas are the algebraic sums of one-half the plus and minus areas found in the tabulation on pages 210 and 211, and the volumes are obtained by applying the prismoidal correction to the volumes in column 4 (*b*) of that table.

Station	End Areas	Volume	m	$\frac{l}{6}\left(\frac{a'-a}{a'+a}\right)$	x	$\frac{vx}{100}$
129	368.5	1,195	30	+3	33	394
128+40	724.8	893	80	−1	79	705
128	484.3	1,711	150	−1	149	2,549
127	439.4	1,074	250	−8	242	2,599
126	157.4					
		Sum = 4,873				Sum = 6,247

The numbers in the fourth column are the distances from the middle sections of the prismoids to the point *A*, at Sta. 129, at which point the free haul begins. Thus, the middle section of the prismoid between Sta. 126 and Sta. 127 is at Sta. 126+50; the distances from this section to Sta. 129 is $(129-126.50)\times 100 = 250$ ft. Similarly, for the prismoid between Sta. 127 and Sta. 128. $m = (129-127.50)\times 100 = 150$ ft.

The value of $\frac{l}{6}\times\frac{a'-a}{a'+a}$ for each prismoid, is given in the fifth column. Thus, for the first prismoid,

$$\frac{100}{6}\times\frac{157.4-439.4}{157.4+439.4} = -8 \text{ ft.}$$

For the second prismoid,

$$\frac{100}{6}\times\frac{439.4-484.3}{439.4+484.3} = -1 \text{ ft.}$$

and similarly for the others.

The numbers in the sixth column are the sums of the corresponding numbers in the fourth and fifth columns; each of these numbers in the sixth column is the distance from the point *A*, to the center of gravity of the corresponding prismoid.

Finally, the overhaul for each prismoid is the product of the volume in the third column by the distance x in the sixth column. These products are written in the seventh column; but, since the distance x is expressed in feet, and the allowance is 1c. per cu. yd. per sta., each product is divided by 100 before writing it in the seventh column. The sum of the numbers in the seventh column is 6,247; the overhaul for the cut is therefore the equivalent of 6,247 cu. yd. overhauled one station.

2. *For the Fill.*—The total volume of the cut is 4,873 cu. yd. Since the shrinkage is 10 %, the volume of this material when placed in the embankment will be 4,873—487 = 4,386 cu. yd. Since the volume of the embankment between Sta. 135 and Sta. 137 is 4,367 cu. yd., the embankment made from the cut practically ends at Sta. 137. Therefore, the point *D*, may be taken as Sta. 137.

The computation of overhaul for fill between Sta. 135, or *B*, and the center of gravity of each prismoid is now computed exactly as in the case of the cut. The results are as follows:

Station	End Area	Volume	m	$\frac{l}{6}\times\frac{a'-a}{a'+a}$	x	$\frac{vx}{100}$
137	769	2,105	150	+6	156	3,284
136	368	2,262	50	−7	43	973
135	854					

Sum = 4,257

The sum of all the values of $\frac{vx}{100}$ is 6,247+4,257 = 10,504. This is the equivalent of 10,504 cu. yd. overhauled one station. At the rate of 1c. per cu. yd. per sta., the allowance for overhaul will be .01×10,504 = \$105.04.

RAILROAD LOCATION

RECONNAISSANCE

The engineering operations preceding the building of a railroad are (1) the reconnaissance, (2) the preliminary survey, and (3) the location.

The *reconnaissance* is a rapid examination of a strip of country lying between the proposed terminals with the following objects in view: (1) To determine the most feasible and economical line between the terminal points; (2) to locate the controlling points, which consist of stream crossings, summits of ridges, and other natural and artificial features of the territory through which the road must necessarily pass in order to come within the limit of permissible cost of construction, and which include such features as the position of towns, manufacturing sites, etc.; (3) to determine the maximum grade and the maximum rate of curvature; (4) to ascertain the kind of material likely to be encountered in the construction of the road, and to determine the effect of the material on the cost of maintenance; (5) to note the resources of the country and its capabilities for future development, and to calculate the probable effect of the building of the road on this development; (6) to obtain a general idea of the approximate cost per mile and of the total cost of the completed road.

For the purpose of determining relative elevations and directions of streams and roads, the engineer should provide himself with an aneroid barometer, a pocket compass, and a hand level. Much useful information can be obtained from existing maps. With this equipment the engineer investigates personally all important points involved and makes comprehensive notes of all topographical features along the route, such as the size and direction of streams, together with their highwater marks; the slope of important waterways that must be crossed; and any other information concerning them that can be secured. Such information as can be obtained regarding the character of the soil, the prevalence of rock, the amount of timber available for construction, the amount of rainfall, etc., should be carefully noted. In addition, the engineer should note the probable quantities of excavation, embankment, and bridging per mile; the prospective fuel supply; the possibilities for business; and all other data from which an approximate estimate of the cost of the proposed railroad can be made.

PRELIMINARY SURVEY

The reconnaissance having been completed and a route selected, the next thing is to make a *preliminary survey*. This consists of an instrumental examination of the route for the following purposes: (1) to determine the relative merits of alternative routes that have been examined on the reconnaissance; (2) to obtain the necessary information for making a map and a profile of the route; (3) to furnish data from which to project the location; and (4) to determine, approximately, the amount of work to be done in the matter of clearing, grading, and bridging, and to furnish data for an approximate estimate of the cost of all materials and labor required for the proposed road.

Preliminary Estimate.—In making a preliminary estimate, great accuracy is not necessary, and no time should be wasted in useless refinements of calculation. The estimate should be high enough to cover all probable cost, and a liberal allowance should be made to cover unforeseen contingencies that may develop during construction. Most experienced engineers make it a rule to add 10 % to a preliminary estimate in order to provide for contingencies.

In estimating for earthwork, the cross-sections may be considered as level cuttings; that is, the cross-section surface may be considered as level unless its slope angle exceeds 10°, in which case a suitable allowance must be made for the slope. The preliminary estimate, which also includes approximate figures for material and labor required for culverts, bridges, trestles, piers, and abutments is then classified and summarized. A sample of a good form of a preliminary estimate of the cost of a proposed railroad follows:

ESTIMATE OF COST—A & B RAILROAD

Clearing 625 A. at $20 per A.	$ 12,500
Earth excavation: 900,000 cu. yd. at 17c.	153,000
Loose-rock excavation: 300,000 cu. yd. at 40c.	120,000
Solid-rock excavation: 200,000 cu. yd. at 80c.	160,000
Overhaul exceeding 600 ft.: 300,000 cu. yd. at 1c.	3,000
Borrowed embankment: 80,000 cu. yd. at 17c.	13,600
Piling: 12,000 lin. ft. at 25c.	3,000
Framed trestles: 300,000 ft. B. M. at $35 per M.	10,500
First-class masonry: 2,800 cu. yd. at $12	33,600
Second-class masonry: 4,200 cu. yd. at $8	33,600
Box culvert masonry: 2,300 cu. yd. at $5	11,500
Dry-rubble masonry: 2,600 cu. yd. at $4	10,400
Concrete masonry: 3,000 cu. yd. at $6	18,000
Riprap: 2,000 sq. yd. at $1.50	3,000
Cast-iron pipe culverts: 40,000 lb. at 3c.	1,200
Vitrified pipe culverts: 1,800 lin. ft. at $1.50	2,700
Total, exclusive of bridges and track	$589,600
Add 10 per cent	58,960
Total cost for grading and trestles	$648,560

LOCATION

The *location* is the operation of fitting the line to the ground in such a manner as to secure the best adjustment of the alinement and grade, consistent with an economical cost of construction. If no topographic map is available, the work of location is done directly on the ground. Ordinarily, however, a topographic party is employed in the preliminary survey and a contour map prepared. The location is then best projected on the map, and it is called a *paper location.*

An example of such location is illustrated in Fig. 1. Here, the line follows the valley of Bear River, and the gradient is determined by the slope of the stream. The gradient adopted is .5%, or .5 ft. per station. The preliminary line is shown dotted, and the located line is drawn full.

Let the grade elevation for Sta. 16 be 155 ft.; the grade elevation for Sta. 17 will, therefore, be 155 ft. + .5 ft. = 155.5 ft.

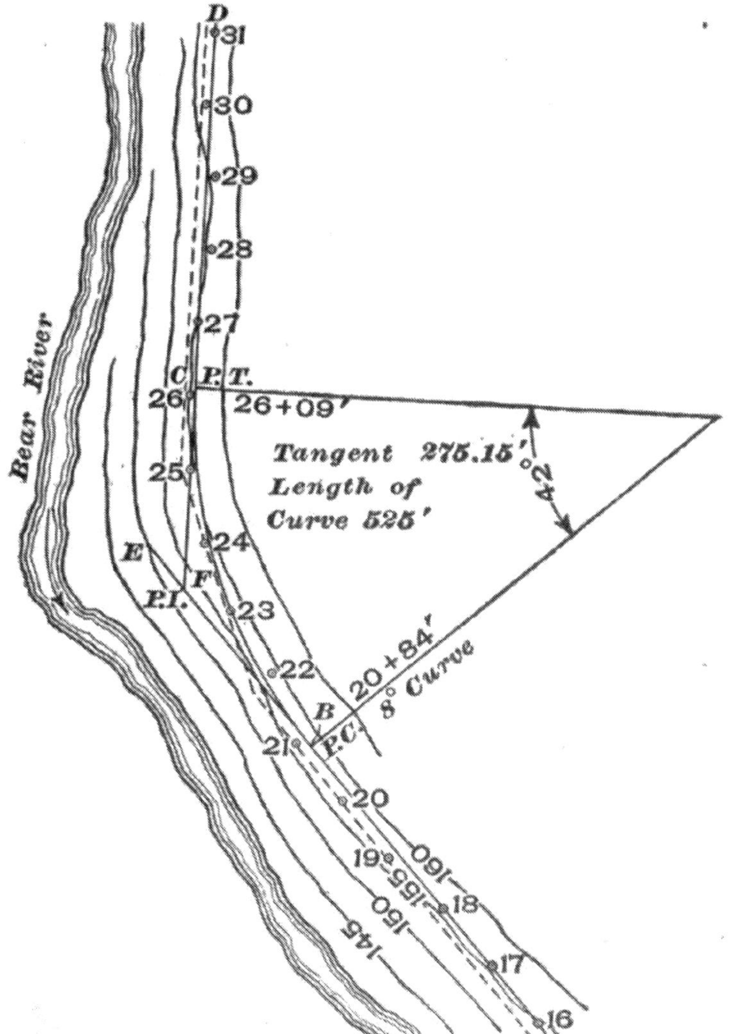
D
31
30
29
28
27
C
P.T.
26
26+09'
Tangent 275.15'
Length of
Curve 525'
42°
25
24
E
F
P.I.
23
22
20+84'
8° Curve
B
21
P.C.
20
19
18
17
16
A
145
150
155
160
Bear River

FIG. 1

The grade elevation for Sta. 18 will be 155.5+.5=156 ft. By the same process, the grade elevation is found for each station shown in the plat; and by means of interpolation between two contour curves, points having the required elevation are located opposite the corresponding stations of the preliminary survey. Each point is marked by a small dot enclosed in a circle. A line joining the points thus designated will be the *grade contour*, or the line where the required gradient meets the surface of the ground. The tangents *AB* and *CD* are then projected so as to conform as closely as practicable to the grade contour, and a suitable curve is inserted for the intersection angle *EFD*. This is most conveniently done by means of a *curved protractor*, an illustration of which is shown in Fig. 2.

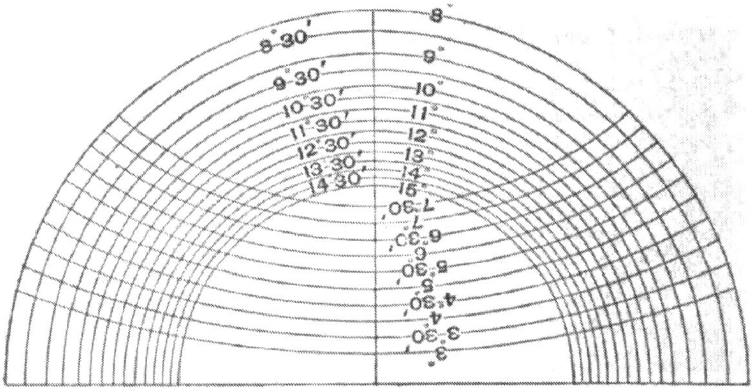

FIG. 2

This instrument, which is made of transparent material, is shifted until there is found a curve that will fit the topography and will close the angle between the tangents, as required.

Curvature.—There is no fixed rule for limiting curvature, but for a permanent track it is desirable to have the curvature as easy as possible. For all ordinary traffic conditions, it is good practice to use such curves as will best conform to existing topographical conditions. Any curve up to 10° will be no obstacle to a speed of 35 mi. per hr., the average speed of passenger trains. This affords a range in curvature that will meet the requirements of most localities.

Compensation for Curvature.—The effect of curvature on a railroad line is to cause a resistance to the movement of trains. When a curve occurs on a gradient, the effect of the curve resistance on ascending trains is practically the same as increasing the gradient. It is customary in fixing the final grades to lighten the grade on a curve an amount sufficient to offset the resistance due to the curvature. This operation is called *compensating for curvature.* The usual rate of compensation for curvature is .03 to .05 ft. per hundred feet per degree of curvature. For example, where the maximum gradient on tangents is 1%, the maximum gradient on a 6° curve, allowing a compensation of .03 ft. per degree, would be 1−(.03 ×6)=.82%. If a compensation of .05 ft. per degree were made, the grade on a 6° curve would be 1−(.05 ×6)=.70%.

Final Grade Lines.—The establishing of final grade lines is illustrated in Fig. 3, where the uncompensated grade is 1.3%, and the compensation for curvature,

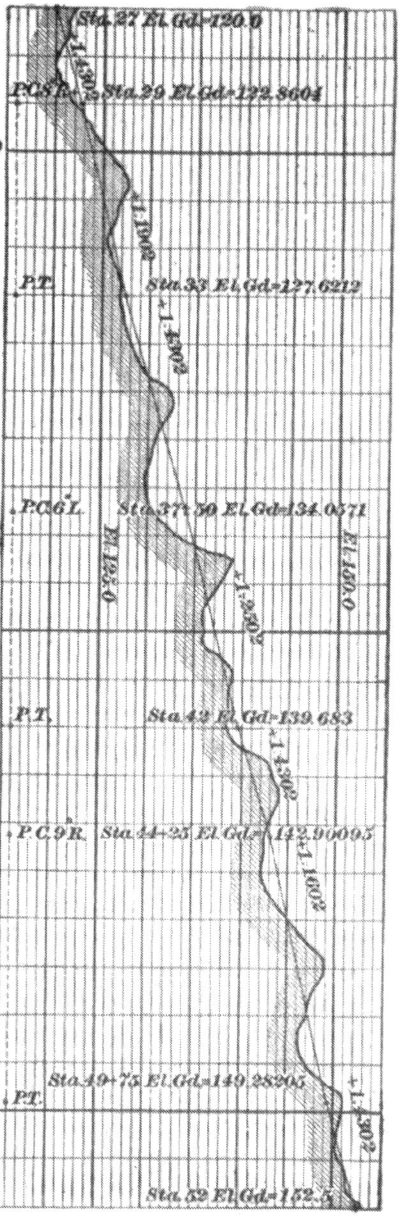

FIG. 3

as shown in the final grade line, is .03 ft. per degree. The location notes for this line are as follows:

Stations	Intersection Angles
52+00	End of grade
49+75 P. T.	
44+25 P. C. 9° R.	49° 30′
42+00 P. T.	
37+50 P. C. 6° L.	27° 00′
33+00 P. T.	
29+00 P. C. 8° R.	32° 00′
27+00	Beginning of grade

The elevation of the grade at Sta. 27 is fixed at 120 ft., and at Sta. 52, at 152.5 ft., giving between these stations an actual rise of 32.5 ft. and an uncompensated grade of 1.3 %. These grade points are marked on the profile with small circles. The total curvature between Sta. 27 and Sta. 52 is 108½°. The resistance due to each degree of curvature being taken as equivalent to an increase of .03 ft. in grade, the total resistance due to 108.5° is equivalent to .03×108.5=3.255 ft. additional rise between Sta. 27 and Sta. 52. Hence, the total theoretical grade between these stations is the sum of 32.5 ft., the actual rise, and 3.255 ft. due to curvature, or 35.755 ft. Dividing 35.755 by 25, the number of stations in the given distance, there results 35.755÷25= +1.4302 ft., as the grade for tangents on this line. The starting point of this grade is at Sta. 27. The P. C. of the first curve is at Sta. 29, giving a tangent of 200 ft. =2 Sta. As the grade for tangents is +1.4302 ft. per sta., the rise in grade between Sta. 27 and Sta. 29 is 1.4302×2 =2.8604 ft. The elevation of grade at Sta. 27 is 120 ft., and the elevation of grade at Sta. 29 is 120+2.8604=122.8604 ft., which is recorded on the profile as shown in the diagram, with the rate of grade, namely, +1.4302, written above the grade line. The first curve is 8°, and, as the compensation per degreee is .03 ft., then, for 8°, or a full station, the compensation is .03×8 = .24 ft. The grade on the curve will therefore be the tangent grade minus the compensation, or 1.4302−.24 = +1.1902 ft. per sta. The P. C. of this curve is at Sta. 29, the P. T. at Sta. 33, making the total length of the curve 400 ft.=4 Sta. The grade on this curve is +1.1902 ft. per sta. and the total rise on the curve is 1.1902×4=4.7608 ft. The elevation of the grade at the P. C. at Sta. 29 is 122.8604; hence, the elevation of grade at the P. T. at Sta. 33 is 122.8604+4.7608=127.6212 ft., which is recorded on the profile together with the grade, namely, +1.1902, written above the grade line. The P. C. of the next curve is at Sta. 37+50, giving an intermediate tangent of 450 ft.=4.5 Sta. The grade for tangents is +1.4302 ft. per sta.; hence, the total rise on the tangent is 1.4302×4.5 =6.4359 ft. Adding 6.4359 ft., to 127.6212 ft., the elevation of grade at Sta. 37+50 is found to be 134.0571 ft., which is recorded on the profile, together with the rate of grade for tangents.

The next curve is 6°, and the compensation in grade per station is .03 ft. × 6=.18 ft. The grade on this curve will therefore be 1.4302−.18=1.2502 ft. per sta. The length of the curve is 450 ft.=4.5 Sta., and the total rise in grade on this curve is +1.2502 ft.×4.5=5.6259 ft. The elevation of the grade at Sta. 37+50, the P. C. of the curve, is 134.0571. The elevation of the grade at Sta. 42, the P. T., is therefore 134.0571 +5.6259=139.683 ft., which is recorded on the profile together with the rate of grade on the 6° curve, namely, +1.2502. The P. C. of the next curve is at Sta. 44+25, giving an intermediate tangent of 225 ft.=2.25 Sta. The total rise on the tangent is therefore, 1.4302×2.25=3.21795 ft. The elevation of grade at the P. T. at Sta. 42 is 139.683; therefore, the elevation of grade at Sta. 44+25 is 139.683+3.21795=142.90095 ft., which is recorded on the profile together with the grade +1.4302.

The last curve is 9°, and the compensation in grade per station is .03×9=.27 ft. The grade on this curve is therefore 1.4302−.27=1.1602 ft. per sta. The length of the curve is 550 ft.=5.5 Sta., and the total rise on the curve is

1.1602×5.5=6.3811 ft. The elevation of grade at Sta. 44 +25, the P. C. of the 9° curve, is 142.90095; hence, the elevation of grade at the P. T., at Sta. 49+75, is 142.90095+6.3811 =149.28205 ft., which is recorded on the profile together with the grade, +1.1602. The end of the line is at Sta. 53, giving a tangent of 225 ft.=2.25 sta. The rise on this tangent is 1.4302×2.25=3.21795 ft., which is added to 149.28205, the elevation of the P. T. at Sta. 49+75. The sum, 152.5 ft., is the elevation of grade at Sta. 52.

The sum of the partial grades should equal the total rise between the extremities of the grade line. The points where the changes of grade occur are marked on the profile with small circles, which are connected by fine lines representing the grade line. These points of change are projected on a horizontal line at the bottom of the profile. The portions of this line that represent curves are dotted, and the portions that represent tangents are drawn full. The P. C. and P. T. of each curve are marked with small circles on this horizontal line, and are lettered as shown in the diagram.

Where the grades are light and the curves have large radii, there will be no need of compensation for curvature. Where the grades exceed .5 % and the curves 5°, compensation should be made.

VERTICAL CURVES

If the grade of the center line of track changes at any point, the two grade lines that intersect at this point form with each other an angle more or less abrupt. If this angle points upwards, it is called a *spur;* if it points downwards, it is called a *sag.*

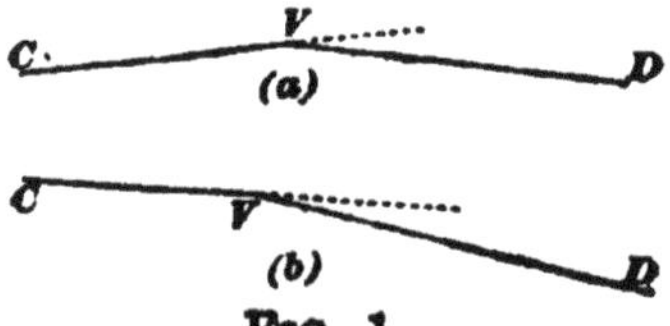

FIG. 1

The angles *CVD* in Fig. 1 (*a*) and (*b*) are spurs; the angles *CVD* in Fig. 2 (*a*) and (*b*) are sags.

Vertical Curve at a Spur.—If *AV* and *BV*, Fig. 3, are two grade lines meeting at *V*, a vertical curve *CMD* must be introduced to join these lines. Between *C* and *D*, the actual grade is established along the vertical curve *CMD*, instead of along

CV and VD. The projections RT and TS of the distances VC and VD from the vertex to the points at which the vertical curve begins and ends are always chosen equal. If K is the middle point of the straight line CD, the vertical curve is always so chosen that it will bisect VK; that is, so that $VM = MK$.

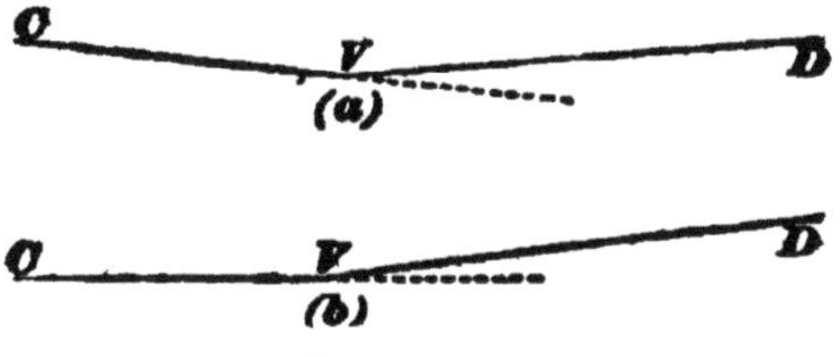

FIG. 2

Let E be the elevation of C, Fig. 3, E' that of D, and H that of V, so that $E = RC$, $E' = SD$, and $H = VT$. Then,

$$VM = \tfrac{1}{2}\left(H - \frac{E+E'}{2}\right)$$

The distance VM is called the *correction in grade* at the point V.

Vertical curves are always made parabolic. It is a property of the parabola that the correction in grade am at any point a is given by the equation,

$$am = VM \times \left(\frac{Ca}{CV}\right)^2$$

The distance $CV = VD$ is always made a whole number of stations; and, to simplify the work, the grade stakes a, b, c,

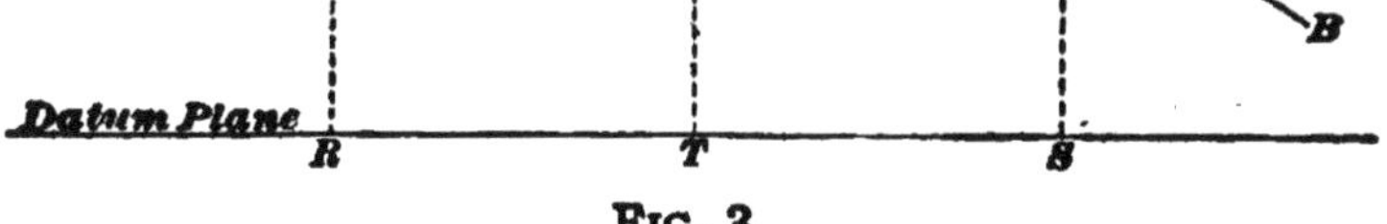

FIG. 3

etc., are so set that they divide the distance CV into a number of equal parts. The corrections in grade at points a', b', and c' along DV are equal to those for the corresponding points along CV. That is, if $Ca = Da'$, then $am = a'm'$; if $Cb = Db'$, then $bm = b'm'$, etc.

EXAMPLE.—A +.4% grade meets a −.5% grade at Sta. 190, the elevation of which is 161.3 ft. If a vertical curve 400 ft. long is inserted, what is the correction in grade and the corrected grade elevation at each station and half station?

SOLUTION.—In this example, $VC = VD = 200$ ft. The elevation of C is $161.3 - 2 \times .4 = 160.5$ ft., $= E$; that of D is $161.3 - 2 \times .5 = 160.3$ ft, $= E'$; that of K is $\frac{1}{2}(E'+E) = \frac{1}{2} \times (160.5 + 160.3) = 160.4$ ft.; and that of V is $H = 161.3$ ft. Substituting these values in the formula for VM,

$$VM = \tfrac{1}{2} \times (161.3 - 160.4) = .45 \text{ ft.}$$

Since, for the first stake, $Ca = 50$ ft. and $CV = 200$ ft., the formula for am gives

$$am = \left(\frac{50}{200}\right)^2 \times VM = \frac{1}{16} \times .45 = .03 \text{ ft.} = a'm'$$

Similarly,

$$bm = \left(\frac{100}{200}\right)^2 \times VM = \frac{1}{4} \times .45 = .11 = b'm'$$

$$cm = \left(\frac{150}{200}\right)^2 \times VM = \frac{9}{16} \times .45 = .25 = c'm'$$

The original and corrected grade elevations are as follows:

Station	*Original Elevation*	*Correction*	*Corrected Elevation*
188	160.50	.00	160.50
+50	160.70	.03	160.67
189	160.90	.11	160.79
+50	161.10	.25	160.85
190	161.30	.45	160.85
+50	161.05	.25	160.80
191	160.80	.11	160.69
+50	160.55	.03	160.52
192	160.30	.00	160.30

Vertical Curve at a Sag.—If two grade lines, AV and VB, Fig. 4, meet so as to form a sag, the vertical curve will evidently be wholly above both grade lines. Using the same notation as before, the correction in grade at the point V will be

$$VM = \frac{1}{2}\left(\frac{E+E'}{2} - H\right)$$

The correction in grade at any point a will be given by the preceding formula for am, as before, but this correction is now to be added to the old elevation of grade at a to obtain the corrected elevation.

EXAMPLE.—The grade of CV, Fig. 4, is -1.2%, that of VD is $+.6\%$, and the elevation of V is $+49.2$ ft. Find the

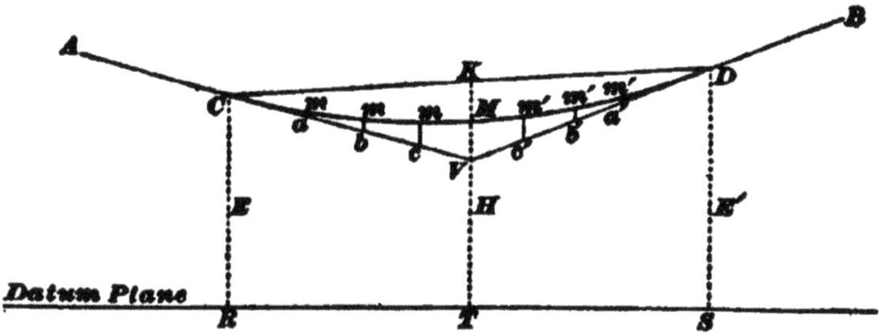

FIG. 4

corrections in grade and the corrected elevations at stakes 100 ft. apart, if the length of the vertical curve is 600 ft.

SOLUTION.—The uncorrected grade elevations are as follows:

Along CV		*Along VD*	
At first stake......	52.8	At fifth stake......	49.8
At second stake....	51.6	At sixth stake.....	50.4
At third stake.....	50.4	At seventh stake, D	51.0
At fourth stake, V.	49.2		

Therefore, $\frac{1}{2}(E+E') = \frac{1}{2}(52.8+51.0) = 51.9$; and, by the preceding formula,

$$VM = \frac{1}{2}(51.9 - 49.2) = 1.35 \text{ ft.}$$

The formula for am may now be applied.

Correction in grade at second stake, 100 ft. from C, is

$$\left(\frac{100}{300}\right)^2 \times 1.35 = \frac{1}{9} \times 1.35 = .15 = \text{correction at sixth stake.}$$

Correction at third stake, $\left(\frac{200}{300}\right)^2 \times 1.35 = \frac{4}{9} \times 1.35 = .60$ = correction at fifth stake.

The corrected elevations will be

At *C*	52.80+ .00=52.80
At second stake	51.60+ .15=51.75
At third stake	50.40+ .60=51.00
At fourth stake	49.20+1.35=50.55
At fifth stake	49.80+ .60=50.40
At sixth stake	50.40+ .15=50.55
At *D*	51.00+ .00=51.00

TRACKWORK

TRACK MATERIALS

Rails.—The illustration shows, in cross-section, the general form of rail adopted by the American Society of Civil Engineers and now used by most railroads: *PQ* is the head; *MN*, the web; and *KL*, the flange, or base. The metal is distributed through the section in the following proportions: head, 42%; web, 21%; flange, 37%. The dimensions indicated in the illustration for the different weights of rails are given in the accompanying table.

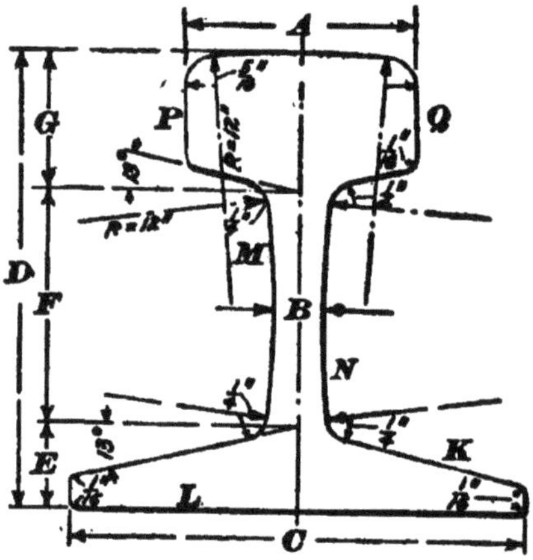

Required Weight of Rail. Rule I, which was first published by the Baldwin Locomotive Works, gives fairly approximate results for light loads; for very heavy loads, however, the weights obtained by it are too large. Rule II agrees more closely with present American practice.

Rule I.—*Divide the greatest load, in pounds, that will be supported by any wheel, by 224; the quotient is the required weight of the rail in pounds per yard.*

WEIGHTS AND DIMENSIONS OF STANDARD RAILS

Rail Part	Weight per Yard, in Pounds												
	40	45	50	55	60	65	70	75	80	85	90	95	100
	Dimensions, in Inches												
A	$1\frac{7}{8}$	2	$2\frac{1}{8}$	$2\frac{1}{4}$	$2\frac{3}{8}$	$2\frac{13}{32}$	$2\frac{7}{16}$	$2\frac{15}{32}$	$2\frac{1}{2}$	$2\frac{9}{16}$	$2\frac{5}{8}$	$2\frac{11}{16}$	$2\frac{3}{4}$
B	$\frac{25}{64}$	$\frac{27}{64}$	$\frac{7}{16}$	$\frac{15}{32}$	$\frac{31}{64}$	$\frac{1}{2}$	$\frac{33}{64}$	$\frac{17}{32}$	$\frac{35}{64}$	$\frac{9}{16}$	$\frac{9}{16}$	$\frac{9}{16}$	$\frac{9}{16}$
C and D	$3\frac{1}{2}$	$3\frac{11}{16}$	$3\frac{7}{8}$	$4\frac{1}{16}$	$4\frac{1}{4}$	$4\frac{7}{16}$	$4\frac{5}{8}$	$4\frac{13}{16}$	5	$5\frac{3}{16}$	$5\frac{3}{8}$	$5\frac{9}{16}$	$5\frac{3}{4}$
E	$\frac{5}{8}$	$\frac{21}{32}$	$\frac{11}{16}$	$\frac{23}{32}$	$\frac{49}{64}$	$\frac{25}{32}$	$\frac{13}{16}$	$\frac{27}{32}$	$\frac{7}{8}$	$\frac{57}{64}$	$\frac{59}{64}$	$\frac{15}{16}$	$\frac{31}{32}$
F	$1\frac{55}{64}$	$1\frac{31}{32}$	$2\frac{1}{16}$	$2\frac{11}{64}$	$2\frac{17}{64}$	$2\frac{3}{8}$	$2\frac{15}{32}$	$2\frac{35}{64}$	$2\frac{5}{8}$	$2\frac{3}{4}$	$2\frac{55}{64}$	$2\frac{61}{64}$	$3\frac{5}{64}$
G	$1\frac{1}{64}$	$1\frac{1}{16}$	$1\frac{1}{8}$	$1\frac{11}{64}$	$1\frac{7}{32}$	$1\frac{9}{32}$	$1\frac{11}{32}$	$1\frac{27}{64}$	$1\frac{1}{2}$	$1\frac{35}{64}$	$1\frac{19}{32}$	$1\frac{41}{64}$	$1\frac{45}{64}$

Rule II.—*The weight of the rail, in pounds per yard, should equal the total number of tons of 2,000 lb. on all the drivers of the heaviest locomotive.*

Required Quantities of Materials.—The six tables that follow show the quantities of materials required in trackwork.

WEIGHT OF RAILS REQUIRED PER MILE OF TRACK

Weight of Rail per Yard	Weight of Track per Mile		Weight of Rail per Yard	Weight of Track per Mile	
Pounds	Tons	Pounds	Pounds	Tons	Pounds
30	47	320	70	110	
35	55		75	117	1,920
40	62	1,920	80	125	1,600
45	70	1,600	85	133	1,280
50	78	1,280	90	141	960
55	86	960	95	149	640
60	94	640	100	157	320
65	102	320			

NUMBER OF RAILS, PAIRS OF ANGLE BARS, AND BOLTS PER MILE OF TRACK

Length of Rail Feet	Number of Rails per Mile	Number of Pairs of Angle Bars	Number of Bolts, Four to Each Joint	Number of Bolts, Six to Each Joint
18	587	587	2,336	3,504
20	528	528	2,112	3,168
21	503	503	2,012	3,018
22	480	480	1,920	2,880
24	440	440	1,760	2,640
25	422	422	1,688	2,532
26	406	406	1,624	2,436
27	391	391	1,564	2,346
28	377	377	1,508	2,262
30	352	352	1,408	2,112
33	320	320	1,280	1,920

NUMBER OF TIES PER MILE

Distance From Center to Center Feet	Number of Ties	Distance From Center to Center Feet	Number of Ties
1½	3,520	2½	2,113
1¾	3,017	2¾	1,921
2	2,640	3	1,761
2¼	2,348		

NUMBER OF TRACK BOLTS IN A KEG OF 200 LB.

Bolts Inches	Size of Nuts Inches	Bolts in Keg	Bolts Inches	Size of Nuts Inches	Bolts in Keg
⅞×4½	1⅜ square	195	⅝×2½	1 square	654
⅞×4	1⅜ square	200	⅞×3¼	1⅜ hexagonal	170
⅞×3¾	1⅜ square	208	¾×3¼	1¼ hexagonal	237
⅞×3½	1⅜ square	216	¾×3½	1¼ hexagonal	228
¾×4	1¼ square	305	¾×4	1¼ hexagonal	220
¾×3¼	1¼ square	329	⅝×3½	1 hexagonal	415
⅝×3½	1 square	576			

RAILROAD SPIKES PER MILE OF TRACK

Rails Used Pounds per Yard	Size Measured Under Head Inches	Average Number per Keg of 200 Lb.	Ties 2 Ft. Between Centers Four Spikes to a Tie	
			Pounds	Kegs
45 to 70	5½×9/16	375	5,870	29⅓
40 to 56	5 ×9/16	400	5,170	26
35 to 40	5 ×½	450	4,660	23⅓
28 to 35	4½×½	530	3,960	20
24 to 35	4 ×½	600	3,520	17⅗
20 to 30	4½×7/16	680	3,110	15½
	4 ×7/16	720	2,910	14½
16 to 25	3½×7/16	900	2,350	11
	4 ×⅜	1,000	2,090	10½
16 to 20	3½×⅜	1,190	1,780	9
	3 ×⅜	1,240	1,710	8½
12 to 16	2½×⅜	1,342	1,575	7⅞

SPACES BETWEEN ENDS OF RAILS

Temperature When Laying Track	Space to be Left Between Ends of Rails Inch	Temperature When Laying Track	Space to be Left Between Ends of Rails Inch
90° above zero	$\frac{1}{16}$	30° above zero	$\frac{1}{4}$
70° above zero	$\frac{1}{8}$	10° above zero	$\frac{5}{16}$
50° above zero	$\frac{3}{16}$	10° below zero	$\frac{3}{8}$

CURVED TRACK

The difference in length between the inner and the outer rail of a curve may be found by either of the following rules:

Rule I.—*Multiply the degree of the curve by the length in stations of 100 ft., and this product by* $1\frac{1}{32}$; *the result will be the difference in length between the inner and the outer rail, in inches.*

Rule II.—*Multiply the distance between the center lines of the rails by the length of the curve, in feet, and divide the product by the radius of the track curve; the quotient will be the required difference in length, expressed in feet.*

For light curves laid to exact gauge, the first rule is the simpler one, but for short curves where the gauge is widened, the second rule should be used.

Curving Rails.—When laying track on curves, in order to have a smooth line, the rails themselves must conform to the curve of the center line. To accomplish this, the rails must be curved. The curving should be done with a rail bender or with a lever, preferably with the former. To guide those in charge of this work, a table of middle and quarter ordinates for a 30-ft. rail for all degrees of curve should be prepared. The middle ordinates in the following table are calculated by the formula

$$m = \frac{c^2}{8R},$$

in which m is the middle ordinate; c, the length of chord, assumed to be of the same length as the rail; and R, the radius of curve. This formula is not theoretically correct; yet the error is so small that it may be ignored in practical work.

In curving rails, the ordinate is measured by stretching a cord from end to end of the rail against the gauge side, as

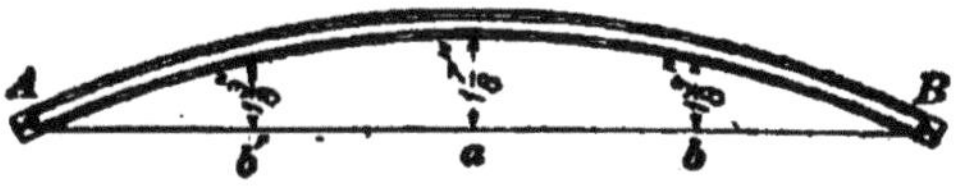

shown in the accompanying illustration. Suppose the rail *AB* is 30 ft. in length, and the curve 8°, then the middle ordinate at *a* should be

$$m = \frac{30^2}{8 \times 716.78} = .157 \text{ ft.} = 1\tfrac{7}{8} \text{ in.}$$

To insure a uniform curve to the rails, the ordinates at the quarter points *b* and *b'* should be tested. In all cases the quarter ordinates should be three-quarters of the middle ordinate. In the illustration, if the rail has been properly curved, the quarter ordinates at *b* and *b'* will be $\frac{3}{4} \times 1\frac{7}{8}$ in. $= 1\frac{13}{32}$, say $1\frac{3}{8}$.

MIDDLE ORDINATES, IN INCHES, FOR CURVING RAILS

Degree of Curve	Length of Rail, in Feet					
	30	28	26	24	22	20
1	$\frac{1}{4}$	$\frac{3}{16}$	$\frac{3}{16}$	$\frac{3}{16}$	$\frac{1}{8}$	$\frac{1}{8}$
2	$\frac{1}{2}$	$\frac{7}{16}$	$\frac{3}{8}$	$\frac{5}{16}$	$\frac{1}{4}$	$\frac{3}{16}$
3	$\frac{11}{16}$	$\frac{5}{8}$	$\frac{9}{16}$	$\frac{7}{16}$	$\frac{3}{8}$	$\frac{5}{16}$
4	$\frac{15}{16}$	$\frac{13}{16}$	$\frac{11}{16}$	$\frac{5}{8}$	$\frac{1}{2}$	$\frac{7}{16}$
5	$1\frac{3}{16}$	$1\frac{1}{16}$	$\frac{7}{8}$	$\frac{3}{4}$	$\frac{5}{8}$	$\frac{9}{16}$
6	$1\frac{7}{16}$	$1\frac{1}{4}$	$1\frac{1}{16}$	$\frac{7}{8}$	$\frac{3}{4}$	$\frac{5}{8}$
7	$1\frac{5}{8}$	$1\frac{7}{16}$	$1\frac{1}{4}$	$1\frac{1}{16}$	$\frac{7}{8}$	$\frac{3}{4}$
8	$1\frac{7}{8}$	$1\frac{5}{8}$	$1\frac{7}{16}$	$1\frac{3}{16}$	1	$\frac{7}{8}$
9	$2\frac{1}{8}$	$1\frac{7}{8}$	$1\frac{5}{8}$	$1\frac{3}{8}$	$1\frac{1}{8}$	$\frac{15}{16}$
10	$2\frac{3}{8}$	$2\frac{1}{16}$	$1\frac{3}{4}$	$1\frac{1}{2}$	$1\frac{1}{4}$	$1\frac{1}{16}$
11	$2\frac{5}{8}$	$2\frac{1}{4}$	$1\frac{15}{16}$	$1\frac{11}{16}$	$1\frac{3}{8}$	$1\frac{1}{8}$
12	$2\frac{13}{16}$	$2\frac{1}{2}$	$2\frac{1}{8}$	$1\frac{13}{16}$	$1\frac{9}{16}$	$1\frac{1}{4}$
13	$3\frac{1}{16}$	$2\frac{11}{16}$	$2\frac{5}{16}$	$1\frac{15}{16}$	$1\frac{5}{8}$	$1\frac{3}{8}$
14	$3\frac{5}{16}$	$2\frac{7}{8}$	$2\frac{1}{2}$	$2\frac{1}{8}$	$1\frac{3}{4}$	$1\frac{1}{2}$
15	$3\frac{9}{16}$	$3\frac{1}{16}$	$2\frac{11}{16}$	$2\frac{1}{4}$	$1\frac{15}{16}$	$1\frac{9}{16}$
16	$3\frac{3}{4}$	$3\frac{1}{4}$	$2\frac{13}{16}$	$2\frac{3}{8}$	$2\frac{1}{16}$	$1\frac{11}{16}$
17	4	$3\frac{1}{2}$	3	$2\frac{9}{16}$	$2\frac{3}{16}$	$1\frac{3}{4}$
18	$4\frac{3}{16}$	$3\frac{11}{16}$	$3\frac{3}{16}$	$2\frac{11}{16}$	$2\frac{5}{16}$	$1\frac{7}{8}$
19	$4\frac{7}{16}$	$3\frac{7}{8}$	$3\frac{3}{8}$	$2\frac{7}{8}$	$2\frac{7}{16}$	2
20	$4\frac{11}{16}$	$4\frac{1}{8}$	$3\frac{9}{16}$	3	$2\frac{9}{16}$	$2\frac{1}{8}$

TURNOUTS

A *turnout* is a contrivance for passing from one track to another. The principal parts are the switch, the frog, and two guard-rails. The *switch*, which is the movable part of the turnout, consists of two switch rails *BA*, *CD*, Fig. 1. The fixed ends *B* and *C* of the switch rails are called the *heels* of the

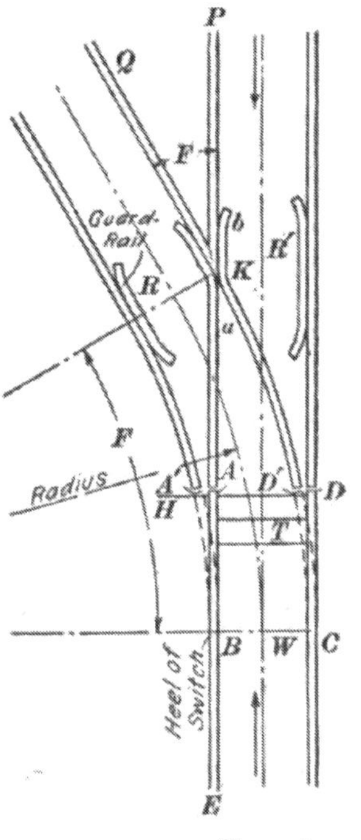

FIG. 1

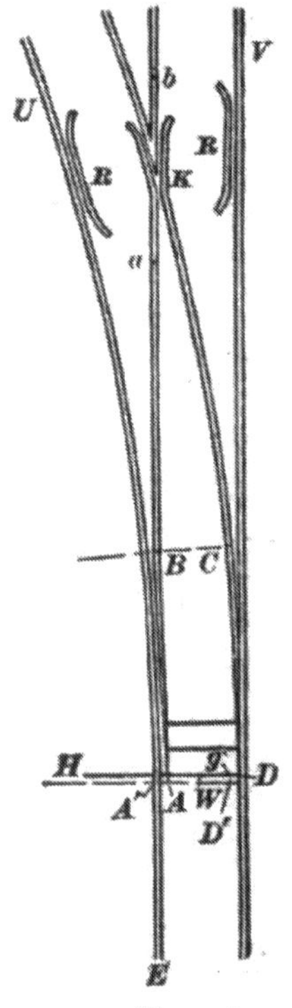

FIG. 2

switch, and the movable ends *A* and *D*, the *toes* of the switch. The cross-tie that supports the toes of the switch is called the *head-block*, and the tie-rod at the toes, the *head-rod*. The distance *AA'* or *DD'* through which the toes move is called the *throw* of the switch. A frog is shown at *K* and two guard-rails at *R*, *R'*.

Switches.—There are two kinds of switches, which differ in the arrangement and form of switch rails, namely, the *stub switch* and the *point switch*. In the stub switch, Fig. 1, a part of each main-track rail is bent over to connect with the side track. In the point switch, Fig. 2, the outer rail *DV* of the main track is spiked rigidly to the ties; the opposite rail *EA'U*, lying partly in the main track and partly in the side track, is also firmly spiked. These two rails are immovable. The two switch rails *BA* and *CD* are planed to thin edges at *A* and *D*. The ends *B* and *C* of these rails are the fixed ends or heels; the thin edges at *A* and *D* are the toes. The head-block is at *H*, and the head-rod at *g*.

The point of the center line at which the turnout begins is called the *point of switch*. In Figs. 1 and 2, *W* is the point of switch. In stub switches, the point of switch is midway between the heels; in point switches, it is midway between the toes and above the head-block.

Frogs and Guard-Rails.—A *frog* is a combination of rails so arranged that the broad tread of the wheel will always have a

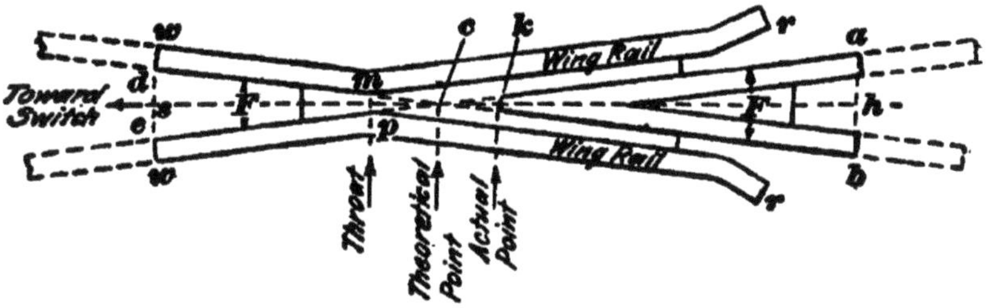

FIG. 3

surface on which to roll, and that the flange of the wheel will have a channel through which to pass. A frog is shown in position on the track at *K*, Fig. 1, and a larger plan of the part at *ab*, Figs. 1 and 2, is shown in Fig. 3.

The wedge-shaped part *akb* of the frog is called the *tongue* of the frog, and its point *k* is called the *actual point of frog*. The actual point of frog is somewhat shortened and rounded. The intersection *c* of the outside edges *ac* and *bc* of the tongue is called the *theoretical point of frog*. When the point of frog is referred to, the theoretical point is usually meant. The bent rails *wr* are called *wing rails;* the narrowest part *mp* of the frog is called the *throat.* The throat of the frog must be wide enough to allow the flanges of the wheels to pass through; it is usually made about 2 in. wide.

Frog Angle and Frog Number.—The angle *acb*, Fig. 3, between the outside edges of the tongue of the frog is called the *frog angle.* This is also equal to the angle *dce* between the outside edges of the tongue produced beyond *c*. The frog angle which is represented by F is also equal to the angle between the two tracks.

The distance *ab* between the gauge lines at the end of the tongue is called the *heel width;* the distance *de*, the *mouth width.* If *sch* is the bisector of the angle F, the distance *ch* is called the *length of frog.*

The ratio of the length to the heel width is called the *frog number*, and is usually denoted by n; that is,

$$n = ch \div ab$$

The relation between n and F is expressed by the formulas

$$n = \tfrac{1}{2} \cot \tfrac{1}{2}F$$

and

$$\cot \tfrac{1}{2}F = 2n$$

Frogs are usually designated by their numbers; thus, a No. 8 frog is one in which $n = 8$.

If the distance *sh* and the widths *ab* and *de*, Fig. 3, are measured on a frog, the frog number n can be determined by the formula

$$n = \frac{sh}{ab + de}$$

Guard-Rails.—Guard-rails, which are usually from 10 to 15 ft. long, are placed opposite the frog on the main track and the switch track, as at R and R' in Figs. 1 and 2. The clear space between the head of the guard-rail and the head of the main or the switch rail should be about 2 in.

FORMULAS AND CALCULATIONS

Radius and Lead of a Turnout for Stub Switches.—Let *RN*, Fig. 4, be the main track and *QP* the turnout. Let *Q* be the point of switch and *K* the point of frog. If a stub switch is employed, the main-track rails will be securely spiked along *YB* and *LD*; the parts *BG* and *DV* of these rails will be movable, so that they may be bent outwards to meet the turnout rails *W* and *Z*. Here, then, the ends *B* and *D* are the heels of the switch, and *G* and *V* are the toes. The head-block is underneath *G* and *V*.

In order to lay out a turnout when the frog angle is given, it is necessary to find the radius r, in terms of the frog angle, and the distance *KB* from the point of frog to the heel of switch, which distance is called the *lead* and is designated by *L*.

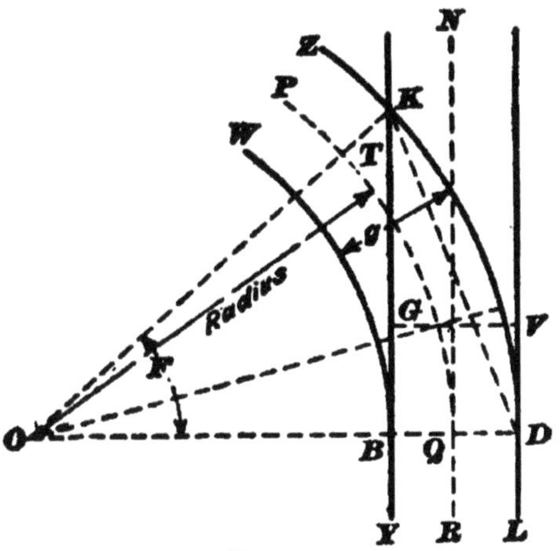

FIG. 4

The formulas for r and L are:

$$r = \tfrac{1}{2}\, g \cot^2 \tfrac{1}{2}F = 2gn^2$$

and

$$L = g \cot \tfrac{1}{2}F = 2gn$$

In these formulas g denotes the gauge. The *standard gauge* of track is 4 ft. 8½ in. = 4.708 ft.

The following table, some parts of which are calculated from the foregoing formulas, can be used in laying out a turnout with a stub switch. The frog number, which is usually given, is stated in the first column; the corresponding frog angle in the second column; and the lead, or *BK*, Fig. 4, in the third column. Then follow columns containing the chord *QT*, Fig. 4, which is equal to $2r \sin \tfrac{1}{2}F$; the radius of the turnout; the corresponding degree of curve, which is equal to $\frac{5{,}730}{r}$; the length l of switch rails *AB*, Fig. 1, obtained by the

DIMENSIONS OF STUB-SWITCH TURNOUTS

Track Circular From Heel of Switch to Point of Frog. Throw = 5½ In.

Frog Number n	Frog Angle F	Lead L	Chord (QT)	Radius	Degree of Curve d	Length of Switch Rails	Distance Ka, Fig. 1
4.0	14° 15′ 00″	37.67	37.38	150.67	38° 46′	11.73	1.50
4.5	12 40 59	42.37	42.12	190.69	30 24	13.19	1.69
5.0	11 25 16	47.08	46.85	235.42	24 32	14.65	1.87
5.5	10 23 20	51.79	51.58	284.85	20 13	16.15	2.06
6.0	9 31 38	56.50	56.30	339.00	16 58	17.64	2.25
6.5	8 47 51	61.21	61.03	397.85	14 26	19.09	2.44
7.0	8 10 16	65.92	65.75	461.42	12 26	20.53	2.62
7.5	7 37 41	70.62	70.47	529.69	10 50	22.03	2.81
8.0	7 9 10	75.33	75.19	602.67	9 31	23.48	3.00
8.5	6 43 59	80.04	79.90	680.36	8 26	24.93	3.19
9.0	6 21 35	84.75	84.62	762.75	7 31	26.43	3.37
9.5	6 1 32	89.46	89.33	849.85	6 45	27.97	3.56
10.0	5 43 29	94.17	94.05	941.67	6 5	29.37	3.75
10.5	5 27 9	98.87	98.76	1,038.19	5 32	30.85	3.94
11.0	5 12 18	103.58	103.47	1,139.42	5 02	32.31	4.12
11.5	4 58 45	108.29	108.19	1,245.36	4 36	33.78	4.31
12.0	4 46 19	113.00	112.90	1,356.00	4 14	35.17	4.50

formula $l = \sqrt{t(2r-t)}$; and the distance *Ka*, Fig. 1, or *cw*, Fig. 3. With different forms of frogs, this distance varies; the engineer should therefore measure it for the different frogs he uses, as it is necessary in determining the length of spiked rail *Aa*, Fig. 1.

Turnout Dimensions for Point Switches.—Let *MN*, Fig. 5, be the center line of the main track and *MJ* that of the turnout. Let *BA* and *CD* be the two switch rails whose fixed ends, or heels, are at *B* and *C*, and whose toes are at *A* and *D*. These

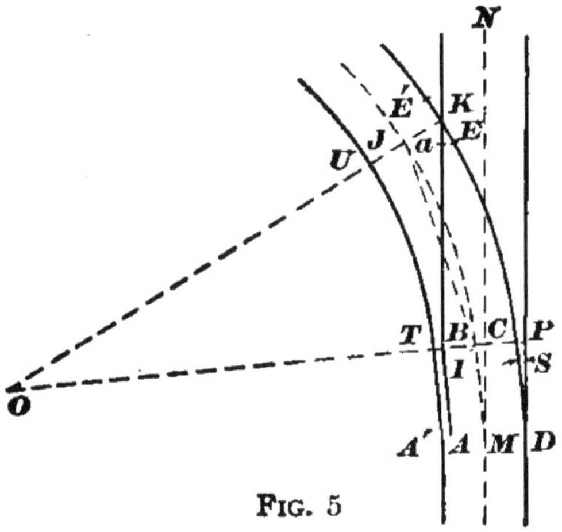

FIG. 5

rails are usually of a uniform length of 15 ft., except for the sharpest curves.

The center line *MIJ* will, when a point switch is used, have a somewhat different position from that which it has when a stub switch is employed. In the stub-switch turnout, the rails *A′TU* and *DCK* are bent to a uniform curve between *M* and *J*; in a point switch, the outer rail is made up of a straight part *DC*, which is the switch rail, and a curved part *CE*, which is tangent to *DC* at *C*. On this account, the lead *A′K* is less with a point switch than with a stub switch.

Since point switches are used on the main line where very accurate work is required, it is necessary to take account of the fact that the short frog rails are not curved, the part *EE′* of the rail being straight.

DIMENSIONS OF POINT-SWITCH TURNOUTS

Turnouts With Straight Point Rails and Straight Frog Rails; Gauge 4 Ft. 8½ In.

Frog Number n	Frog Angle F	Switch Angle S (CDP)	Lead L $(A'K)$	Radius r	Degree of Curve d	Chord (JI)	Length of Switch Rails (CD)	Length of Straight Frog Rail (KE)
4.0	14° 15′ 00″	3° 40′	32.20	125.21	47° 05′	23.09	7.5	1.50
4.5	12 40 49	3 40	34.29	159.25	36 36	25.03	7.5	1.69
5.0	11 25 16	2 45	41.85	197.65	29 22	29.88	10.0	1.87
5.5	10 23 20	2 45	44.16	240.44	24 00	32.03	10.0	2.06
6.0	9 31 39	1 50	56.00	288.09	19 59	38.66	15.0	2.25
6.5	8 47 51	1 50	58.84	340.19	16 54	41.34	15.0	2.44
7.0	8 10 16	1 50	61.65	397.65	14 27	43 98	15.0	2.62
7.5	7 37 41	1 50	64.36	460.00	12 29	46.50	15.0	2.81
8.0	7 9 10	1 50	67.04	527.91	10 52	48.99	15.0	3.00
8.5	6 43 59	1 50	69.60	600.94	9 33	51.38	15.0	3.19
9.0	6 21 35	1 50	72.20	681.16	8 25	53.80	15.0	3.37
9.5	6 1 32	1 50	74.70	767.11	7 28	56.11	15.0	3.56
10.0	5 43 29	1 50	77.04	858.14	6 41	58.28	15.0	3.75
10.5	5 27 9	1 50	79.51	959.00	5 59	60.57	15.0	3.94
11.0	5 12 18	1 50	81.82	1,065.52	5 23	62.69	15.0	4.12
11.5	4 58 45	1 50	84.09	1,180.16	4 51	64.78	15.0	4.31
12.0	4 46 19	1 50	86.16	1,299.93	4 24	66.67	15.0	4.50

In computing the dimensions of a point-switch turnout, the usual data are the length $AB = DC$ of the switch rail, the angle CDP between the outer switch rail and the main rail. This angle is called the *switch angle*, and will be represented by S. The frog number or the frog angle must also be known, as well as the length of the straight part EE'. It is then required to determine the radius OI of the center line of a turnout whose outer rail shall be tangent to the switch rail DC at C and to the frog rail EE' at E, and to find the lead $A'K$ of this turnout.

The formulas for computing these quantities are so complicated that, in practice, tables giving the various dimensions of point switches are always employed.

The accompanying table contains all the dimensions necessary for laying out a point switch when the frog number is known. It contains the frog angle, the switch angle CDP, Fig. 5, the lead $A'K$, the radius OI of the center line of the turnout, the degree of curve of this center line, the chord JI, the length $AB = CD$ of the switch rails, and the length $KE = Ka$ of the straight frog rail.

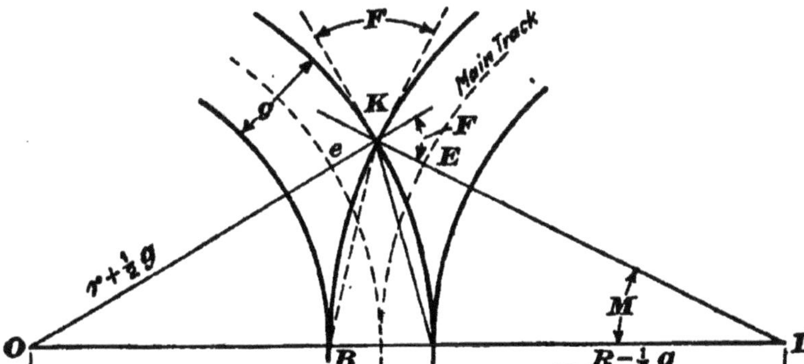

FIG. 6

Turnouts from the Outer Side of a Curved Track.—A turnout from the outer side of a curved track is shown in Fig. 6. The radius $DE = R$ of the main track, Fig. 6, the frog angle F, or frog number n, and the gauge g are usually known; from these the lead $BK = L$, and the radius $Oe = r$ of the center line of the turnout must be computed. The angle M, Fig. 6, must first be found by the formula

$$\tan \tfrac{1}{2}M = \frac{g}{2R} \cot \tfrac{1}{2}F = \frac{gn}{R}$$

Then, the lead must be determined by the formula

$$L = 2\,(R + \tfrac{1}{2}g) \sin \tfrac{1}{2}M$$

Finally, r is given by the formula

$$r + \tfrac{1}{2}g = \frac{R + \tfrac{1}{2}g}{\sin\,(F - M)} \sin M$$

When r has been found, the degree of curve is given by the formula

$$d = \frac{5{,}730}{r}$$

If the main-track curve is not very sharp, this value of d may be obtained by subtracting the degree of curve of the main track from that obtained from the sixth column of the table for stub switches. The lead L may also be taken from the table.

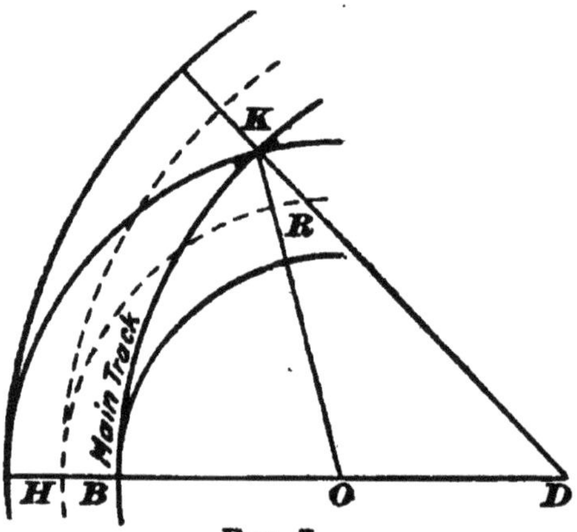

FIG. 7

If the curvature of the main track is very sharp, or if the frog angle is very small, the turnout may curve in the same direction as the main track; in which case, the degree of curve taken from the stub-switch table will be less than the degree of curve of the main track. The difference between the two degrees of curve will still be equal to the degree of curve of the turnout.

If the degrees of curve are equal, the turnout rails will be straight.

Turnout to the Inner Side of Curved Track.—A turnout to the inner side of a curved track is shown in Fig. 7. The radius OR of the turnout is always less than the radius DH of the main track. The degree of curve of the center line of the turnout and the lead BK are found as follows:

Rule I.—*Take from the table for a stub switch, or from the table for a point switch, the value of the degree of curve corresponding to the given frog number. Add this to the degree of curve of the main track. The sum is the degree of curve of the turnout.*

Rule II.—*Take the value of the lead from the table for a stub switch, or from the table for a point switch, corresponding to the given frog number. This will be the value of the desired lead BK, Fig. 7.*

CONNECTING CURVES

A *connecting curve* is a curve introduced to connect a turnout with a side track. Thus, in Fig. 8, the two parallel straight tracks are connected by the turnout ME and the curved track ED. The values of n and g, and the distance a, usually taken as 13 ft., must be known; then the radius $r' = O_1D = O_1E$, and distance KT may be computed by the formulas

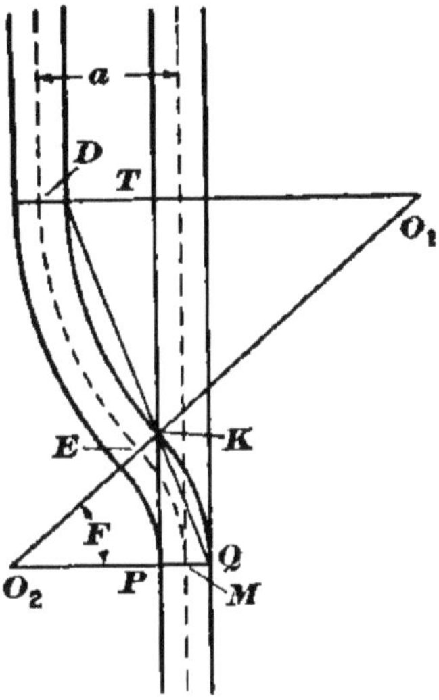

FIG. 8

$$r' = 2\,(a-g)\,n^2 + \tfrac{1}{2}a$$

and
$$KT = \frac{a-g}{g} \times L$$

L is the lead PK of the turnout, and, in such cases as this, is always to be taken from the table for a stub switch, even when the point switch is inserted, because in deriving the formula for KT, QK and ME are assumed to be circular arcs.

CROSS-OVERS

A *cross-over* is a stretch of track that connects two parallel tracks, and enables a train to pass from one track to the other. Thus, in Fig. 9, if *UV* and *U'V'* are two parallel tracks, the track *RZR'* is a cross-over. This cross-over consists of two equal turnouts *Rm* and *R'm'*, whose frog angles at *K* and *K'* are equal, and a reversed curve *mZm'* connecting the ends of these turnouts, *Z* being the point of reversal.

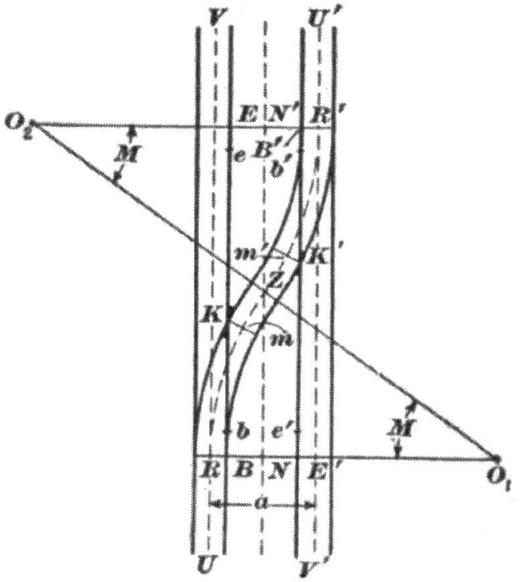

FIG. 9

Cross-Over Between Two Parallel Straight Tracks.—To lay out the cross-over, it is necessary to know the radius *r*, the central angle *M*, and the distances $BE = B'E'$. The radius *r* may be taken from the table for stub switches. Then,

$$\sin M = \sqrt{\frac{a}{r}\left(1 - \frac{a}{4r}\right)}$$

and

$$BE = 2r \sin M$$

When the tracks are less than 30 ft. apart, the value of $\frac{a}{4r}$

may be dropped. The formulas for sin M and BE then become, respectively,

$$\sin M = \sqrt{\frac{a}{r}}$$

and

$$BE = 2\sqrt{ar}$$

The preceding formulas apply only to stub switches; to apply them to point switches, proceed as follows: Having located one frog point K of the point-switch turnout, measure back from K the lead KB for a stub-switch turnout taken from the table, and from the point R of the center line opposite B run in the curve RmZ to the point of reversal. Then, measure off the distance $BE = 2\sqrt{ar}$, and from the point B' opposite to E lay off the stub-switch lead $B'K'$ to locate the second point of frog K'. Then run in the center-line curve $R'Z$. The two frog points and the reversed curve mZm' are thus located. Finally, measure back from K and K' the distances $Kb = K'b'$ equal to the lead for point switches, to locate the toes of the point switches at b and b', and complete the location of these switches as explained under Laying Out Turnouts.

It is evident that the whole length of the cross-over when point switches are employed is $be = b'e' = BE - 2 \times Bb = 2\sqrt{ar} - 2 \times Bb$. Therefore,

$$be = b'e' = 2\sqrt{ar} - 2 \times (\text{lead of stub switch} - \text{lead of point switch})$$

A stake is usually driven at Z, midway between the inner rails and midway between the points N and N', and the turnout curves are continued to this point. This is more accurate than to attempt to determine the point of reversal by the use of the central angle M.

Another Form of Cross-Over Between Two Parallel Straight Tracks.—A second form of cross-over is shown in Fig. 10. In this form, the ends of the two equal turnouts are connected by a straight track $KTK'T'$. The cross-over with a reversed curve, Fig. 9, is much shorter than this straight-track cross-over, and thus requires less length of track and occupies less room. The straight-track form is, however, to be preferred; it is less wearing on the rolling stock because it gives the wheel trucks a better opportunity to adjust themselves to the reversion of curvature.

In order to lay out a straight-track cross-over, it is only necessary to compute the distance $BE = B'E'$, Fig. 10, in addition to the usual dimensions of the two turnouts, which may be done by taking the lead L from the stub-switch table and applying the formula:

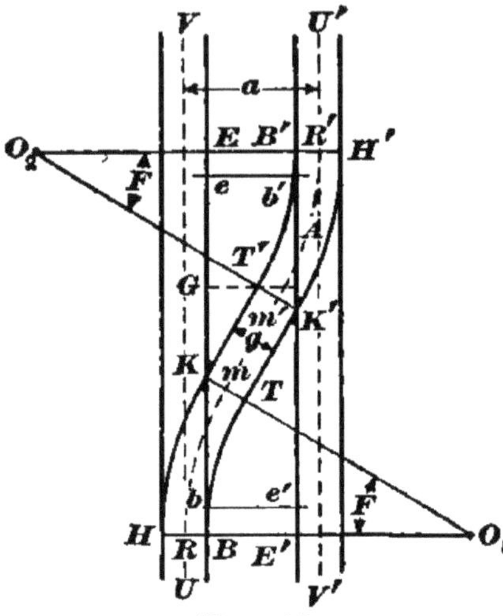

FIG. 10

$$BE = 2L - \frac{a}{4n} + (a - 2g)n$$

The turnout Rm having been put in place, the distance BE is laid off and the heels B' and H' of the second turnout are located opposite the point E. This turnout is then laid out as far as m', and finally the straight rails KT' and $K'T$ are laid adjoining the ends of the two turnouts.

The only modification of the work for a point switch arises from the fact that the lead $Kb = K'b'$ of the point switch is less than that of the stub switch. The whole length of cross-over is, for a point switch,

$$be = 2L' - \frac{a}{4n} + (a - 2g)n$$

Here L' is the lead taken from the table for point switches.

LAYING OUT TURNOUTS

To Lay Out a Stub Switch.—Having decided on the position of the end b, Fig. 11, of the frog rail, measure the total length of the frog and deduct it from the length of the rail to be cut, marking with red chalk on the flange of the rail the point at which the rail is to be cut. From Fig. 3,

$$n = \frac{ch}{ab}$$

and $$ch = n \times ab$$

To calculate the distance from the heel to the theoretical point of frog, the width of the frog at the heel is measured and multiplied by the frog number. For example, if the width of the frog at the heel is 8½ in., and a No. 8 frog is to be used, the theoretical distance from the heel to the point of frog is $8.5 \times 8 = 68$ in. $= 5$ ft. 8 in. Measure off this distance from the point marking the heel of the frog; this will locate the point of frog, which should be distinctly marked with red chalk on the flange of the rail. It is a common practice to make a distinct mark on the web of the main-track rail, directly opposite to the point of frog. This point, being under the head of the rail, is protected from wear and the weather. The heel of the turnout is then located by measuring back the lead from the point of frog. Next, make a chalk-mark on both main-track rails on a line marking the center of the head-block. A more permanent mark is made with a center punch. Stretch a cord touching these marks, and drive a stake on each side of the track, with a tack in each. This line should be at right angles to the center line of the track, and the stakes should be sufficiently far from the track not to be disturbed when putting in switch ties. Next, cut the switch ties to proper length; draw the spikes from the track ties, three or four at a time, and remove the ties from the track, replacing them with switch ties, and tamping the latter securely in place. When all the long ties are tamped, cut the main-track rail for the frog, being careful that the amount cut off is just equal to the length of the frog. If, by increasing or decreasing the length of the lead 5%, the cutting of a rail can be avoided, this should be done, especially for frogs above No. 8.

Full-length rails (30 ft.) should be used for moving or switch rails, and care should be taken to leave a joint of proper width at the head-chair. The head-chairs should be spiked to the head-block so that the main-track rails will be in perfect line. From 8 to 10 ft. of the switch rails should be spiked to the ties. The tie-rods are placed between the switch ties, which should not be more than 15 in. from center to center of tie. The connection-rod should be attached to the head-rod and switch stand. With these connections made, the switch stand is easily placed to give the proper throw of the switch.

It is common practice to fasten the switch stand to the head-block with track spikes, but a better fastening is made with bolts. The stand is first properly placed, the holes are marked and bored, and the bolts passed through from the under side of the head-block. This obviates all danger of movement of the switch stand in fastening, which is liable to occur when spikes are used, and insures a perfect throw.

The use of track spikes is admissible when holes are bored to receive them, in which case a ⅝-in. auger should be used for standard track spikes. The switch stand should, when possible, be placed facing the switch, so as to be seen from the engineer's side of the engine—the right-hand side.

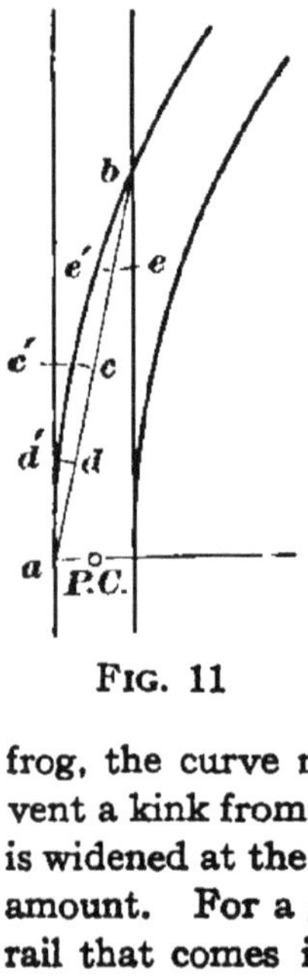

FIG. 11

Next stretch a cord from the heel *a*, Fig. 11, to the point *b*, of the frog. This cord will take the position of the chord of the arc of the outer rail of the turnout curve. Mark the middle point *c* and the quarter points *d* and *e*, and at these points lay off the offsets *dd'*, *cc'*, and *ee'*. Add to these offsets the distance from the gauge line to the outside of the rail flange, and mark the points on the switch ties. Spike the rail to these marks and place the other at easy track gauge from it. Spike the rails of the turnout, as far as the point of frog, to exact gauge, unless the gauge has been widened owing to the sharpness of the curve. Beyond the point of frog, the curve may be allowed to vary a little in gauge to prevent a kink from showing opposite the frog. In case the gauge is widened at the frog, increase the guard-rail distance an equal amount. For a gauge 4 ft. 8½ in., place the side of the guard-rail that comes in contact with the car wheels at 4 ft. 6¾ in. from the gauge line of the frog. This gives a space of 1¾ in. between the main rail and the guard-rail. In case the gauge is widened ¼ or ½ in., increase the guard-rail distance an equal amount.

When the turnout curve is very sharp, it will be necessary to curve the switch rails, to avoid an angle at the head-block.

The rails should be carefully curved before being laid, and great pains should be taken to secure a perfect line.

To Lay Out a Point Switch.—The frog point K, Fig. 12, having been located exactly as for a stub switch, the lead KB is next laid off from K to the toe of switch B, and the positions of B and D are marked on the main-track rails. From D, the length DN of the switch rail, which is usually 15 ft., is then measured forwards to N, and the position of N is marked on the web or flange of the rail. The heel M is usually $5\frac{3}{4}$ in. from the point N. The point I is located on a line perpendicular to MD and at a distance $\frac{1}{2}g$ from M. The point J is similarly located from the point H. As a check on the work, the length of the chord JI should have the value given in the table for point switches.

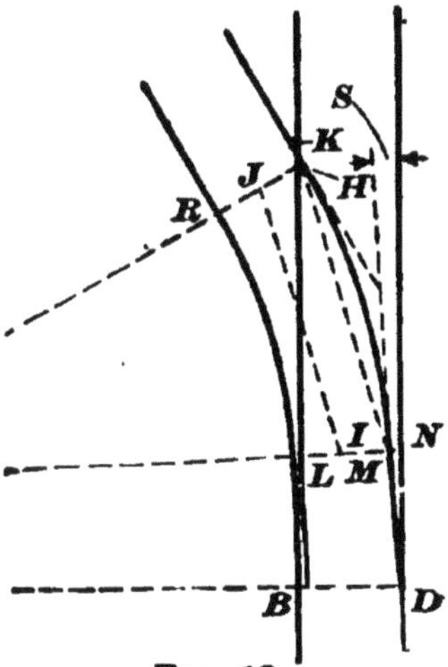

FIG. 12

Switch ties of the requisite number and length should be prepared and placed in the track in proper order. As in the case of stub switches, all long switch ties should be in place before the rail is cut for placing the frog; also, the ends M and L of the rails, with which the switch points connect, should be exactly even; otherwise the tie-rods will be skewed, and the switch will not work or fit well. The tie-rods should next be fastened in position, care being taken to place them in their proper order, the head-rod being numbered 1. Each rod is marked with a center punch, the number of punch marks corresponding to the number of the rod.

The switch rails are now coupled with the rails LK and MK, and the sliding plates are then placed in position and securely spiked to the ties. The head-rod is then connected with the switch stand, and the switch is closed, giving a clear main track. The stand is then adjusted for this position of the switch, and bolted fast to the head-block. Next, rail BR is crowded against the switch point so as to insure a close fit, and secured in place with a rail brace at each tie. The laying of the rails of the turnout is then continued.

The rail *MH* is to be bent and spiked in place by laying off offsets from the chord *MH* exactly as explained for stub switches. The rail between *M* and *H* usually consists of two pieces of plain rail bent to the proper curve. The outer rail of the main track is not disturbed.

Switch Timbers.—Every first-class railroad has its own standards for switches, which include the necessary switch timbers. The number of ties and their lengths may be determined by the following rules:

Rule I.—*To find the number of ties required for any switch lead, reduce to inches the distance from the head-block to the last long tie behind the frog, and divide this distance by the number of inches from center to center of ties; the quotient will be the number of ties required.*

Rule II.—*Measure the length of the tie next the head-block and the length of the last long tie behind the frog. Find the difference, in inches, between them. Divide this difference by the number of ties in the switch lead; the quotient will be the increase in length per tie from the head-block toward the frog to have the ends of the tie in proper line on both sides of the track.*

MECHANICS

FALLING BODIES

When the center of gravity of a moving body passes over equal distances in equal intervals of time, the body has a *uniform motion;* otherwise, the motion is *variable.* The *velocity* in a uniform motion is constant and is equal to the distance traversed by the center of gravity of the body in a unit of time, as feet per second, miles per hour, etc. When, in a variable motion, the velocity increases or decreases uniformly with the time, the motion is designated, respectively, as *uniformly accelerated* or *uniformly retarded,* and the rate of increase or decrease is called *acceleration* or *retardation,* being equal to the amount that the velocity increases or decreases in a unit of time. A body falling under the action of gravity is a case of uniformly accelerated motion, the acceleration being equal to 32.16 ft. per sec. and being usually denoted by g.

Let t = number of seconds that the body falls;
v = velocity, in feet per second, at the end of the time t;
h = distance that the body falls during the time t.

Then,

$$v = gt = \frac{2h}{t} = \sqrt{2gh} = 8.02\sqrt{h},$$

$$h = \frac{vt}{2} = \frac{gt^2}{2} = \frac{v^2}{2g} = .015547\, v^2$$

and

$$t = \frac{v}{g} = \frac{2h}{v} = \sqrt{\frac{2h}{g}} = .24938\sqrt{h}$$

CENTRIFUGAL FORCE

Let F = centrifugal force, in pounds;
W = weight of revolving body, in pounds;
r = distance from the axis of motion to the center of gravity of the body, in feet;
v = velocity, in feet per second.

Then,

$$F = \frac{Wv^2}{gr}$$

When the track on a bridge is curved, the moving cars exert on the bridge a lateral thrust, equal to F, that has to be taken by the lateral bracing of the bridge. In applying the preceding formula, W is to be taken as the maximum weight of the live load for which the chords of the bridge are designed; v is usually expressed in miles per hour and $r = \frac{5{,}730}{D}$, D being the degree of curvature. The formula then becomes

$$F = .00001167\, v^2 DW$$

For curves of 4° or under, v is usually taken as 60 mi. per hr., and usually for D exceeding 4°, $v = 60 - 2D$.

W is to be assumed as acting 5 ft. above the base of the rail. The overturning moment due to the force F is therefore

$5 \times F$ ft.-lb.; and, if d is the distance, in feet, from center to center of rails, the vertical force on the outer rail due to this overturning moment is $\frac{5F}{d}$ lb.

WORK

Work is the overcoming of resistance through a distance. The unit of work is the *foot-pound;* that is, it equals 1 lb. raised vertically 1 ft. The amount of work done is equal to the resistance in pounds multiplied by the distance in feet through which it is overcome. If a body is lifted, the resistance is the weight, or the overcoming of the attraction of gravity, the work done being the weight W, in pounds, multiplied by the height h of the lift, in feet, or Wh ft.-lb.

Power is the amount of work performed in a unit of time. One H. P. is 550 ft.-lb. of work in 1 sec., 33,000 ft.-lb. in 1 min. or 1,980,000 ft.-lb. in 1 hr. In the metric system, 1 H. P. is 75 meter kilograms per second, usually written 75 m. Kg. sec.

Kinetic energy is the capacity of a moving body to perform work. If the moving body has a weight W and a velocity v, the work that it is capable of doing in being brought to rest is $\frac{Wv^2}{2g}$. A body falling through a height of h ft. acquires during its fall a velocity of $v = \sqrt{2gh}$; its kinetic energy is therefore,

$$\frac{W\left(\sqrt{2gh}\right)^2}{2g} = Wh$$

EXAMPLE 1.—What is the horsepower of a stream of water discharging 12 cu. ft. per sec. through a height of 125 ft.?

SOLUTION.—The kinetic energy per second, is $62.5 \times 12 \times 125$ ft.-lb., 62.5 being the weight of 1 cu. ft. of water. The horsepower is, therefore,

$$\frac{62.5 \times 12 \times 125}{550} = 170.5$$

EXAMPLE 2.—What is the kinetic energy per second of a jet of water whose area of cross-section is .1 sq. ft. and whose velocity is 10 ft. per sec.?

SOLUTION.—In this case, $W = 62.5 \times .1 \times 10 = 62.5$ lb. The kinetic energy is therefore,

$$\frac{62.5 \times 10^2}{2g} = \frac{6{,}250}{64.32} = 97.2 \text{ ft.-lb. per sec.}$$

COMPOSITION AND RESOLUTION OF FORCES

The *resultant* of two or several forces acting on a body is the single force that, if acting alone, would produce the same effect as the several forces combined. The latter forces are called *components* with respect to the resultant.

Composition of forces is the process of finding the resultant when the components are known, and the converse process of finding the components when the resultant is given, is called *resolution of forces.*

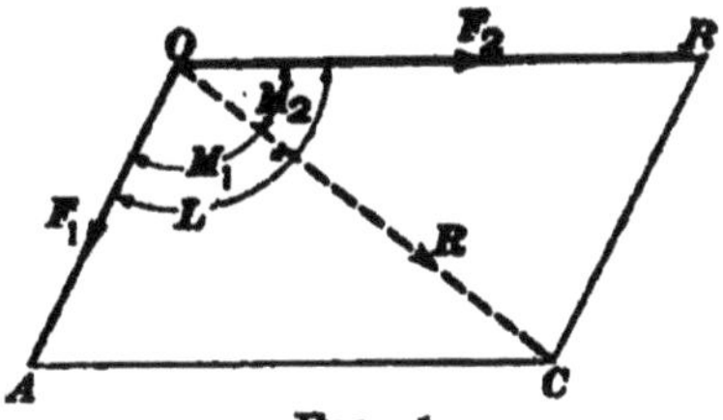

FIG. 1

Parallelogram of Forces.—If two forces, as F_1 and F_2, Fig. 1, are represented in magnitude and direction by two lines, as OA and OB, their resultant R will be represented in magnitude and direction by the diagonal OC of the parallelogram $OACB$ which is constructed by drawing BC and AC parallel to OA and OB, respectively, and joining the intersection C with O.

The resultant R can also be determined analytically; its magnitude by the formula $R = \sqrt{F_1^2 + F_2^2 \times 2F_1F_2 \cos L}$, and the angles M_1 and M_2 that R makes with F_1 and F_2, respectively, may be found by the formulas,

$$\sin M_1 = \frac{F_2 \sin L}{R}$$

and

$$\sin M_2 = \frac{F_1 \sin L}{R}$$

For *rectangular components*, $L = 90°$. The formulas then become:

$$R = \sqrt{F_1^2 + F_2^2}$$

$$\sin M_1 = \frac{F_2}{R}$$

$$\sin M_2 = \frac{F_1}{R}$$

Resolution of Forces.—A given force may have an innumerable number of combinations of components. The problem is, however, determinate when the directions of the components are given.

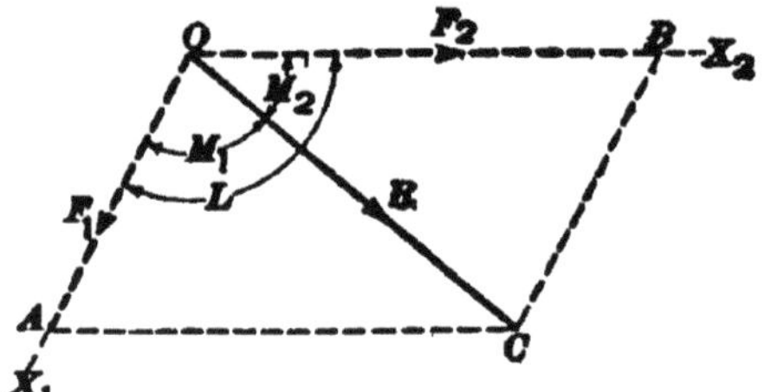

FIG. 2

Let OC, Fig 2, represent in magnitude and direction the force R acting at O, and let it be required to find its components in the directions OX_2 and OX_1. Draw from C, lines parallel to these directions meeting OX_1 at A and OX_2 at B. Then, OA and OB are the required components F_1 and F_2. They may be determined also analytically by the formulas,

$$F_1 = \frac{R \sin M_2}{\sin (M_1 + M_2)}$$

and

$$F_2 = \frac{R \sin M_1}{\sin (M_1 + M_2)}$$

When F_1 and F_2 are perpendicular to each other, then $M_1 + M_2 = 90°$

and

$$F_1 = R \sin M_2$$
$$F_2 = R \sin M_1$$

MOMENTS OF FORCES

The *moment* of a force about a point is the product obtained by multiplying the magnitude of the force by the perpendicular distance from the point to the line of action of the force. In Fig. 3, the moment of F about the point C is Fp; and about the point C_1 it is Fp_1.

The point to which a moment is referred, or about which a moment is taken, is called the *center of moments*, or *origin of moments*. The perpendicular p or p_1 from the origin of moments on the line of action of the force is called the *lever arm* or simply the *arm*, of the force with respect to the origin.

A moment is expressed in foot-pounds, inch-tons, etc., according to the units to which the force and its arm are referred.

The moment is either positive or negative, depending on the direction in which the force tends to cause rotation. It is positive for clockwise motion, and negative for counter-clockwise motion. Thus, the moment of F about C is positive and the moment about C_1 is negative, because, if the arms p and p_1 were bars, the force would tend to rotate p in a clockwise direction, and p_1 in a counter-clockwise direction.

CENTER OF GRAVITY

The *center of gravity* of a figure or a body is that point upon which the figure or the body will balance no matter in what position it may be placed, provided it is acted upon by no other force than gravity.

If a plane figure is alike, or symmetrical, on both sides of a center line, the latter line is termed an *axis of symmetry*, and the center of gravity lies in this line. If the figure is symmetrical about any other

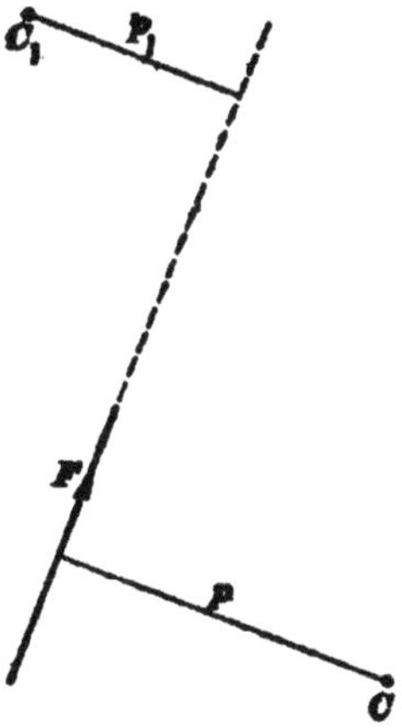

Fig. 3

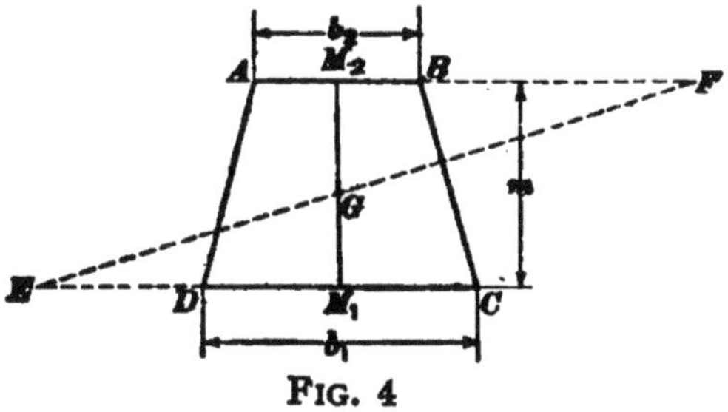

Fig. 4

axis, the intersection of the two axes will be the center of gravity of the section; thus, the center of gravity of a *parallelogram* is at the intersection of the diagonals and that of a *circle* or an *ellipse* is at the geometrical center of the figure. The center of gravity of a *triangle* lies on a line drawn from a vertex to the middle point of the opposite side, and at a distance from that side equal to one-third the length of the line; or it is at the intersection of lines drawn from the vertexes to the middle points of the opposite sides.

To find the center of gravity of a *trapezoid*, Fig. 4, lay off $BF = DC$ and $DE = AB$; the center of gravity is at the intersection of EF with $M_1 M_2$, the line joining the middle points of the parallel sides. GM_1 can also be determined by the formula

$$GM_1 = \frac{m(b_1+2b_2)}{3\,(b_1+b_2)}$$

The center of gravity of any *quadrilateral* may be determined as follows: First divide it, with a diagonal, into two triangles and join with a straight line the centers of gravity of the two triangles; then, with the second diagonal, divide the figure into two other triangles and join the centers of gravity of these triangles with a straight line. The center of gravity of the quadrilateral is at the intersection of the lines joining the centers of gravity of the two sets of triangles.

For an *arc of a circle*, the center of gravity lies on the radius drawn to the middle point of the arc (an axis of symmetry) and at a distance from the center equal to the length of the chord multiplied by the radius and divided by the length of the arc.

For a *semicircle*, the distance from the center $= \frac{2r}{\pi} = .6366\ r$, when $r =$ the radius.

For the area included in a *half circle*, the distance of the center of gravity from the center is

$$\frac{4r}{3\pi} = .4244r$$

For a *circular sector*, the distance of the center of gravity from the center equals two-thirds of the length of the chord multiplied by the radius and divided by the length of the arc.

For a *circular segment*, let A be its area and C the length of its chord; then the distance of the center of gravity from the center of the circle is equal to $\frac{C^3}{12A}$.

The center of gravity of any ***irregular plane figure*** can be determined by applying the following principle: The static moment of any plane figure with regard to a line in its plane—that is, the product of its area A by the distance D of its center of gravity from that line—is equal to the algebraic sum of the static moments of the separate parts into which the figure may be divided, with regard to the same axis, or

$$AD = a_1d_1 + a_2d_2, \text{ etc.},$$

in which, a_1 a_2, etc., denote the areas of the subdivided parts of the figure, and d_1, d_2, etc. are the distances of their respective centers of gravity from the reference line. Solving this equation for the value of D,

$$D = \frac{a_1d_1 + a_2d_2 + \text{etc.}}{A}$$

The figure whose center of gravity is required is divided into separate parts whose centers of gravity are easily ascertained, usually into rectangles or triangles. A suitable axis is then assumed with reference to which the expressions a_1d_1, a_2d_2, etc. are found, and their sum is divided by $A = a_1 + a_2 +$ etc., the quotient giving D. The center of gravity of the whole figure lies, therefore, on a line parallel to the assumed axis and distant D from it. In a similar manner, another line containing the center of gravity is obtained, the intersection of the two lines giving its exact position.

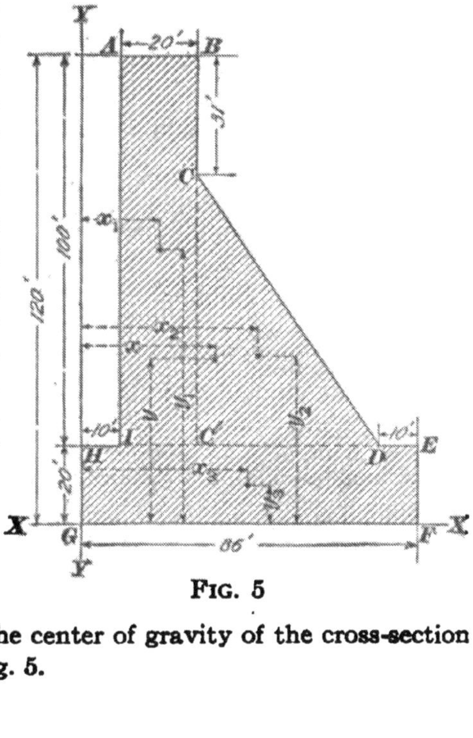

FIG. 5

EXAMPLE 1.—Find the center of gravity of the cross-section of the dam shown in Fig. 5.

SOLUTION.—Divide the section into the rectangles $ABC'L$ and $HEFG$ and the triangle CDC', and assume the lines $X-X$ and $Y-Y$ as reference lines. Then,

$$y=\frac{a_1y_1+a_2y_2+a_3y_3}{a_1+a_2+a_3}$$

and
$$x=\frac{a_1x_1+a_2x_2+a_3x_3}{a_1+a_2+a_3}$$

From the illustration, $a_1=100\times20=2{,}000$, $y_1=70$, $x_1=20$; $a_2=\frac{69\times46}{2}=1{,}587$, $y_2=43$, $x_2=45.33$; $a_3=86\times20=1720$, $y_3=10$, and $x_3=43$. Substituting these values,

$$y=\frac{2{,}000\times70+1{,}587\times43+1{,}720\times10}{2{,}000+1{,}587+1{,}720}=42.48$$

and
$$x=\frac{2{,}000\times20+1{,}587\times45.33+1{,}720\times43}{5{,}307}=35.03$$

EXAMPLE 2.—Find the center of gravity of the bridge chord section shown in Fig. 6.

SOLUTION.—The center of gravity is on the line YY, which is an axis of symmetry. To find the distance y, divide the section into angles and plates and take moments about XX. The areas and centers of gravity of the angles might be located by the preceding principles or taken from a manufacturer's handbook. They are: for the 4″×4″×½″ angle, area = 3.75 sq. in. and distance from center of gravity to back of angle = 1.18 in.; for the 3½″×3½″×½″ angle, area = 3.25 sq. in. and distance from center of gravity to back of angle = 1.06 in. Hence, the moment of the bottom angles is $2\times3.75\times1.18=8.85$ and that of the top angles is $2\times3.25\,(15-1.06)=90.61$. The moment of the two web plates is $2\times15\times\frac{1}{2}\times7.5=112.5$, and that of the cover-plate, $24\times\frac{1}{2}\times15.25=183.00$. The sum of the moments is, $8.85+90.61+112.5+183.00=394.96$. The sum of the areas is $2\times3.25+2\times3.75+24\times\frac{1}{2}+2\times15\times\frac{1}{2}=41$ sq. in. Then, $y=394.96\div41=9.63$ in.

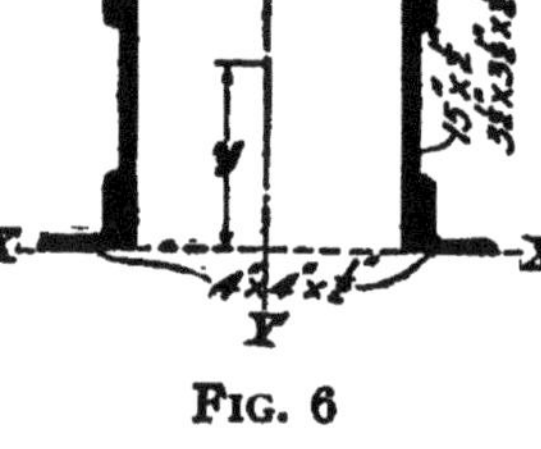

FIG. 6

Center of Gravity of Solids.—For a solid having three axes of symmetry, all perpendicular to each other, like a sphere, cube, right parallelopiped, etc., the point of intersection of these axes is the center of gravity.

For a cone or pyramid, draw a line from the apex to the center of gravity of the base; the required center of gravity is one-fourth the length of this line from the base, measured on the line.

For two bodies, the larger weighing W lb., and the smaller P lb., the center of gravity will lie on the line joining the centers of gravity of the two bodies and at a distance from the larger body equal to $\frac{Pa}{P+W}$, where a is the distance between the centers of gravity of the two bodies.

For any number of bodies, first find the center of gravity of two of them, and consider them as one weight whose center of gravity is at the point just found. Find the center of gravity of this combined weight and a third body. So continue for the rest of the bodies, and the last center of gravity will be the center of gravity of the whole system of bodies.

To find the center of gravity mechanically, suspend the object from a point near its edge and mark on it the direction of a plumb-line from that point; then suspend it from another point and again mark the direction of a plumb-line. The intersection of these two lines will be directly over the center of gravity.

MOMENT OF INERTIA

The *moment of inertia* of a plane surface about a given axis is the sum of the products obtained by multiplying each of the elementary areas, into which the surface may be conceived to be divided, by the square of its distance from the axis.

The moment of inertia is usually designated by the letter I. The value of the moment of inertia used in calculating the strength of beams and columns is usually taken about the neutral axis of the figure, which, with the exception of reinforced-concrete sections, is passing through the center of gravity of the figure.

MOMENTS OF INERTIA, ETC.

Name of Section.		I	$\frac{I}{c}$	r^2
Solid circular		$\frac{\pi d^4}{64}$	$\frac{\pi d^3}{32}$	$\frac{d^2}{16}$
Hollow circular		$\frac{\pi(d^4-d_1^4)}{64}$	$\frac{\pi(d^4-d_1^4)}{32d}$	$\frac{d^2+d_1^2}{16}$
Solid square		$\frac{d^4}{12}$	$\frac{d^3}{6}$	$\frac{d^2}{12}$
Hollow square		$\frac{d^4-d_1^4}{12}$	$\frac{d^4-d_1^4}{6d}$	$\frac{d^2+d_1^2}{12}$
Solid rectangular		$\frac{bd^3}{12}$	$\frac{bd^2}{6}$	$\frac{b^2}{12}$
Hollow rectangular	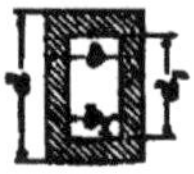	$\frac{bd^3-b_1d_1^3}{12}$	$\frac{bd^3-b_1d_1^3}{6d}$	$\frac{b^3d-b_1^3d_1}{12(bd-b_1d_1)}$
Solid triangular		$\frac{bd^3}{36}$	$\frac{bd^2}{24}$	$\frac{d^2}{18}$
Solid elliptical		$\frac{\pi bd^3}{64}$	$\frac{\pi bd^2}{32}$	$\frac{b^2}{16}$
Hollow elliptical		$\frac{\pi}{64}(bd^3-b_1d_1^3)$	$\frac{\pi(bd^3-b_1d_1^3)}{32d}$	$\frac{b^3d-b_1^3d_1}{16(bd-b_1d_1)}$
I-beam	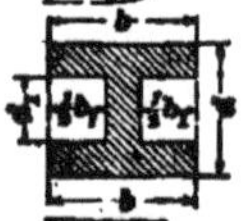	$\frac{bd^3-b_1d_1^3}{12}$	$\frac{bd^3-b_1d_1^3}{6d}$	$\frac{b^3d-b_1^3d_1}{12(bd-b_1d_1)}$
Cross with equal arms (approximate)				$\frac{d^2}{22.5}$
Angle with equal arms (approximate)				$\frac{d^2}{25}$

Formulas for the values of I about an axis passing through the center of gravity of the section are given for various forms of sections in the accompanying table. For any other section, it can be computed by means of the following principles:

Principle I.—*The moment of inertia of a section about any axis is equal to the algebraic sum of the moments of inertia about the same axis, of the separate parts of which the figure may be conceived to consist.*

Principle II.—*The moment of inertia of any figure about an axis not passing through the center of gravity, is equal to the moment of inertia about a parallel axis through the center of gravity, plus the product of the entire area of the section by the square of the distance between the two axes.*

EXAMPLE 1.—Find the moment of inertia about the neutral axis XX of the Bethlehem I column section having dimensions as shown in Fig. 1.

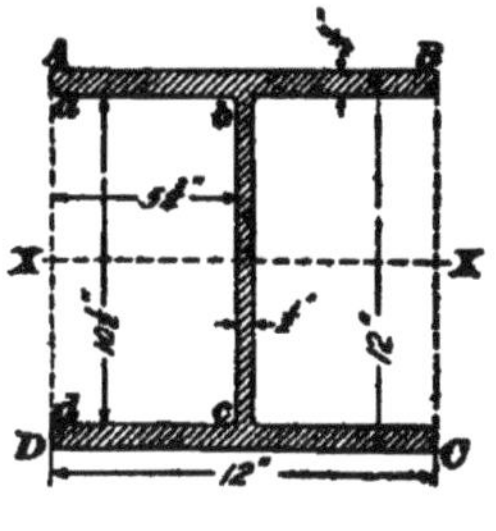

FIG. 1

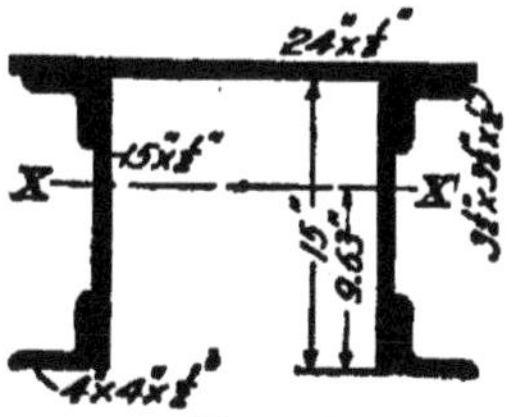

FIG. 2

SOLUTION.—Conceive the section to consist of the square $ABCD$ minus twice the rectangle $abcd$. Then, by applying principle I and the formulas of the table for moments of inertia,

$$I = \frac{12^4}{12} - \frac{2 \times 5.75 \times 10.5^3}{12} = 618.6$$

NOTE.—This result can be obtained directly by the I beam formula, given in the same table.

EXAMPLE 2.—Find the moment of inertia of the section shown in Fig. 2 about the neutral axis parallel to the cover-plate.

SOLUTION.—The neutral axis passes through the center of gravity, which has been found to be 9.63 in. from the back of the bottom angles. The distances of the centers of gravity of the subdivisions of this section from the axis XX, Fig. 2, are:

For the cover-plate 15.25−9.63.............. = 5.62
For the web-plates 9.63−7.50................ = 2.13
For the $3\frac{1}{2}''\times3\frac{1}{2}''\times\frac{1}{2}''$ ∟'s, 15.00−1.06−9.63. = 4.31
For the $4''\times4''\times\frac{1}{2}''$ ∟'s, 9.63−1.18......... = 8.45

The moments of inertia of the respective parts about their own neutral axes parallel to XX are:

For the cover-plate................ $\dfrac{24\times(\frac{1}{2})^3}{12} = .25$

For the web-plates.............. $\dfrac{2\times\frac{1}{2}\times15^3}{12} = 281.25$

From a steel manufacturer's handbook, the value of I for a $3\frac{1}{2}''\times3\frac{1}{2}''\times\frac{1}{2}''$ ∟ is found to be 3.64; and for a $4''\times4''\times\frac{1}{2}''$ ∟ it is 5.56. Applying principle II, the moment of inertia of the entire section is, $I = .25+24\times\frac{1}{2}\times5.62^2+281.25+2\times15\times\frac{1}{2}\times 2.13^2+2\times3.64+2\times3.25\times4.31^2+2\times5.56+2\times3.75\times8.45^2 = 1{,}403.22$.

RADIUS OF GYRATION

Let I denote the moment of inertia of any section and a its area; then, the relation between I and a is expressed in the formula, $I = ar^2$, in which r is a constant depending on the shape of the section and is called the *radius of gyration* of the section referred to the same axis as I. Then,

$$r = \sqrt{\frac{I}{a}}$$

EXAMPLE 1.—What is the radius of gyration of the section shown in Fig. 1 about the axis XX?

SOLUTION.—The moment of inertia of this section has been found to be 618.6 and its area is $2\times12\times\frac{3}{4}+10.5\times\frac{1}{2} = 23.25$ sq. in. Substituting in the formula,

$$r = \sqrt{\frac{618.6}{23.25}} = 5.16$$

EXAMPLE 2.—Determine the distance b in the strut made up of two latticed channels, as shown in Fig. 3, so that the radii of gyration about the axes XX and YY will be equal.

SOLUTION.—Let I_x, r_x, I_y, r_y be, respectively, the moments of inertia and radii of gyration of a single [about the axes XX and YY; a its area and CG, its center of gravity, then, from the figure, $b=d-c$, and $I_x=ar_x^2$; also, $I_y=ar_y^2+ad^2$. Hence, by the condition of the problem, $ar_x^2=ar_y^2+ad^2$, or $r_x^2=r_y^2+d^2$. Whence, $d=\sqrt{r_x^2-r_y^2}$. The values of r_x, r_y, and c for any [may be taken from a steel manufacturer's handbook. For instance, for a 15″ [of 33 lb. $r_x=5.62$, $r_y=.912$, and $c=.794$; hence, $d=\sqrt{5.62^2-.912^2}=5.546$, and $b=d-c=5.546-.794=4.752$.

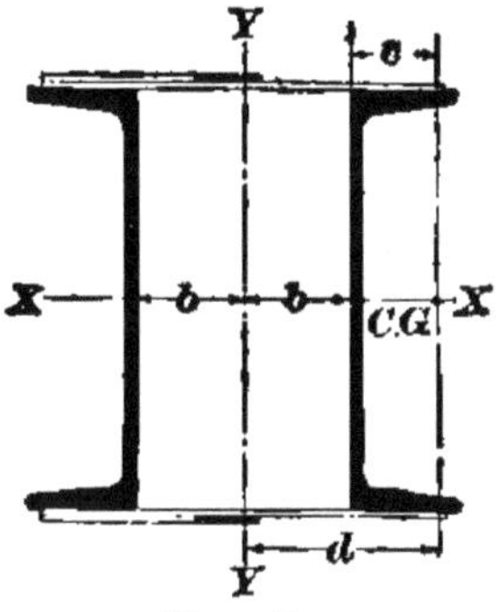

FIG. 3

A practical rule giving good approximate results for a channel column or strut is to subtract r_y from r_x; the result is b. Applying this rule for the 15″ [of 33 lb. column or strut, $b=5.62-.912=4.708$.

SECTION MODULUS AND MOMENT OF RESISTANCE

The expression $\frac{I}{c}$, in which I is the moment of inertia and c the distance of the outermost fiber of the section from the neutral axis, is called the *section modulus*. For a given material, this quantity is a measure of the capacity of the section to resist bending. Multiplied by the unit stress to which the outermost fibers are subjected under given loads, the product gives the amount of bending moment the section is resisting, and is therefore called *moment of resistance*. If f is the unit stress that certain loads develop in the outermost fibers of the section, the moment of resistance is

$$M_r=\frac{I}{c}f$$

EXAMPLE 1.—What is the section modulus of a 20-in. I beam at 75 lb. whose moment of inertia is 1,268.9?

SOLUTION.—Since the neutral axis passes through the center of the section, the distance c is in this case equal to one-half the depth; that is, $\frac{20}{2} = 10$. The section modulus is therefore

$$\frac{I}{c} = \frac{1,268.9}{10} = 126.9$$

EXAMPLE 2.—When subjected to loads perpendicular to the cover-plates the outermost fibers of the section shown in Fig. 2, are stressed to 16,000 lb. per sq. in., What is the resisting moment of the section?

SOLUTION.—The moment of inertia of the section has been found to be 1,403.22 and the outermost fibers are 9.63 in. from the neutral axis; hence, the section modulus is equal to $\frac{1,403.22}{9.63} = 145.7$; this multiplied by 16,000 gives 2,331,200 in.-lb.

Formulas for obtaining directly the section moduli of sections frequently used are given in the table of Moments of Inertia, etc. For rolled-steel sections, they are given in manufacturers' handbooks.

FRICTION

Friction is the resistance that a body meets from the surface on which it moves. It depends on the degree of roughness of the surfaces in contact, and is directly proportional to the perpendicular pressure between the surfaces. It is independent of the extent of the surfaces in contact, so long as the normal pressure remains the same. It is generally greater between surfaces of the same material than between those of different materials, and greater between soft bodies than hard ones.

Coefficient of Friction.—The ratio between the resistance to the motion of a body due to friction and the perpendicular pressure between the surfaces is called the *coefficient of friction.* When the coefficient of friction between two surfaces is known, the frictional resistance is obtained by multiplying the normal pressure by the coefficient.

EXAMPLE.—What is the resistance per linear foot of a retaining wall against sliding when the normal pressure on the foundation is 10,000 lb. per lin. ft. of wall and the coefficient of friction of the masonry on the foundation is .65?

SOLUTION.—The frictional resistance is 10,000×.65 = 6,500 lb. The coefficient of friction of the wheels of suddenly stopping engines and cars on the rails is usually assumed at .20. The rails on bridges or trestles will transfer to the bridge or trestle tower the frictional forces produced by the brakes in order to stop the cars, causing stresses that have to be provided for in the structure.

EXAMPLE.—What is the longitudinal force on a bridge caused by the sudden stopping of a car weighing 60,000 lb.?

SOLUTION.— 60,000×.20 = 12,000 lb.

Angle of Friction.—When a body, as B in the accompanying illustration, weighing W lb. is placed on an inclined plane making an angle a with the horizontal, the normal pressure is $N = W \cos a$; and, if the coefficient of friction is denoted by f, the frictional resistance against sliding down of the body is $F = fN = fW \cos a$. This force acts in a direction opposite to that of the force $P = W \sin a$. When the angle a is such that F just balances, or is equal to, P, so that the slightest force will cause the body to slide, the angle is then called the *angle of friction*. The tangent of that angle is equal to the coefficient of friction, or $f = \tan a$.

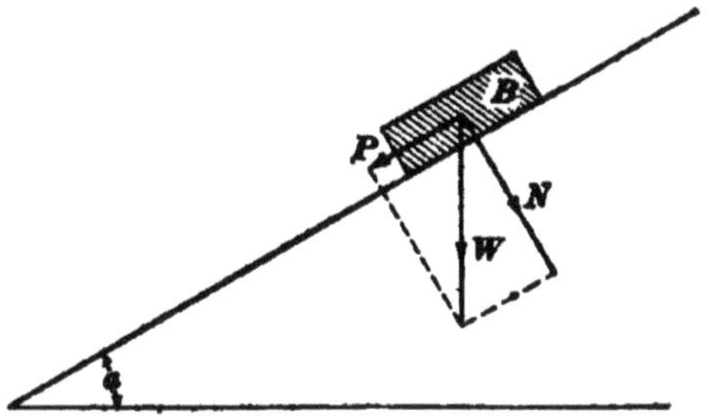

Angle of Repose.—On a sloping bank of loose material, such as sand, earth, etc., when the angle of slope is such that the particles are on the point of moving, the angle is called the *angle of repose*. It is the same as the angle of friction of the material on itself. The slope is then called the *slope of repose*, or the *natural slope of the material*, for it is the slope that the material will assume when subject to gravity only.

EXAMPLE.—The coefficient of friction of dry sand on itself is .65; what is its angle of repose?

SOLUTION.—The angle of repose is the same as the angle of friction, whose tangent equals the coefficient of friction;

consequently, .65 = tan a, and from a table of natural tangents $a = 33°$.

The accompanying tables give coefficients of friction and angles of repose of a number of materials.

COEFFICIENTS OF FRICTION AND ANGLES OF REPOSE FOR MASONRY MATERIALS

Material	Coefficient of Friction	Angle of Repose Degrees
Fine-cut granite, on same, dry	.60	31
Fine-cut granite, on rough-pointed granite, dry	.65	33
Rough-pointed granite, on same, dry	.70	35
Well-dressed soft limestone, on same, dry	.75	37
Concrete blocks, on same, dry	.65	33
Concrete blocks, on fine-cut granite, dry	.60	31
Common brick, on same, dry	.65	33
Common brick, on well-dressed soft limestone, dry	.65	33
Common brick, on well-dressed hard limestone, dry	.60	31
Common brick, on same, with slightly damp mortar	.75	37
Hard brick, on same, with slightly damp mortar	.70	35
Hard limestone, on same, with slightly damp mortar	.65	33
Common brick, on same, with fresh mortar	.50	27
Well-dressed granite, on same, with fresh mortar	.50	27
Granite, roughly worked, on dry sand and gravel	.50 to .60	27 to 31
Granite, roughly worked, on wet sand	.35 to .45	19 to 24
Granite, roughly worked, on dry clay	.50	27
Granite, roughly worked, on moist clay	.35	19

Rolling Friction.—The friction between the circumference of a rolling body and the surface upon which it rolls is known as *rolling friction*. It is due to the compressibility of substances, the weight of the rolling body causing a small depression in the supporting surface and a flattening of the roller. Its magnitude

COEFFICIENTS OF FRICTION, ANGLES OF REPOSE, AND WEIGHTS OF EARTHS

Material	Coefficient of Friction	Angle of Repose Degrees	Weight Pounds per Cubic Foot
Mixed earth, dry	.70	35	95
Mixed earth, damp	.80	39	115
Mixed earth, wet	.40	22	115
Sand, dry	.65	33	110
Sand, wet	.05	3	125
Loam, dry	.70	35	75 to 100
Loam, wet	.50	27	90 to 120
Clay, dry	1.00	45	100
Clay, wet	.30	17	125

COEFFICIENTS AND ANGLES OF FRICTION FOR MISCELLANEOUS MATERIALS

Materials	Coefficient of Friction	Angle of Friction Deg.	Angle of Friction Min.
Cast iron on cast iron	.15	8	32
Cast iron on brass	.15	8	32
Cast iron on oak	.49	26	6
Wrought iron on wrought iron	.14	7	58
Wrought iron on cast iron	.19	10	46
Wrought iron on brass	.17	9	39
Wrought iron on mahogany	.18	10	12
Wrought iron on oak	.62	31	47
Steel on cast iron	.20	11	19
Steel on brass	.15	8	32
Steel on ice	.014	0	48
Yellow copper on cast iron	.19	10	46
Yellow copper on oak	.62	31	48
Brass on cast iron	.22	12	25
Brass on brass	.20	11	19
Brass on wrought iron	.16	9	6
Bronze on cast iron	.21	11	52
Bronze on wrought iron	.16	9	5
Bronze on bronze	.20	11	19
Oak on oak	.48	25	38
Oak on elm	.25	14	3
Oak on cast iron	.37	20	19
Leather on oak	.33	18	16
Leather belt on oak drum	.27	15	7
Leather belt on cast iron	.56	29	15
Leather packing	.56	29	15

depends on the materials of the roller and supporting surface, and is proportional to the normal pressure exercised by the roller on the rolling surface. It depends also on the diameter of the roller, being less for large rollers than for small ones. On highways with soft compressible surfaces, the resistance is also affected by the width of the wheel tires, being greater for narrow tires than for wide ones.

ROLLING FRICTION FOR DIFFERENT ROADWAY SURFACES

Character of Roadway Surface	Rolling Friction			
	In Pounds per Gross Ton			Mean In Terms of Load
	Maximum	Minimum	Mean	
Earth, ordinary	300	125	200	$\frac{1}{11}$
Earth, dry and hard	125	75	100	$\frac{1}{22}$
Gravel, common	147	140	143	$\frac{1}{16}$
Gravel, hard rolled			75	$\frac{1}{30}$
Macadam, ordinary	140	60	90	$\frac{1}{25}$
Macadam, good	80	41	60	$\frac{1}{37}$
Macadam, best	64	30	50	$\frac{1}{45}$
Cobblestone, ordinary			140	$\frac{1}{16}$
Cobblestone, good			75	$\frac{1}{30}$
Granite block, ordinary			90	$\frac{1}{25}$
Granite block, good	80	45	56	$\frac{1}{40}$
Granite block, best	40	25	34	$\frac{1}{66}$
Belgian block, ordinary			56	$\frac{1}{40}$
Belgian block, good	50	26	38	$\frac{1}{60}$
Plank	56	32	44	$\frac{1}{50}$
Wooden block, in good condition	40	20	30	$\frac{1}{75}$
Asphalt	39	15	22	$\frac{1}{100}$

The accompanying table gives the maximum, minimum, and mean values of the coefficient of rolling friction for different roadway surfaces. They are expressed in pounds required to overcome the resistance on a level road of a gross ton (2,240 lb.). The mean value is also expressed as a ratio between the frictional resistance and the load.

STRENGTH OF MATERIALS

DEFINITIONS OF TERMS

Stress is the cohesive force by which the particles of a body resist the external load that tends to produce an alteration in the form of the body. It is always equal to the effective external force acting upon the body; thus, a bar subjected to a direct pulling force of 1,000 lb. endures a stress of 1,000 lb. *Unit stress* is the stress or load per unit of area, usually taken per square inch of section. For instance, if the bar mentioned above is 1 in.×2 in. in section, the unit stress of the bar will be 1,000÷2 (sectional area)=500 lb. *Tensile stress* is produced when the external forces tend to stretch a body, or pull the particles away from one another. A rope by which a weight is suspended is an example of a body subjected to tensile stress. *Compressive stress* is produced when the forces tend to compress the body, or push the particles closer together. A post or column of a building is subjected to compressive stress. *Shearing stress* is produced when the forces tend to cause the particles in one section of a body to slide over those of the adjacent section. A steel plate acted on by the knives of a shear, and a beam carrying a load, are subjected to shearing stress. Tension, compression, and shear are called *simple* or *direct stresses*, to distinguish them from bending and torsion.

The amount of alteration in form of a body produced by a stress is called *deformation*, or *strain*. It may be tensile deformation, compressive deformation, or shearing deformation, according as the stress producing it is tensile, compressive, or shearing. The *rate of deformation*, also called *unit deformation*, is the deformation of a body, subjected to tension or compression, per unit of length. If an iron bar 6 ft. long is subjected to a force that elongates it 1 in., the rate of deformation will be 1 in.÷72 (length of the bar in inches)=.0139 in.

The *modulus* or *coefficient of elasticity* is the ratio between the stresses and corresponding deformations for a given material, which may have a somewhat different modulus of

AVERAGE ULTIMATE STRENGTHS OF METALS IN POUNDS PER SQUARE INCH

Kind of Metal	Compression	Tension	Elastic Limit	Shearing	Modulus of Rupture	Modulus of Elasticity
Aluminum:						
Aluminum, commercial	12,000	15,000	6,500	12,000		11,000,000
Aluminum, nickel		40,000	22,000			
Brass, Bronze, and Copper:						
Brass, cast	(30,000)	24,000	6,000	36,000	20,000	9,000,000
Brass wire, annealed **(softened by** reheating)		50,000				
Brass wire, unannealed		80,000	16,000			14,000,000
Bronze, aluminum	120,000	75,000				
Bronze, gun metal	(20,000)	32,000	10,000		53,000	10,000,000
Bronze, manganese	120,000	60,000	30,000			
Bronze, phosphor		50,000	24,000			14,000,000
Bronze, Tobin		66,000	40,000			4,500,000
Copper, bolts	30,000	30,000				
Copper, cast	(40,000)	24,000	6,000	30,000	22,000	10,000,000
Copper wire, annealed (softened by **reheating)**		36,000				15,000,000
Copper wire, unannealed		60,000	10,000			18,000,000

Cast and Wrought Iron:						
Iron, cast	80,000	15,000	6,000	18,000	30,000	12,000,000
Iron chains		35,000				
Iron, corrugated					40,000	
Iron wire, annealed (softened by reheating)		60,000				15,000,000
Iron wire, unannealed		80,000	27,000			25,000,000
Iron, wrought, shapes	46,000	48,000	26,000	40,000	44,000	27,000,000
Iron, wrought, rerolled bars	48,000	50,000	27,000	40,000	48 000	26,000,000
Lead:						
Lead, cast		2,000	1,000			1,000,000
Lead pipe		1,600				
Cast and Structural Steel:						
Steel, castings	70,000	70,000	40,000	60,000	70,000	30,000,000
Steel, structural, soft	56,000	56,000	30,000	48,000	54,000	29,000,000
Steel, structural, medium	64,000	64,000	33,000	50,000	60,000	29,000,000
Steel wire, annealed (softened by reheating)		80,000	40,000			29,000,000
Steel wire, unannealed		120,000	60,000			30,000,000
Steel wire, crucible		180,000	80,000			30,000,000
Steel wire, for suspension bridges		200,000	90,000			30,000,000
Steel wire, special tempered		300,000				
Tin and Zinc:						
Tin, cast	(6,000)	3,500	1,800		4,000	4,000,000
Zinc, cast	(20,000)	5,000	4,000		7,000	13,000,000

NOTE.—Compression values enclosed in parentheses indicate loads producing 10% reduction in original lengths.

AVERAGE ULTIMATE STRENGTH OF WOODS, IN POUNDS PER SQUARE INCH

Kind of Timber	Extreme Fiber Stress	Modulus of Elasticity	Compression With Grain	Shearing With Grain
Douglas fir.......	5,000	1,380,000	4,400	500
Hemlock.........	3,500	900,000	4,000	250
Long-leaf, or Georgia, pine..	7,000	1,500,000	5,000	500
Short-leaf pine...	6,000	1,200,000	4,200	400
Western, or ponderosa, pine....	4,500	850,000	3,100	
White oak.......	7,000	1,240,000	5,000	800
White pine.......	4,000	870,000	3,500	300

elasticity for tension, compression, and shear. If k is the increase per unit of length of a material subjected to tensile stress and s the unit stress producing this elongation, the modulus of elasticity of the material for tension is

$$E = \frac{s}{k}$$

For example, if a wrought-iron bar subjected to a unit tensile stress of 10,000 lb. per sq. in. is stretched .0003625 in. per inch of length, the modulus of elasticity of the wrought iron for tension is

$$E = \frac{10{,}000}{.0003625} = 27{,}586{,}200 \text{ lb. per sq. in.}$$

It should be observed that E must be expressed in the same units as the unit stress s; in this example, in pounds per square inch.

If the total length of a bar is L, its sectional area A, the total stress to which the bar is subjected P, and the total deflection produced K, then the modulus of elasticity of the material of the bar will be

$$E = \frac{PL}{AK}$$

In this formula, L and K must be referred to in the same unit of length, and A in the corresponding unit of area. Thus, if L is in inches, K also must be in inches, and A must be in square inches.

EXAMPLE.—A steel rod 10 ft. long and 2 sq. in. in cross-section is stretched .12 in. by a weight of 54,000 lb. What is the tension modulus of elasticity of the material?

SOLUTION.—To apply the formula, the stress $P = 54{,}000$ lb.; $L = 10$ ft. $= 120$ in.; $A = 2$ sq. in.; and $K = .12$ in. Therefore,

$$E = \frac{54{,}000 \times 120}{2 \times .12} = 27{,}000{,}000 \text{ lb. per sq. in.}$$

The relation $E = p \div l$ is true only when equal additions of stress cause equal increases of strain. Previous to rupture, this condition ceases to exist, and the material is said to be strained beyond the *elastic limit*, which, therefore, is that degree of stress within which the modulus of elasticity is nearly constant and equal to the unit stress divided by the unit strain.

The *ultimate strength* of a given material in tension, compression, or shear is that unit stress which is just sufficient to break it, and is equal to the maximum stress causing rupture divided by the original area of the cross-section. The preceding tables show the average ultimate strengths, in pounds per square inch, of both metals and woods.

Working stress is the maximum unit stress to which the parts of a structure are to be subjected.

Factor of safety is the ratio of the ultimate strength to working stress. The factor of safety required for a structure depends on the material and on the character of the loads applied—that is, whether the loads are quiescent or such that cause impact and vibrations. For stone and brick, a factor of safety of from 10 to 30 is used; for timber, from 8 to 15; for cast iron, from 6 to 20; for reinforced concrete, from 4 to 6; and for structural steel, from 3 to 6.

It is obvious that structures subjected to loads causing impact should be designed for a higher factor of safety than those having to carry static loads. When a structure, as a bridge, carries both dead and live loads, the modern practice favors the specifying of one working unit stress for both kinds of loads, and providing for the effect of vibration by increasing the live-load stress or bending moment by an amount I determined from a so-called impact formula. The formula most in use for railroad bridges is

$$I = \frac{300}{L+300} S$$

in which S = maximum live-load stress or bending moment in the member, and L = length, in feet, of single track that must be loaded in order to obtain the value S.

SIMPLE, OR DIRECT, STRESS

Formula for Simple Stress.—If P is an external force producing tension, compression, or shear uniformly distributed over an area A, and s is the unit working stress, then $P = sA$ is the fundamental formula for designing parts of structures subjected to a simple, or direct, stress. When designing members that are in tension, A must be taken as the net area of the section. This is determined by deducting from the gross section the greatest number of pin, bolt, or rivet holes that can be cut by a plane at right angles to the section. Rivet holes are usually taken ⅛ in. larger than the diameter of the rivet.

Important Applications of Formulas for Direct Stress. 1. Tension members and short compression members of roof

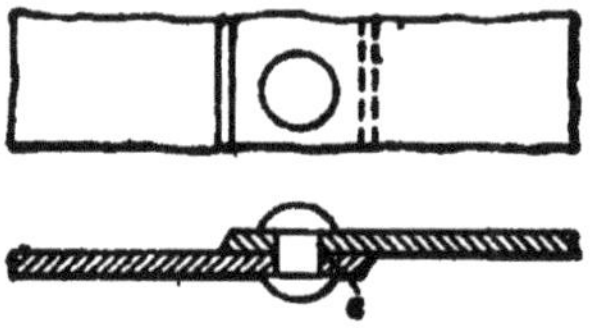

FIG. 1

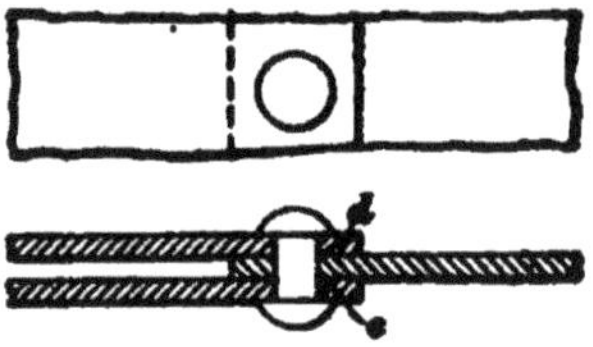

FIG. 2

or bridge trusses are examples of simple stress, and their sections are determined by the preceding formula.

EXAMPLE.—A tension member of a roof truss is made of two 3½″×3½″×½″ angles connected by one line of rivets ¾ in. in diameter. What stress will it carry at 16,000 lb. per sq. in.?

SOLUTION.—The gross sectional area of a 3½″×3½″×½″ angle is 3.25 sq. in. The deduction for one rivet hole is (¾+⅛)

$\times \frac{1}{2} = .5$. The net area is $3.25 - .5 = 2.75$. The carrying capacity of the angle is therefore $2.75 \times 16{,}000 = 44{,}000$ lb.

2. Riveted joints also are examples of simple stress. In the joint shown in Fig. 1, the rivet is in *single shear*, because there is only one section e of the rivet subjected to a shearing stress. The amount R that one rivet will carry being equal to the area of the cross-section of the rivet multiplied by the unit shearing stress, or $R = sA$, the number n of rivets required to transfer a stress T by single shear is

$$n = \frac{T}{R} = \frac{T}{As}$$

In Fig. 2, the rivet is subjected to shear on two sections, d and e, and it is said to be in *double shear*. The amount of stress that one rivet can carry in double shear is twice that of one in single shear, and, using the preceding notation,

$$n = \frac{T}{2R}$$

The *bearing value* of a rivet is the compressive stress induced by the rivet in bearing on the plate, and is also calculated by the simple-stress formula, $P = sA$, P being the value of a rivet in bearing, s the unit working stress in bearing, and A the bearing area, which, as it is customary to assume, is the thickness of the plate multiplied by the diameter of the rivet. In calculating the required number of rivets, both the shearing and the bearing value of one rivet are determined and the critical value (the smaller) used.

The following tables give the shearing and bearing values of rivets, in pounds, for different values of the working stress.

3. *Strength of Cylindrical Shells and Pipes With Thin Walls.* When a cylinder is subjected to internal pressure, the tensile stress developed in the walls or shell of the cylinder is called *circumferential stress*, or *hoop tension*. Let s be the intensity of this stress; d, the internal diameter of the cylinder; p, the intensity of pressure on the inner surface of the cylinder; and t the thickness of the shell.

Then, $$t = \frac{pd}{2s}$$

and $$s = \frac{pd}{2t}$$

SHEARING AND BEARING VALUES OF RIVETS, IN POUNDS

Diam. of Rivet Inch	Area of Rivet Square Inch	Shear Values at 6,000 Lb. per Sq. In.		Bearing Values for Different Thicknesses of Plate, in Inches, at 12,000 Lb. per Sq. In.										
		Single Shear	Double Shear	$\frac{1}{4}$	$\frac{5}{16}$	$\frac{3}{8}$	$\frac{7}{16}$	$\frac{1}{2}$	$\frac{9}{16}$	$\frac{5}{8}$	$\frac{11}{16}$	$\frac{3}{4}$	$\frac{13}{16}$	$\frac{7}{8}$
$\frac{1}{2}$	.1963	1,180	2,360	1,500	1,880	2,250	2,630	3,000						
$\frac{5}{8}$	.3068	1,840	3,680	1,880	2,340	2,810	3,280	3,750	4,220	4,690				
$\frac{3}{4}$	.4418	2,650	5,300	2,250	2,810	3,380	3,940	4,500	5,060	5,630	6,190	6,750		
$\frac{7}{8}$	.6013	3,610	7,220	2,630	3,280	3,940	4,590	5,250	5,910	6,560	7,220	7,880	8,530	9,190
1	.7854	4,710	9,420	3,000	3,750	4,500	5,250	6,000	6,750	7,500	8,250	9,000	9,750	10,500

Diam. of Rivet Inch	Area of Rivet Square Inch	Shear Values at 7,500 Lb. per Sq. In.		Bearing Values for Different Thicknesses of Plate, in Inches, at 15,000 Lb. per Sq. In.										
		Single Shear	Double Shear	$\frac{1}{4}$	$\frac{5}{16}$	$\frac{3}{8}$	$\frac{7}{16}$	$\frac{1}{2}$	$\frac{9}{16}$	$\frac{5}{8}$	$\frac{11}{16}$	$\frac{3}{4}$	$\frac{13}{16}$	$\frac{7}{8}$
$\frac{1}{2}$	.1963	1,470	2,940	1,880	2,340	2,810	3,280	3,750						
$\frac{5}{8}$	.3068	2,300	4,600	2,340	2,930	3,520	4,100	4,690	5,280	5,860				
$\frac{3}{4}$	.4418	3,310	6,630	2,810	3,520	4,220	4,920	5,630	6,330	7,030	7,720	8,440		
$\frac{7}{8}$	.6013	4,510	9,020	3,280	4,100	4,920	5,740	6,560	7,380	8,200	9,030	9,850	10,670	11,480
1	.7854	5,890	11,780	3,750	4,690	5,630	6,560	7,500	8,440	9,380	10,310	11,250	12,190	13,130

SHEARING AND BEARING VALUES OF RIVETS, IN POUNDS

Diam. of Rivet Inch	Area of Rivet Square Inch	Shear Values at 9,000 Lb. per Sq. In.		Bearing Values for Different Thicknesses of Plate, in Inches, at 18,000 Lb. per Sq. In.										
		Single Shear	Double Shear	1/4	5/16	3/8	7/16	1/2	9/16	5/8	11/16	3/4	13/16	7/8
1/2	.1963	1,770	3,530	2,250	2,810	3,380	3,940	4,500						
5/8	.3068	2,760	5,520	2,810	3,520	4,220	4,920	5,630	6,330	7,030				
3/4	.4418	3,980	7,950	3,380	4,220	5,060	5,910	6,750	7,590	8,440	9,280	10,130		
7/8	.6013	5,410	10,820	3,940	4,920	5,910	6,890	7,880	8,860	9,840	10,830	11,810	12,800	13,780
1	.7854	7,070	14,140	4,500	5,630	6,750	7,880	9,000	10,130	11,250	12,380	13,500	14,630	15,750

Diam. of Rivet Inch	Area of Rivet Square Inch	Shear Values at 11,000 Lb. per Sq. In.		Bearing Values for Different Thicknesses of Plate, in Inches, at 22,000 Lb. per Sq. In.										
		Single Shear	Double Shear	$\frac{1}{4}$	$\frac{5}{16}$	$\frac{3}{8}$	$\frac{7}{16}$	$\frac{1}{2}$	$\frac{9}{16}$	$\frac{5}{8}$	$\frac{11}{16}$	$\frac{3}{4}$	$\frac{13}{16}$	$\frac{7}{8}$
$\frac{1}{2}$	.1933	2,160	4,320	2,750	3,440	4,130	4,820	5,500						
$\frac{5}{8}$	.3068	3,370	6,750	3,440	4,300	5,160	6,020	6,880	7,740	8,600				
$\frac{3}{4}$	.4418	4,860	9,720	4,130	5,160	6,190	7,220	8,250	9,280	10,320	11,340	12,380		
$\frac{7}{8}$	.6013	6,610	13,230	4,810	6,020	7,220	8,430	9,630	10,840	12,040	13,240	14,440	15,640	16,840
1	.7854	8,640	17,280	5,500	6,880	8,250	9,630	11,000	12,380	13,750	15,130	16,500	17,880	19,250

The first formula serves to compute the thickness when p, d, and s (working stress) are given; and the second one is used to compute the intensity of stress when the intensity of pressure p and the dimensions of the cylinder are given.

EXAMPLE.—What should be the thickness of walls of a cast-iron water pipe, inside diameter 24 in., to resist a water pressure of 200 lb. per sq. in., using a unit working stress of 2,000 lb.

SOLUTION.—Here, $d=24$, $p=200$, and $s=2{,}000$. Substituting in the formula for t,

$$t=\frac{200\times 24}{2\times 2000}=1.2 \text{ in.}$$

4. *Temperature Stresses.*—If a bar subjected to change of temperature is constrained so that it can neither expand nor contract, the constraint exerts on it a force sufficient to prevent the deformation. This causes in the bar a corresponding stress called *temperature stress.* It is compressive when the change of temperature is a rise, and tensile when a fall.

COEFFICIENT OF EXPANSION FOR A NUMBER OF SUBSTANCES

Name of Substance	Linear Expansion	Surface Expansion	Cubic Expansion
Cast iron	.00000617	.00001234	.00001850
Copper	.00000955	.00001910	.00002864
Brass	.00001037	.00002074	.00003112
Silver	.00000690	.00001390	.00002070
Wrought iron	.00000686	.00001372	.00002058
Steel (untempered)	.00000599	.00001198	.00001798
Steel (tempered)	.00000702	.00001404	.00002106
Zinc	.00001634	.00003268	.00004903
Tin	.00001410	.00002820	.00003229
Mercury	.00003334	.00006668	.00010010
Alcohol	.00019259	.00038518	.00057778
Gases			.00203252
Concrete	.0000065	.000013	.0000195

Let T be the stress induced in a bar, whose area is a, by a rise or fall of $t°$; let, also, c be the coefficient of expansion and E the modulus of elasticity of the material. Then,

$$T=ctaE$$

EXAMPLE.—A wrought-iron bar 1.5 in. square has its ends fastened to firm supports. What is the stress produced in it by a change of 50° in its temperature?

SOLUTION.—Here, $E=25,000,000$; $a=1.5\times1.5=2.25$ sq. in., and $t=50$; and, according to the accompanying table, $c=.00000686$. Substituting in the formula, $T=.00000686\times50\times2.25\times25,000,000=19,294$ lb.

BEAMS

A body resting upon supports and liable to transverse stress is called a *beam*. Beams are designated by the number and location of the supports, and may be simple, cantilever, fixed, or continuous. A *simple beam* is one that is supported at each end, the distance between its supports being the *span*. A *cantilever* is a beam that has one or both ends overhanging the support; or a beam that has one end firmly fixed and the other end free. A *fixed beam* is one that has both ends firmly secured. A *continuous beam* is one which rests upon more than two supports.

Reactions.—The loads acting on a beam are balanced by the reactions or supporting forces; their sum must therefore be equal

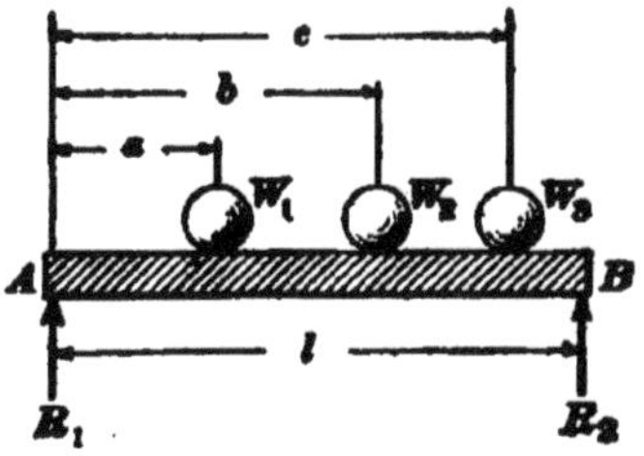

FIG. 1

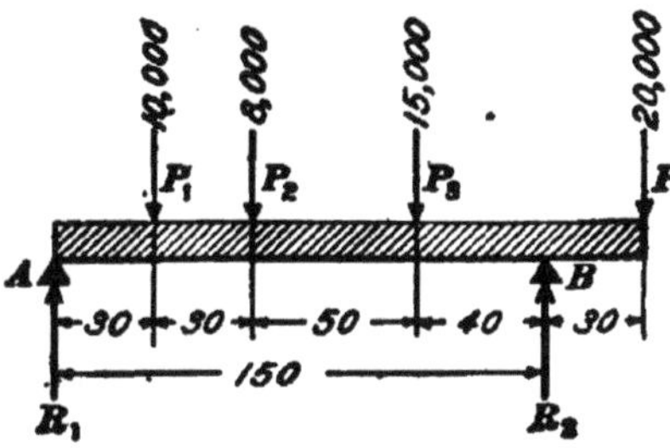

FIG. 2

to the sum of the loads. To find any reaction, as R_2, at B, Fig. 1, take moments of all the external forces about the other support A and divide their sum by the span. With reference to Fig. 1,

$$R_2=\frac{W_1a+W_2b+W_3c}{l}$$

The reaction R_1 can be found in a similar manner by taking moments about the support B. Their sum R_1+R_2 must be equal to the sum of loads $W_1+W_2+W_3$.

EXAMPLE.—Find the reactions of a cantilever bridge loaded as shown in Fig. 2.

SOLUTION.—Substituting given values in the formula and noting that the moment of P_4 about B is of opposite sign to the moments of the other loads,

$$R_1=\frac{10{,}000\times 120+8{,}000\times 90+15{,}000\times 40-20{,}000\times 30}{150}$$

$$=12{,}800 \text{ lb.}$$

and

$$R_2=\frac{10{,}000\times 30+8{,}000\times 60+15{,}000\times 110+20{,}000\times 180}{150}$$

$$=40{,}200 \text{ lb.}$$

The sum of the loads is 10,000+8,000+15,000+20,000 =53,000. The sum of the reactions is 40,200+12,800=53,000.

External Shear and Bending Moment.—The forces acting on a beam tend, on the one hand, to shear its fibers vertically and, on the other hand, to bend it, producing compressional stresses in the fibers on one side of the neutral axis and tensional on the other side. The tendency to shear the fibers vertically is determined by the external shear, and that of bending by the bending moment.

The *external shear* at any section of a beam is the algebraic sum of all the external forces (loads and reactions) on one side of the section. Forces acting upwards are considered positive, and those acting downwards, negative. For brevity, external shear is often called simply *shear*, but it must not be confused with shearing stress at the section. The external shear is equal to either reaction minus the sum of the loads between that reaction and the section considered. The maximum shear is always equal to the greater reaction. For a simple beam with a uniformly distributed load, the maximum shear is at the supports, and is equal to one-half the load, or to the reaction; the shear changes at every point of the loaded length, the minimum shear being zero at the center of the span. The maximum shear in a simple beam having a single load concentrated at the center is equal to one-half the load, and is uniform throughout the beam. Where a beam supports several concentrated loads, changes in the amount of shear occur only at the points where the loads are applied. The external shear is resisted by the internal shear, or shearing stress, of the beam, which is numerically equal to the external shear. If the external shear is denoted by V, and the area of the cross-section by A, the average intensity of shearing stress in the section is $\frac{V}{A}$. This shearing stress is not uniformly distributed, and in beams of rectangular cross-section, the maximum intensity of shearing stress is $\frac{3V}{2A}$. Hence, a rectangular beam must be so designed that this value will not exceed the working shearing strength of the material. In metallic beams with thin webs (plate girders), the shearing stress may be considered as uniformly distributed over the cross-section of the web. There is, also, at every horizontal or longitudinal section of the beam, a horizontal shearing stress the intensity of which at any point is equal to the intensity of the vertical shearing stress at that point.

Although the maximum intensity of shearing stress, both horizontal and vertical, in wooden beams is usually small, the shearing strength of wood along the grain is also small. As the horizontal external shear usually acts along the grain, the safe load for a wooden beam may depend on its shearing strength and not on its bending strength. For instance, the safe load for a beam 4 in.×12 in. and 4 ft. long is 16,000 lb., uniformly distributed, when based on a fiber strength of 1,000 lb. per sq. in. Such a load will produce a shearing stress per unit of area equal to $\frac{3\times 8{,}000}{2\times 48}=250$ lb. per sq. in., which exceeds the working shearing stress for the wood along the grain by about 100 lb. per sq. in.

The ***bending moment*** at any section of a loaded beam is equal to the algebraic sum of the moments of all the external forces (loads and reactions) to the right or left of the section about that section. For example, the bending moments at several points on the beam shown in Fig. 3 are as follows: At $W_1 = R_1a$; at $W_2 = R_1(a+b) - W_1b$; at $W_3 = R_1(a+b+c) - [W_2c + W_1(b+c)]$, or R_2d.

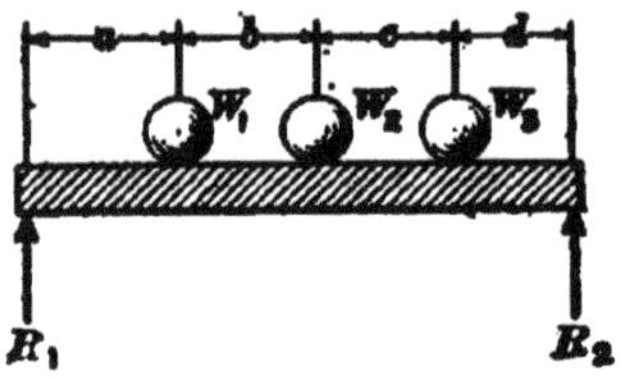

FIG. 3

The bending moment varies, depending on the shear, and attains a maximum value at the point where the shear changes sign. If the loads are concentrated at several points, the maximum bending moment will be under the load at which the sum of all the loads between one support up to and including the load in question first becomes equal to, or greater than, the reaction at the support. Hence, to find the maximum bending moment in any simple beam:

Rule.—*Compute the reactions and determine the point where the shear changes sign. Calculate the moment about this point of either reaction, and of each load between the reaction and the point, and subtract the sum of the latter moments from the former.*

EXAMPLE.—What is the maximum bending moment of the beam loaded as shown in Fig. 4?

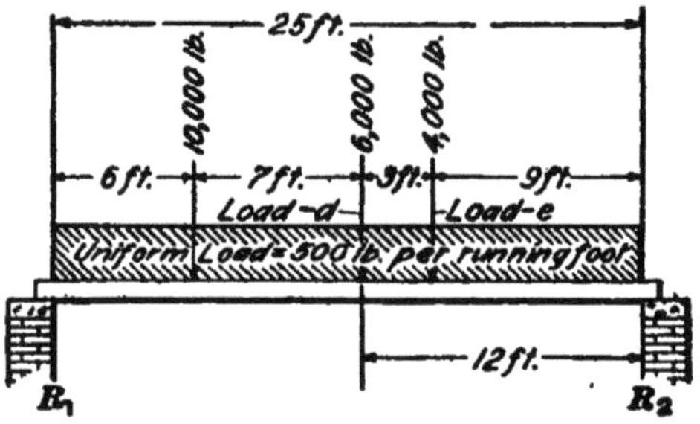

FIG. 4

SOLUTION.—The reactions due to the uniform load are equal to half of the load; those due to the concentrated loads are computed by the principle given under Reactions. Both added give $R_1 = 18,170$ lb. and $R_2 = 14,330$ lb. Beginning at R_1 and subtracting the loads in succession, it is found that the shear just to the left of the load d is $18,170 - 16,500$; and just to right of the load d it becomes negative. Hence, the shear

FORMULAS FOR MAXIMUM SHEAR AND BENDING MOMENTS OF BEAMS

Case	Method of Loading	Maximum Shear	Maximum Moment
I		W	Wl
II		W	$\frac{Wl}{2}$
III		W	$\frac{Wl}{3}$
IV		W	$\frac{2Wl}{3}$
V		W	$\frac{Wl}{2}$
VI		W	$\frac{Wl}{2}$
VII		W	$\frac{Wl}{2}$
VIII	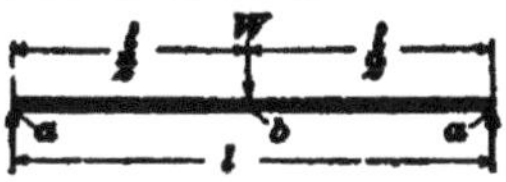	$\frac{W}{2}$	$\frac{Wl}{4}$
IX	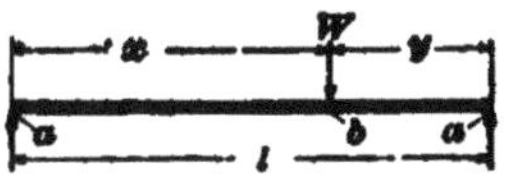	$\frac{Wy}{l}$ or $\frac{Wx}{l}$	$\frac{Wxy}{l}$
X		$\frac{W}{2}$	$\frac{Wx}{2}$

TABLE—(*Continued*)

Case	Method of Loading	Maximum Shear	Maximum Moment
XI		$\frac{W}{2}$	$\frac{Wl}{8}$
XII		$\frac{W}{2}$	$\frac{Wl}{12}$
XIII	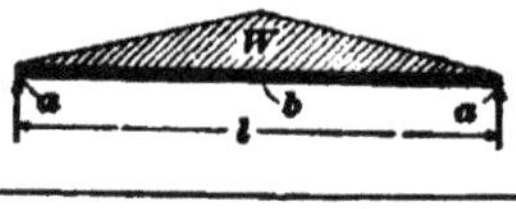	$\frac{W}{2}$	$\frac{Wl}{6}$
XIV	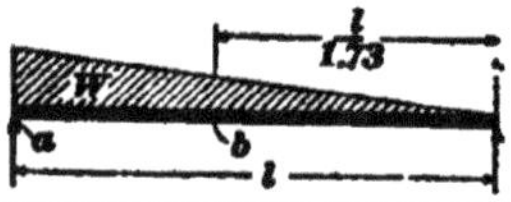	$\frac{2W}{3}$	$\frac{52Wl}{405}$
XV	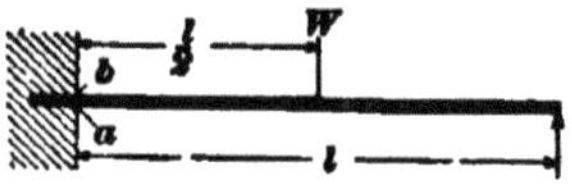	$\frac{11}{16}W$	$\frac{3}{16}Wl$
XVI		$\frac{5}{8}W$	$\frac{Wl}{8}$
XVII		$\frac{W}{2}$	$\frac{Wl}{8}$
XVIII		$\frac{W}{2}$	$\frac{Wl}{12}$
XIX		$\frac{W}{2}$	$\frac{Wx}{2}$
XX	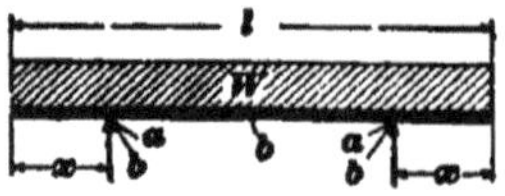	$\frac{Wx}{l}$ or $W\left(\frac{l-2x}{2l}\right)$	$\frac{Wx^2}{2l}$ or $\frac{W}{2}\left(\frac{l}{4}-x\right)$

changes sign under the load *d* and the bending moment is maximum at that point. It is equal to $18{,}170\times13-10{,}000\times7-\frac{13^2\times500}{2}=123{,}960$ ft.-lb.

Formulas for the maximum bending moments and shears for beams loaded and supported in different ways are given in the accompanying table.

For a beam supporting moving loads, the maximum bending moment occurs:

1. For a single load, when the load is at the middle of the span.

2. For two equal loads, under either load, when the two loads are on opposite sides of the center and one of the loads is at a distance from the center equal to one-fourth the distance between the loads.

3. For two unequal loads, under the heavier load, when that load and the center of gravity of the two loads are equidistant from the center of gravity of the beam.

Example.—A beam 24 ft. long supports two moving loads 6 ft. apart. The left-hand load is 8,000 lb., and the right-hand load is 4,000 lb. Find the maximum bending moment.

Solution.—The center of gravity of the loads is 2 ft. from the left-hand load. The maximum bending moment occurs under the heavy load, and obtains when the latter is 1 ft. to the left of the center of the beam. The left reaction is, then, $\frac{12{,}000\times11}{24}=5{,}500$ lb., and the maximum bending moment is $5{,}500\times11=60{,}500$ ft.-lb.

Designing of Beams.—In every section of a carrying beam there is induced an internal moment called the *moment of resistance*, which is equal to the bending moment at that section. As previously explained, the resisting moment is equal to $\frac{I}{c}f$; and, if the maximum bending moment is denoted by M, $M=\frac{I}{c}f$; whence, $\frac{M}{f}=\frac{I}{c}$, which is the fundamental formula for the designing of beams; f is the working stress in flexure, which is the modulus of rupture divided by a suitable factor of safety. The ***modulus of rupture***, also called the ***ultimate strength of flexure***, is the extreme fiber stress that a material subjected to bending can withstand. Its value is intermediate between the ultimate strength in compression and tension. In the sixth column of the table on pages 276 and 277 are given the average values of the modulus of rupture for several kinds of metal.

When a beam is to be designed to carry certain loads, the maximum bending moment is determined and divided by f. The latter is usually given or is found by dividing the modulus of rupture of the material by a suitable factor of safety. The problem then reduces itself to the finding of a section that has a value of $\frac{I}{c}$, the section modulus, equal to $\frac{M}{f}$. For rolled-steel sections, the value of $\frac{I}{c}$ can be taken from a manufacturers' handbook. For a rectangular section,

$$\frac{I}{c}=\frac{bd^2}{6}$$

b being the breadth and d the depth of the section. Since the expression contains two unknown quantities b and d, a value for either one may be assumed and substituted, and the formula solved for the other. If a built-up beam is used, the section has to be found by trial; a suitable section is first assumed and its section modulus is computed by the principles given under the heading Moment of Inertia; if necessary, it is modified until it is equal to $\frac{M}{f}$.

EXAMPLE.—Design both a rolled-steel I beam and a solid wooden beam 10 ft. long, each to carry a uniform load of 250 lb. per ft. in addition to a central load of 2,000 lb., assuming for wood a working stress of 1,000 lb. per sq. in. and for steel 15,000 lb. per sq. in.

SOLUTION.—The maximum bending moment occurs at the middle of the beam and is equal to the sum of the moments due to the uniform load and the central load. Expressed in inch-pounds,

$$M=\frac{2{,}000\times 120}{4}+\frac{250\times 10\times 120}{8}=97{,}500 \text{ in.-lb.}$$

For a steel beam, $\frac{M}{f}=\frac{97{,}500}{15{,}000}=6.5$. From a manufacturer's handbook, a 6-in. I beam at 12.25 lb. has a section modulus of 7.3 and can therefore be used. For a wooden beam, $\frac{M}{f}$ $=\frac{97{,}500}{1{,}000}=97.5=\frac{bd^2}{6}$. Assuming that $b=6$ in., $d=\sqrt{97.5}$ $=10$ in. nearly.

Stiffness.—In designing a beam, it sometimes becomes necessary to ascertain the amount that it will deflect under given loads. This, for instance, is the case when designing supports for machinery parts or joists for plastered ceilings, in which latter case the deflection should not exceed $\frac{1}{360}$ of the span. The accompanying table gives deflection formulas for the most usual cases. In these formulas l is the span, in inches; W, the total load acting on the beam; I, the moment of inertia of the cross-section of the beam; and E, the modulus of elasticity of the material.

EXAMPLE 1.—A timber simple beam 10 ft. long, and having a width of 4 in. and a depth of 12 in., carries a uniform load of 400 lb. per ft. What is the deflection?

SOLUTION.—According to the table, the deflection for a uniformly distributed load is $\frac{5\,Wl^3}{384\,EI}$. In this case, $l=10\times 12$ $=120$; $W=400\times 10=4{,}000$; $E=1{,}500{,}000$; and $I=\frac{4\times 12^3}{12}$ $=576$. Substituting in the formula,

$$\text{Deflection}=\frac{5\times 4{,}000\times 120^3}{384\times 1{,}500{,}000\times 576}=.1 \text{ in.}$$

FORMULAS FOR DEFLECTION OF BEAMS

Case	Method of Loading	Deflection Inches
I		$\frac{Wl^3}{3\,EI}$
II	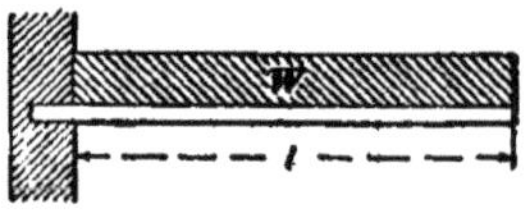	$\frac{Wl^3}{8\,EI}$
III	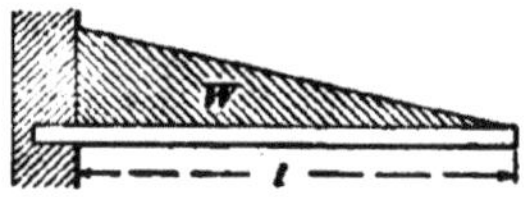	$\frac{Wl^3}{15\,EI}$
IV	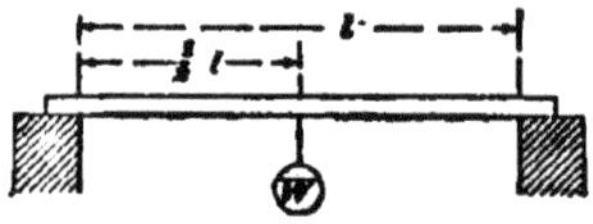	$\frac{Wl^3}{48\,EI}$
V	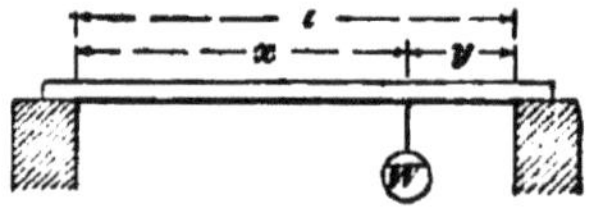	$\frac{Wxy(2l-x)\sqrt{3x(2l-x)}}{27\,lEI}$
VI	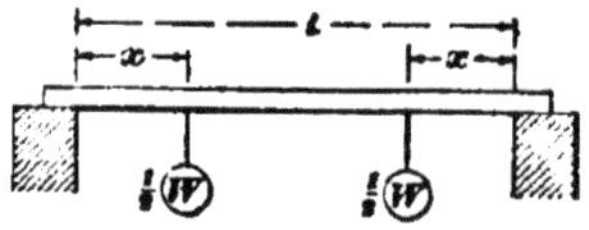	$\frac{Wx}{48\,EI}(3l^2-4x^2)$
VII	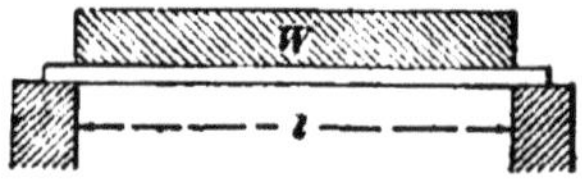	$\frac{5\,Wl^3}{384\,EI}$
VIII		$\frac{3\,Wl^3}{320\,EI}$

TABLE—(*Continued*)

Case	Method of Loading	Deflection Inches
IX	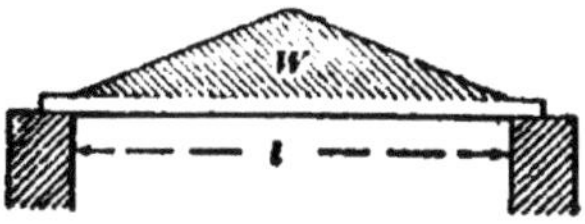	$\frac{Wl^3}{60\,EI}$
X	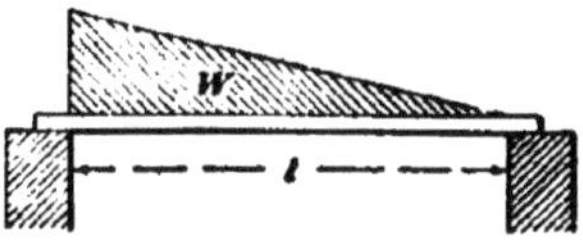	$\frac{47\,Wl^3}{3{,}600\,EI}$
XI	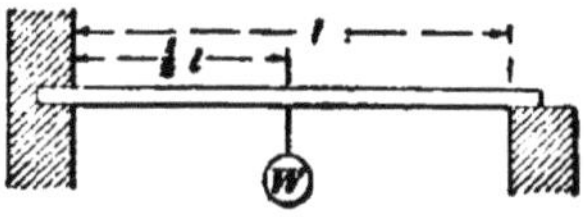	$\frac{3\,Wl^3}{322\,EI}$
XII	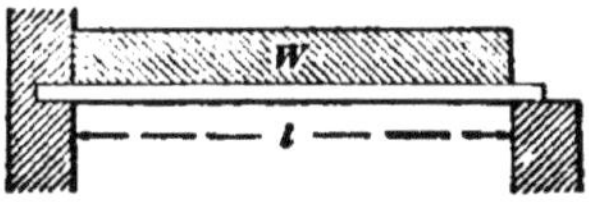	$\frac{5\,Wl^3}{926\,EI}$
XIII	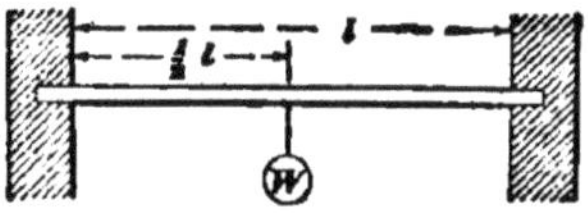	$\frac{Wl^3}{192\,EI}$
XIV		$\frac{Wl^3}{384\,EI}$
XV		For overhang: $\frac{Wx}{12\,EI}(3\,xl-4\,x)^2$ For part between supports: $\frac{Wx}{16\,EI}(l-2\,x)^2$

COLUMNS

The strength of a compression member depends on the ratio of its length to its least lateral dimension, or, what is the same thing, on the *ratio of slenderness*; that is, the ratio of its length to its radius of gyration.

For compression members whose ratio of slenderness does not exceed 30, the formula $s=\frac{P}{A}$, for simple stress, may be used.

When this ratio exceeds 30, but is not more than 150, s should be reduced by *Rankine's formula*,

$$s=\frac{s_u}{1+\frac{k_1 l^2}{r^2}}$$

in which s_u is the ultimate strength in compression, which should be divided by a suitable factor of safety; l, the length; and r, the radius of gyration. Both l and r are expressed in the same unit. The values of k_1 which depend on the material of the column and the condition of its ends—that is whether fixed or round—are given in the following table:

VALUES OF k_1 (RANKINE'S FORMULA)

Material	Both Ends Flat or Fixed	One End Round	Both Ends Round
Cast iron.......	$\frac{1}{5,000}$	$\frac{1.78}{5,000}$	$\frac{4}{5,000}$
Wrought iron...	$\frac{1}{36,000}$	$\frac{1.78}{36,000}$	$\frac{4}{36,000}$
Steel...........	$\frac{1}{25,000}$	$\frac{1.78}{25,000}$	$\frac{4}{25,000}$
Wood..........	$\frac{1}{3,000}$	$\frac{1.78}{3,000}$	$\frac{4}{3,000}$

When the value of $\frac{l}{r}$ exceeds 150, *Euler's formula*, which is given later, should be used.

The *straight-line formula* is more convenient for determining the value of s, and is now in extensive use. It is only approximate, giving values of s that differ somewhat from those obtained by Rankine's formula; but the difference is on the side of safety. For the same notation as before, the straight-line formula is

$$s = s_u - k\frac{l}{r}$$

The values of s_u and k are given in the accompanying table, in which will also be found the limit of $\frac{l}{r}$ within which the formula may be used. When $\frac{l}{r}$ exceeds this limit, Euler's formula, which follows, should be used.

CONSTANTS FOR THE STRAIGHT-LINE AND EULER'S FORMULAS

	Medium Steel		Wrought Iron		Cast Iron
	Flat Ends	Pin Ends	Flat Ends	Pin Ends	Flat Ends
s_u	52,500	52,500	42,000	42,000	80,000
k	179	220	128	157	438
limit of $\frac{l}{r}$	195	159	218	178	122
$nE\pi^2$	666 m	444 m	666 m	444 m	395 m

EXAMPLE.—What is the ultimate strength per square inch of a medium-steel column 25 ft. long both ends of which are fixed and the radius of gyration of which is 2.5?

SOLUTION.—By the straight-line formula,

$$s = 60{,}000 - 179 \times \frac{25 \times 12}{2.5} = 38{,}520 \text{ lb. per sq. in.}$$

Using Rankine's formula,

$$S = \frac{60{,}000}{1 + \frac{(25 \times 12)^2}{25{,}000 \times 2.5^2}} = 38{,}070 \text{ lb. per sq. in.}$$

Euler's Formula.—Structural members in compression whose ratio of slenderness exceeds 150 should preferably not be used. Sometimes, however, long columns cannot be avoided, and when $\frac{l}{r}$ exceeds the limits for which the preceding formulas may be applied, Euler's formula should be used. This formula is as follows:

$$\frac{P}{A}=\frac{n\pi^2E}{\left(\frac{l}{r}\right)^2}$$

in which E is the modulus of elasticity of the material and n is a constant depending on the end condition, having the value of 1 for columns with both ends pivoted and 4 for columns with both ends fixed. The preceding table gives the values of $n\pi^2E$, expressed in millions of pounds.

Formula for Wooden Columns.—The formula for determining the strength of wooden columns having flat or square ends was deduced from exhaustive tests of full-size specimens, made at the Watertown Arsenal, Mass., and may be expressed as follows:

$$S=U-\frac{Ul}{100d},$$

in which S is the ultimate strength of column, per square inch of section; U, the ultimate compressive strength of the material, per square inch; l, the length of the column, in inches; d, the dimension of the least side of the column, in inches.

This formula may be applied to all wooden columns, the length or height of which is not under 10 times nor over 45 times the dimension of the least side. In other words, $\frac{l}{d}$ should not be less than 10 nor more than 45. If the length is less than 10 times the least side, the direct compressive strength of the material per square inch, multiplied by the sectional area of the column, in square inches, will give the strength of the column. If the length is over 45 times the least side, Rankine's formula should be used.

COMBINED STRESSES

Bending Combined with Compression or Tension.—Assume that P is the axial force acting on the beam; M, the maximum bending moment to which the beam is subjected; A, the cross-sectional area of the beam; I, its moment of inertia; and c, the distance from the neutral axis of the most distant fiber, having the same kind of stress (tension or compression) as that caused by P. Then, the working stress should not exceed

$$s=\frac{P}{A}+\frac{Mc}{I}$$

In case of compression, s should, in addition, be reduced by one of the compression formulas previously given.

The preceding formula for s is the one commonly used in practice, but it is only approximate. When more accurate results are required, the following formula should be used,

$$s=\frac{P}{A}+\frac{Mc}{I\pm k\frac{Pl^2}{E}}$$

Here, l is the span; E, the modulus of elasticity, and k, a constant having the following values:

	Value of k
For a cantilever loaded at end	$\frac{1}{2}$
For a cantilever loaded uniformly	$\frac{1}{4}$
For a beam supported at both ends and loaded at center	$\frac{1}{12}$
For a beam supported at both ends and loaded uniformly	$\frac{5}{48}$
For a beam fixed at both ends and loaded at center	$\frac{1}{16}$
For a fixed beam uniformly loaded	$\frac{1}{32}$

The minus sign before k is for the case when the direct stress is compressive, and the plus sign, when it is tensile.

STRENGTH OF ROPES AND CHAINS

Ropes.—If C is the circumference of a rope in inches and P the working load in pounds, then, for hemp and manila rope,

$$P = 10C^2$$

This formula gives a factor of safety of from 7½ for manila or tarred hemp rope to about 11 for the best three-strand hemp rope.

For iron-wire rope of seven strands, nineteen wires to a strand,

$$P = 600C^2$$

and for the best steel-wire rope of seven strands, nineteen wires to the strand,

$$P = 1{,}000\ C^2$$

The last two formulas are based on a factor of safety of 6.

Chains.—If P is the safe load in pounds and d the diameter of link in inches, then, for open-link chains made from a good quality of wrought iron,

$$P = 12{,}000\ d^2$$

and for stud-link chains,

$$P = 18{,}000\ d^2$$

Chain Cables.—The strength of a chain link is less than twice that of a straight bar of a sectional area equal to that of one side of the link. A weld exists at one end and a bend at the other, each requiring at least one heat, which produces a decrease in the strength. The report of the committee of the U. S. Testing Board, on tests of wrought-iron and chain cables, contains the following conclusions:

"That beyond doubt, when made of American bar iron, with cast-iron studs, the studded link is inferior in strength to the unstudded one.

"That, when proper care is exercised in the selection of material, the strength of chain cables will vary by about 5% to 17% of the resistance of the strongest. Without this care the variation may rise to 25%.

"That with proper material and construction the ultimate resistance of the chain may be expected to vary from 155% to 170% of that of the bar used in making the links, and show an average of about 163%.

"That the proof test of a chain cable should be about 50% of the ultimate resistance of the weakest link."

From a great number of tests of bars and unfinished cables, the committee considered that the average ultimate resistance

ULTIMATE RESISTANCE AND PROOF TESTS OF CHAIN CABLES

Diam. of Bar Inches	Average Resist. = 163% of Bar Pounds	Proof Test Pounds	Diam. of Bar Inches	Average Resist. = 163% of Bar Pounds	Proof Test Pounds
1	71,172	33,840	$1\frac{9}{16}$	162,283	77,159
$1\frac{1}{16}$	79,544	37,820	$1\frac{5}{8}$	174,475	82,956
$1\frac{1}{8}$	88,445	42,053	$1\frac{11}{16}$	187,075	88,947
$1\frac{3}{16}$	97,731	46,468	$1\frac{3}{4}$	200,074	95,128
$1\frac{1}{4}$	107,440	51,084	$1\frac{13}{16}$	213,475	101,499
$1\frac{5}{16}$	117,577	55,903	$1\frac{7}{8}$	227,271	108,058
$1\frac{3}{8}$	128,129	60,920	$1\frac{15}{16}$	241,463	114,806
$1\frac{7}{16}$	139,103	66,138	2	256,040	121,737
$1\frac{1}{2}$	150,485	71,550			

and proof tests of chain cables made of the bars, whose diameters are given, should be such as are shown in the accompanying table.

MASONRY

MATERIALS OF CONSTRUCTION

The materials employed in the construction of masonry are stone, brick, terra cotta, and the cementing materials used in the manufacture of mortars, namely, lime, cement, and sand.

STONE

Strength of Stone.—In ordinary buildings and engineering structures, stones are generally under compression. Occasionally, they are subjected to cross-stresses, as in lintels over wide openings. They are never subjected to direct tension. As a general rule, a stone should not be subjected to a greater compressive stress than one-tenth of the ultimate crushing strength, as found by experiment.

The resistance to crushing varies within wide limits, owing to the great variety in the structure of the stones; the method of preparing and finishing the test pieces also affects the results; hence, the great variations found in the values given by different experiments. The accompanying table shows the average resistance of the principal building stones to crushing and to rupture when used as beams.

CRUSHING STRENGTH AND MODULUS OF RUPTURE OF BUILDING STONE

Stone	Crushing Strength Pounds per Square Inch	Modulus of Rupture Pounds per Square Inch
Granite	15,000	1,800
Sandstone	10,000	1,200
Limestone	13,000	1,500
Marble	14,000	2,160

Absorptive Power of Stone.—The absorptive power of a stone is a very important property, a low absorption generally indicating a good quality. The accompanying table gives the average percentage of water absorbed by stones.

ABSORPTIVE POWER OF STONE

Stone	Absorptive Capacity Per Cent.
Granites	.066 to .155
Sandstones	.410 to 5.480
Limestones	.200 to 5.000
Marbles	.080 to .160
Trap	.000 to .019

Durability of Stone.—The following rough estimate, based on observations made in the city of New York, indicates the number of years a sound stone may be expected to last without being discolored or disintegrated to such an extent as to require repairs:

Name of Stone	*Life of Stone* *Years*
Coarse brownstone....................	5 to 15
Compact brownstone..................	100 to 200
Limestone............................	20 to 40
Granite..............................	75 to 200
Marble...............................	40 to 200

BRICK

Size and Weight.—The dimensions of bricks vary considerably. The standard adopted by the National Brickmakers' Association is, for common clay brick, $8\frac{1}{4}$ in.×4 in.×$2\frac{1}{4}$ in., and for face or pressed brick (clay) $8\frac{3}{8}$ in.×$4\frac{1}{8}$ in.×$2\frac{1}{4}$ in. The weight of a common clay brick is about $4\frac{1}{2}$ lb.; that of a pressed-clay, enameled brick, about 7 lb. Enameled and glazed bricks are made in two sizes: English size, 9 in.×3 in. ×$4\frac{1}{2}$ in.; American size, $8\frac{3}{8}$ in.×$2\frac{1}{4}$ in.×$4\frac{1}{8}$ in. The usual dimensions for firebricks are 9 in.×$4\frac{1}{2}$ in.×$2\frac{1}{2}$ in.; various sizes and forms are made to suit the required work. The dimensions of the lime-sand bricks are $8\frac{3}{8}$ in.×$4\frac{1}{8}$ in.×$2\frac{11}{16}$ in. The weight varies between 5 and 6 lb.

WEIGHT AND STRENGTH OF BRICK

Kind of Brick	Weight Pounds per Cubic Foot	Crushing Strength Pounds per Square Inch
Best pressed-clay...........	150	5,000 to 15,000
Common hard-clay..........	125	5,000 to 8,000
Soft-clay....................	100	450 to 600
Lime-sand.................	120	3,600 to 7,600
Firebrick..................	120	1,000 to 1,500

The accompanying table gives the approximate weight and resistance to crushing of brick.

Requisites for Good Brick.—Bricks of good quality should be of regular shape, with parallel surfaces, plane faces, and sharp square edges. They should be of uniform texture; burnt hard; and thoroughly sound, free from cracks and flaws. They should emit a clear ringing sound when struck a sharp blow. A hard well-burned brick should not absorb more than one-tenth of its weight of water; it should have a specific gravity of 2 or more. The crushing strength of a brick laid flat should be at least 6,000 lb. per sq. in. The modulus of rupture should be at least 1,000 lb. per sq. in.

CEMENTING MATERIALS

Lime.—Common lime, commercially called *quicklime*, is manufactured by calcining, or burning, at a temperature of from 1,400° to 2,000° F., stones composed of pure or very nearly pure carbonate of lime. The product is practically pure oxide of calcium. It is prepared for use, converting it into calcium hydrate, by the addition of water. This process is called *slaking*. The quantity of water required in slaking lime is about one-third the volume of the lime.

Lime weighs about 66 lb. per bu., or about 53 lb. per cu. ft. One barrel of lime, weighing 230 lb., will make about 2¼ bbl., or .3 cu. yd. of stiff paste. In 1-to-3 mortar, 1 bbl. of unslaked lime will make about 6¾ bbl. of mortar; or 1 bbl. of lime paste will make about 3 bbl. of mortar. For a 1-to-2 mortar, use is made of about 1 bbl. of quicklime to 5 or 5½ bbl. of sand.

Hydraulic Cements.—The hydraulic cements are divided into three main classes; namely, Portland cement, natural cement, and pozzuolana. These cements differ from the limes by not slaking after calcination.

Portland cement is the product resulting from the process of grinding an intimate mixture of calcareous (containing lime) and argillaceous (containing clay) materials, calcining (heating) the mixture until it starts to fuse, or melt, and grinding the resulting clinker to a fine powder.

Natural cement is made by calcining natural argillaceous or silicious limestones at a heat just below fusion and grinding the product to powder.

Pozzuolana, or ***puzzolan***, ***cement*** is a material resulting from grinding together, without subsequent calcination, an intimate mixture of slaked lime, and a puzzolanic substance, such as blast-furnace slag or volcanic scoria. The only variety of puzzolan cement employed extensively in American practice is *slag cement.* This cement is made by grinding together a mixture of blast-furnace slag and slaked lime. The slag used for this purpose is granulated, or quenched, in water as soon as it leaves the furnace, which operation drives off most of the dangerous sulphides and renders the slag puzzolanic.

AVERAGE WEIGHTS OF HYDRAULIC CEMENTS

Kind of Cement	Net Weight of Bag Pounds	Net Weight of Barrel Pounds	Weight per Cubic Foot Pounds	
			Packed	Loose
Portland	94	376	100–120	70–90
Natural......	94	282	75–95	45–65
Slag..........	82½	330	80–100	55–75

Portland cement may be distinguished by its dark color, heavy weight, slow rate of setting, and greater strength. Natural cement is characterized by lighter color, lighter weight, quicker set, and lower strength. Slag cement is somewhat similar to Portland, but may be distinguished from it by its lilac-bluish color, by its lighter weight, and by the greater fineness to which it is ground.

Portland cement is adaptable to any class of mortar or concrete construction, and is unquestionably the best material for all such purposes. Natural and slag cements, however are cheaper, and under certain conditions, may be substituted for the more expensive Portland cement. All heavy construction, especially if exposed, all reinforced-concrete work, sidewalks, concrete blocks, foundations of buildings, piers, walls, abutments, etc., should be made with Portland cement. In second-class work, as in rubble masonry, brick sewers, unimportant work in damp or wet situations, or in heavy work in which the working loads will not be applied until long after completion, natural cement may be employed to advantage. Slag cement is best adapted to heavy foundation work that is immersed in water or at least continually

REQUIREMENTS FOR HIGH-GRADE CEMENTS

Requirements	Portland Cement	Natural Cement	Slag Cement
Specific gravity:			
Not less than	3.1	2.8	2.7
Fineness:			
Residue on No. 100 sieve, not over	8%	10%	3%
Residue on No. 200 sieve, not over	25%	30%	10%
Time of Setting:			
Initial, not less than	20 min.	10 min.	20 min.
Hard, not less than	1 hr.	30 min.	1 hr.
Hard, not more than	10 hr.	3 hr.	10 hr.
Tensile strength per sq. in.:			
7 da., neat, not less than	500 lb.	125 lb.	350 lb.
28 da., neat, not less than	600 lb.	225 lb.	450 lb.
7 da., 1–3 quartz, not less than	170 lb.	50 lb.	125 lb.
28 da., 1–3 quartz, not less than	240 lb.	110 lb.	200 lb.
Soundness:			
Normal pats in air and water for 28 da. to be	sound and hard	sound and hard	sound and hard
Boiling test to be	sound and hard		sound and hard
Analysis:			
Magnesia, *MgO*, not over	4%		4%
Anhydrous sulphuric acid, SO_3 not over	1.75%		
Sulphur, *S*, not over			1.3%

damp. This kind of cement should never be exposed directly to dry air, nor should it be subjected either to attrition or impact.

The preceding tables give the average weights of hydraulic cements and the various requirements for high-grade cements.

Sand.—Dry sand weighs from 80 to 115 lb. per cu. ft. Moist sand occupies more space and weighs less per cubic foot than dry sand.

The voids of ordinary sand range from one-fourth to one-half of the volume. The more uneven the grains in size, the smaller the percentage of voids.

The fineness of sand is measured by determining the percentage passing through five sieves, the first having 400 meshes, the second 900, the third 2,500, the fourth 6,400, and the fifth 28,900 per sq. in. When the grains range from $\frac{1}{8}$ to $\frac{1}{16}$ in., the sand is called *coarse;* when from $\frac{1}{16}$ to $\frac{1}{24}$ in., *fine;* and when from $\frac{1}{30}$ to $\frac{1}{80}$ in., *very fine.* When it is composed of sizes varying within these limits it is termed *mixed sand.*

MORTARS

Lime mortar is ordinarily composed of 1 part of slaked lime to 4 parts of sand. This kind of mortar should not be used in foundation work below the water-line, or in continually damp situations; neither should it be used in freezing weather.

MATERIALS REQUIRED PER CUBIC YARD OF MORTAR

Kind of Mixture	Portland Cement Barrels	Loose Sand Cubic Yards
1-1	4.95	.65
1-2	3.28	.88
1-3	2.42	1.01
1-4	1.99	1.06
1-5	1.62	1.11
1-6	1.34	1.15
1-7	1.18	1.17
1-8	1.05	1.18

Portland-cement mortar is composed of Portland cement and sand in proportions that vary from 1 part of cement and 1 part of sand to 1 part of cement and 6 parts of sand, this variation being according to the strength of the mortar desired. The common proportion for ordinary masonry is 1 part of cement to 3 parts of sand. For pointing face joints, 1 part of cement to either 1 or 2 parts of sand is used.

Natural-cement mortar is usually composed of 1 part of cement and 2 parts of sand. This proportion is found to possess sufficient adhesion and resistance to crushing for ordinary masonry above ground.

In the preceding table are given the quantities of materials required to produce 1 cu. yd. of compacted mortar. The proportions are by volume, a cement barrel being assumed to contain 3.6 cu. ft.

Mortar Impervious to Water.—Both lime and cement mortar absorb water; consequently, they disintegrate under the action of frost. Impermeability of the mortar may be increased by carefully grading the sand and increasing the amount of cement. The addition of a small amount of lime tends to reduce the volume and number of the voids and thus aids in reducing the permeability. Practically impermeable mortar may be made by adding to the mortar a mixture of alum and soap. The proportions usually employed are $\frac{3}{4}$ lb. of pulverized alum to each cubic foot of sand, and $\frac{3}{4}$ lb. of potash soap to each gallon of water. The alum and soap combine and form compounds of alumina and fatty acids that are insoluble in water. The strength of the mixture is but little inferior to the strength of the mortar of the same proportions.

Strength of Mortar.—The strength that mortar should possess is of three kinds; namely, *compressive*, *cohesive*, and *adhesive*. The degree to which it should possess any one of these depends on the position in which it is employed. In ashlar masonry, resistance to compression is all that is required; in uncoursed rubble masonry and in brick masonry, it must possess adhesiveness, or the capacity of adhering to the surface of the stones or brick in order to prevent their displacement. In masonry of all classes that may have to develop transverse stresses, it must possess cohesiveness or tensile strength.

The tensile and the compressive strength of a given mortar depend on the adhesive strength of the cementing medium and on the character of the aggregate. Coarse and fine sand in the proportion of about 4 parts of coarse grains ($\frac{3}{100}$ to $\frac{2}{10}$ in. in diameter) and 1 part of very fine grains (less than $\frac{2}{100}$ in. in diameter) usually produce the strongest mortar. Screenings from broken stone usually produce stronger mortars than sand, because of their greater density. Mixtures of sand and screenings often produce stronger mortar than either material alone. With the same aggregate, the strongest and most impermeable mortar is that containing the largest percentage of cement in a given volume of the mortar. With the same percentage of cement in a given volume of mortar, the strongest, and usually the most impermeable, mortar is that which has the greatest density, that is, which in a unit volume has the largest percentage of solid materials.

In the accompanying table is given a fair average of the tensile strength that may be expected from mortars of Portland and natural cements that are made in the field and with a sand of fair quality but not especially prepared.

The strength of Portland-cement mortar increases up to about 3 mo.; after that period, it remains practically constant for an indefinite time. Natural-cement mortar, on the

TENSILE STRENGTH OF CEMENT MORTARS

Proportions		Tensile Strength, in Pounds per Square Inch					
		Portland Cement			Natural Cement		
Cement Parts	Sand Parts	7 da.	28 da.	3 mo.	7 da.	28 da.	3 mo.
1	1	450	600	610	160	245	280
1	2	280	380	395	115	175	215
1	3	170	245	280	85	130	165
1	4	125	180	220	60	100	135
1	5	80	140	175	40	75	110
1	6	50	115	145	25	60	90
1	7	30	95	120	15	50	75
1	8	20	70	100	10	45	65

other hand, continues to increase in strength for 2 or 3 yr., its greatest strength being about 25% in excess of that attained in 3 mo. The strength of slag-cement mortar averages about three-quarters of that of Portland-cement mortar.

The compressive strength of cement mortar is about eight times its tensile strength, and the strength of mortar in cross-breaking and shear may be taken at about one and one-half to two times the tensile strength.

The adhesion of 1–2 Portland-cement mortar, 28 da. old to sandstone averages about 100 lb. per sq. in.; to limestone, 75 lb.; to brick, 60 lb.; to glass, 50 lb.; and to iron or steel, 75 to 125 lb. Natural-cement mortars have nearly the same adhesive strength as those made with Portland cement.

CONCRETE

Concrete consists of cement, water, sand, and large or small fragments of broken stone, gravel, or cinder. The plastic cement, either by itself or with the sand, is called the *matrix* and the hard material the *aggregate*.

Cement for Concrete.—The cement used for concrete work is almost exclusively hydraulic cement, generally Portland cement. Natural cement is not so strong and reliable as Portland. It sets more quickly, but takes longer to obtain its ultimate strength. It is used where economy demands it, but should never be placed under water. In civil-engineering work it is seldom employed, except in the form of mortar. A very good substitute for Portland cement in concrete for use under water is pozzuolana cement. This cement never gets very hard, but it withstands the action of sea-water even better than Portland cement. It will, however, soon fail if subjected to much attrition and wear.

Water for Concrete.—The wetter the concrete is the easier it will be put in place, but mixtures that are too wet are not so strong as medium mixtures. The quantity of water that will make the best mixture is such that after the concrete has been put in place and rammed, it will quake like jelly when struck with a spade, and water will come to the surface. If the concrete is wetter than this, the water will have a slight chemical effect on the cement, and, moreover, the sand and cement will tend to separate from the broken stone.

In cinder concrete, owing to the porosity of the cinders, it is necessary to use a little more water, so that the cement will be liquid enough to fill the little cavities in each cinder. This precaution is indispensable when the concrete is to be used with steel, as otherwise the steel will be corroded by the action of air reaching it through the pores in the cinders.

Sand for Concrete.—The sand used for concrete should be sharp and free from loam and chemical salts, particularly salts of a hygroscopic nature. The sand should not be too fine. An investigation made by A. S. Cooper on the effect that the size of the grains of sand has on the strength of mortar led him to the conclusion that, up to a certain limit, mortars become stronger as the grains of sand used become larger. However, the amount of cement required to fill the voids between the grains of sand is an item of importance, and increases with the size of the grains themselves. It is, therefore, customary to use sand with some coarse grains in it, but with enough smaller grains to fill the voids between the larger ones.

Aggregates.—When concrete is to be used in a place where it may have to withstand the action of fire, it is necessary that the aggregate be of such nature that it will not disintegrate and crumble away. Limestone and marble chips are objectionable as aggregates, as the action of heat causes them to swell, crack, and crumble to dust. Trap rock, cinder, and broken brick are among the best aggregates for concrete that is to be exposed to the action of fire. It should be borne in mind, however, that broken brick will soon soften in concrete placed under water.

Limestone is unsafe to use in reinforced-concrete work, unless special care is taken to see that the steel is well protected from the stone by a layer of cement. Another material that is considered injurious to steel, if the latter is not coated with cement, is cinders; their damaging effect is not due so much to the sulphur in them, as commonly claimed, as to their porosity. However, in certain proportions in which the cinder is not so predominant—as in a mixture of 1 part of cement, 2 parts of sand, and 3 parts of cinder—the corrosive effect on the steel is inconsiderable if the concrete is properly mixed.

Proportioning Ingredients.—The proper proportion of ingredients for the best concrete is such that there will be enough cement in the mixture to bind all the materials together, and that the materials will be of such various sizes that all voids will be filled. When a concrete is made of cement, sand, and stone, and the stone is of such a size that it will pass through a 3-in. ring, but will not pass through a 2½-in. ring, the concrete is weaker and requires more cement than one made with stone graded from 3 in. down. When the stone is graded in size, the smaller-sized stones fill the voids between the larger stones, and thus reduce the amount of cement required. The grading of the stone also makes the concrete stronger. Some engineers specify that the stone must pass through a ring 2 in. in diameter, more engineers specify a 2½-in. ring, and even a 3-in. ring is not uncommon. For very thin walls, and for small work, such as concrete blocks, it is necessary, of course, that the size of broken stones shall not be too large to place them in the mold. It can, however, be stated as a general proposition that the larger the stones, the stronger will be the concrete.

Usual Proportions of Concrete.—For reinforced concrete and more important concrete work, such as piers and dams, a 1-2-4 mixture is generally used. In columns, even a richer mixture is sometimes required. For less important work, a 1-2-6 mixture is commonly used; and for rubble concrete, even a 1-4-8 mixture is sometimes employed.

Methods of Measuring Concrete Ingredients.—Cement is bought by the barrel, but it is usually shipped by the bag. Four bags of Portland cement make a barrel. Natural cement comes in the same-sized bags, or in larger bags making 3 bg. to a barrel. An ordinary box car holds from 400 to 600 bg. The purchaser is charged with the bags by the manufacturer, unless they are of paper, but he gets a rebate for those he returns. A barrel of Portland cement weighs 375 lb.; a barrel of natural cement, 300 lb.

Cement is usually measured by the barrel the way it comes from the manufacturer, or as 4 bg. to the barrel, while broken stone and sand are measured loose in a barrel. Portland cement, after it is taken out of its original packing and stirred up, fills a larger volume than when packed. It is, therefore, necessary to state just how the cement is to be measured; and, as said before, it is customary to measure it by the barrel compact. A cement barrel contains about 3.6 cu. ft.

Fuller's Rule for Quantities.—If c is the number of parts of cement; s, the number of parts of sand; g, the number of parts of gravel or broken stone; C, the number of barrels of Portland cement required for 1 cu. yd. of concrete; S, the number of cubic yards of sand required for 1 cu. yd. of concrete; and G, the number of cubic yards of stone or gravel required for 1 cu. yd. of concrete. Then,

$$C=\frac{11}{c+s+g}$$

$$S=\frac{3.8}{27}Cs$$

and

$$G=\frac{3.8}{27}Cg$$

If the broken stone is of uniformly large size with no smaller stone in it, the voids will be greater than if the stone is graded. Therefore, 5% must be added to each value found by the preceding formulas.

EXAMPLE.—If a 1–2–4 mixture be considered, what will be: (*a*) the number of barrels of cement, (*b*) the number of cubic yards of sand, and (*c*) the number of cubic yards of stone required for 1 cu. yd. of concrete?

SOLUTION.—(*a*) Here, $c=1$, $s=2$, and $g=4$. Substituting these values in the formula for C,

$$C=\frac{11}{1+2+4}=1.57$$

(*b*) Substituting the values of C and s in the formula for S,

$$S=\frac{3.8}{27}\times 1.57\times 2=.44$$

(*c*) Substituting the values of C and g in the formula for G,

$$G=\frac{3.8}{27}\times 1.57\times 4=.88$$

Table of Concrete Quantities.—The following table, which gives the quantities of ingredients for concrete of various proportions, has been prepared by Edwin Thacher. As will be observed, he takes into account the difference in the character and size of the stone or gravel used.

QUANTITIES OF INGREDIENTS FOR CONCRETE OF VARIOUS PROPORTIONS

Proportion of Ingredients			Ingredients Required for 1 Cu. Yd. of Rammed Concrete											
			Stone, 1 In. and Under—Dust Screened Out			Stone, 2½ In. and Under—Dust Screened Out			Stone, 2½ In. With Most Small Stone Screened Out			Gravel, ¾ In. and Under		
Cement	Sand	Stone	Cement Bbl.	Sand Cu. Yd.	Stone Cu. Yd.	Cement Bbl.	Sand Cu. Yd.	Stone Cu. Yd.	Cement Bbl.	Sand Cu. Yd.	Stone Cu. Yd.	Cement Bbl.	Sand Cu. Yd.	Gravel Cu. Yd.
1	1.0	2.0	2.57	.39	.78	2.63	.40	.80	2.72	.41	.83	2.30	.35	.74
1	1.0	2.5	2.29	.35	.88	2.34	.36	.89	2.41	.37	.92	2.10	.32	.80
1	1.0	3.0	2.06	.31	.94	2.10	.32	.96	2.16	.33	.98	1.89	.29	.86
1	1.0	3.5	1.84	.28	.98	1.88	.29	1.00	1.88	.29	1.05	1.71	.26	.91
1	1.5	2.5	2.05	.47	.78	2.09	.48	.80	2.16	.49	.82	1.83	.42	.73
1	1.5	3.0	1.85	.42	.84	1.90	.43	.87	1.96	.45	.89	1.71	.39	.78
1	1.5	3.5	1.72	.39	.91	1.74	.40	.93	1.79	.41	.96	1.57	.36	.83
1	1.5	4.0	1.57	.36	.96	1.61	.37	.98	1.64	.38	1.00	1.46	.33	.88
1	1.5	4.5	1.43	.33	.98	1.46	.33	1.00	1.51	.35	1.06	1.34	.31	.91
1	2.0	3.0	1.70	.52	.77	1.73	.53	.79	1.78	.54	.81	1.54	.47	.73
1	2.0	3.5	1.57	.48	.83	1.61	.49	.85	1.66	.50	.88	1.44	.44	.77
1	2.0	4.0	1.46	.44	.89	1.48	.45	.90	1.53	.47	.93	1.34	.41	.81
1	2.0	4.5	1.36	.42	.93	1.38	.42	.95	1.43	.43	.98	1.26	.38	.86
1	2.0	5.0	1.27	.39	.97	1.29	.39	.98	1.33	.39	1.03	1.17	.36	.89
1	2.5	3.5	1.45	.55	.77	1.48	.56	.79	1.51	.58	.81	1.32	.50	.70

1	2.5	4.0	1.35	.52	.82	1.38	.53	.84	1.42	.54	.87	1.24	.47	.75
1	2.5	4.5	1.27	.48	.87	1.29	.49	.88	1.33	.51	.91	1.16	.44	.80
1	2.5	5.0	1.19	.46	.91	1.21	.46	.92	1.26	.48	.96	1.10	.42	.83
1	2.5	5.5	1.13	.43	.94	1.15	.44	.96	1.18	.44	.99	1.03	.39	.86
1	2.5	6.0	1.07	.41	.97	1.07	.41	.98	1.10	.41	1.03	.98	.37	.89
1	3.0	4.0	1.26	.58	.77	1.28	.58	.78	1.32	.60	.80	1.15	.52	.72
1	3.0	4.5	1.18	.54	.81	1.20	.55	.82	1.24	.57	.85	1.09	.50	.75
1	3.0	5.0	1.11	.51	.85	1.14	.52	.87	1.17	.54	.89	1.03	.47	.78
1	3.0	5.5	1.06	.48	.89	1.07	.49	.90	1.11	.51	.93	.97	.44	.81
1	3.0	6.0	1.01	.46	.92	1.02	.47	.93	1.06	.48	.97	.92	.42	.84
1	3.0	6.5	.96	.44	.95	.98	.44	.96	1.00	.45	1.01	.88	.40	.87
1	3.0	7.0	.91	.42	.97	.92	.42	.98	.94	.42	1.05	.84	.38	.89
1	3.5	5.0	1.05	.56	.80	1.07	.57	.82	1.11	.59	.85	.96	.50	.76
1	3.5	5.5	1.00	.53	.84	1.02	.54	.85	1.06	.56	.89	.92	.48	.78
1	3.5	6.0	.95	.50	.87	.97	.51	.89	1.00	.53	.92	.88	.46	.80
1	3.5	6.5	.92	.49	.91	.93	.49	.92	.96	.51	.95	.83	.44	.82
1	3.5	7.0	.87	.47	.93	.89	.47	.95	.91	.49	.98	.80	.43	.85
1	3.5	7.5	.84	.45	.96	.86	.45	.98	.86	.47	1.01	.76	.41	.87
1	3.5	8.0	.80	.42	.97	.82	.43	1.01	.81	.45	1.04	.73	.39	.89
1	4.0	6.0	.90	.55	.82	.92	.56	.84	.95	.58	.87	.83	.51	.77
1	4.0	6.5	.87	.53	.85	.88	.53	.87	.91	.55	.90	.80	.49	.79
1	4.0	7.0	.83	.51	.89	.84	.51	.90	.87	.53	.93	.77	.47	.81
1	4.0	7.5	.80	.49	.91	.81	.50	.93	.84	.51	.96	.73	.44	.83
1	4.0	8.0	.77	.47	.93	.78	.48	.95	.81	.49	.98	.71	.43	.86
1	4.0	8.5	.74	.45	.95	.76	.46	.98	.78	.47	1.01	.68	.42	.88
1	4.0	9.0	.71	.43	.97	.73	.44	1.01	.75	.45	1.04	.65	.40	.89
1	5.0	9.0	.66	.50	.90	.67	.52	.93	.70	.53	.96	.61	.46	.83
1	5.0	10.0	.62	.47	.95	.63	.48	.96	.65	.50	1.00	.57	.43	.87

AVERAGE ULTIMATE STRENGTH OF CONCRETE MADE FROM PORTLAND CEMENT, SAND, AND CRUSHED STONE

Proportion of Ingredients			Tension Pounds per Sq. In.				Compression Pounds per Sq. In.				Shear Pounds per Sq. In.			
Cement	Sand	Stone	7 Da.	1 Mo.	3 Mo.	6 Mo.	7 Da.	1 Mo.	3 Mo.	6 Mo.	7 Da.	1 Mo.	3 Mo.	6 Mo.
1	2.0	4	160	210	240	250	1,600	2,150	2,400	2,500	200	269	300	313
1	2.5	5	143	195	225	235	1,430	1,950	2,250	2,350	179	244	281	294
1	3.0	6	125	180	210	220	1,250	1,800	2,100	2,200	156	225	263	275
1	3.5	7	110	166	196	208	1,100	1,660	1,960	2,000	138	208	245	260
1	4.0	8	98	152	182	195	980	1,520	1,820	1,950	123	190	228	214
1	4.5	9	85	140	169	184	850	1,400	1,690	1,840	106	175	211	230
1	5.0	10	75	126	155	172	750	1,260	1,550	1,720	94	158	194	215
1	5.5	11	65	112	142	160	650	1,120	1,420	1,600	81	140	178	200
1	6.0	12	60	100	130	150	600	1,000	1,300	1,500	75	125	163	188

Strength and Weight of Plain Concrete.—The average ultimate strength of concrete in tension, compression, and shear is given in the accompanying table for different proportions of mixture, the aggregate of which is broken stone. Concrete made of gravel is 75% as strong and concrete made with cinders is about 65% as strong.

As the strength of concrete increases with age, it is necessary for the engineer to know when the concrete will be loaded. It is customary to assume a factor of safety based on the strength of the concrete after 6 mo. The engineer must be careful that the concrete, in the first few months after being laid, is not subjected to too great stresses. For general work, a factor of safety of 5 on concrete 6 mo. old is recommended. This will give the required strength for the first few months, and yet will not be wasteful of material at any time. A factor of safety of 4 on concrete 6 mo. old may be used for steady loads, such as earth fills, water pressure, etc.

The *weight of concrete* depends mainly on the kind of aggregate used. It averages about 140 to 150 lb. per cu. ft., for broken-stone and gravel concrete, and 110 to 115 for cinder concrete.

REINFORCED CONCRETE

FORMULAS FOR RECTANGULAR BEAMS

Reinforced concrete is concrete in which steel or iron is embedded in order to increase the strength of the former.

Fundamental Principles.—Many theories have been advanced as a basis for the design of reinforced-concrete beams, and it is not yet known which is most nearly correct. The formulas that follow are based on the so-called *straight-line theory*, which has been almost universally adopted in the United States and has been recommended by a Joint Committee composed of members of the leading engineering societies of this country. This theory is based on the following assumptions, and principles derived from these assumptions:

1. A plane section of a beam remains plane after it has been subjected to bending.

2. For one and the same material, the unit stresses at different points of a beam subjected to bending are proportional to their distances from the neutral axis.

3. The unit stresses in steel and concrete at points equidistant from the neutral axis are proportional to their respective moduli of elasticity.

4. The concrete is assumed to take only compressional stresses, all the tensional stresses being carried by the steel.

5. The internal stresses in the section of a reinforced-concrete beam subjected to bending form a couple consisting of the resultant of all compressional stresses taken by the concrete, on one hand, and the tensional stresses taken by the steel, on the other hand.

It is also assumed that the value of the ratio of the moduli of elasticity of steel and concrete (usually denoted by n) is constant within the limits of the working stresses of the materials. This value of n greatly varies with the qualities of the material and labor employed in the manufacture of the concrete, and is usually specified by city ordinances.

The reinforced-concrete tables given later are computed for $n=12$ and $n=15$, which are prevalent in the present engineering practice.

Definitions.—The *economic steel ratio* is that ratio of the area of steel to the area of concrete at which both the steel and concrete can be stressed to their maximum allowable limit at the same time, and is denoted by p_e. If a lower ratio is used, the stress in the concrete will not reach its limit without overstressing the steel, and if a higher ratio is employed the full strength of the steel cannot be utilized without overstressing the concrete. The economic steel ratio, or as it is also called the *critical value* of steel, is not a fixed quantity; it depends on the ratio of the allowable maximum unit stresses of steel and concrete.

The *stress ratio* is the ratio of the stresses actually produced in the steel and concrete by a given external moment. When n is constant, the value of the stress ratio depends only on the amount of steel used. For the critical value of steel, the stress ratio equals the ratio of the allowable maximum unit stress in steel and that in concrete.

Notation.—The accompanying illustration shows a section of a reinforced-concrete beam. The following notation is used for the different elements involved in its design:

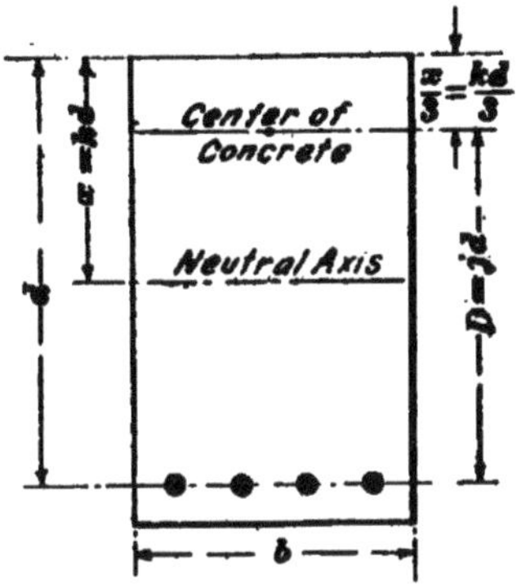

b = width of beam, in inches;

d = effective depth = distance of steel from top of beam;

$x = kd$ = distance of neutral axis from top of beam;

$k=\frac{x}{d}$;

D = arm of stress couple = distance between center of steel and center of concrete;

$j=\frac{D}{d}$;

A = total area of steel;

p = steel ratio $=\frac{A}{bd}$;

p_e = economic steel ratio;

M = bending moment;

f_s and f_c = stresses in steel and concrete, respectively, actually produced by the bending moment M;

F_s and F_c = maximum allowable unit stresses in steel and concrete, respectively;

M_s and M_c = working moment of resistance of steel and concrete for the unit stresses F_s and F_c, respectively;

r = stress ratio $= f_s : f_c$;

r_e = stress ratio when economic percentage of steel is used, which is equal to $F_s : F_c$;

E_s and E_c = moduli of elasticity of steel and concrete, respectively;

$n=\frac{E_s}{E_c}$;

C_s = section-modulus coefficient for steel = jp;

C_c = section-modulus coefficient for concrete = $\frac{kj}{2}$;

R = coefficient whose values are given later in tabular form.

Formulas.—Following are the formulas for rectangular reinforced-concrete beams:

$$A = pbd \qquad (1)$$

$$k = \sqrt{p^2n^2+2pn}-pn \qquad (2)$$

$$j = 1-\frac{k}{3} \qquad (3)$$

$$C_s = jp \qquad (4)$$

$$C_c = \frac{kj}{2} \qquad (5)$$

$$M = C_sbd^2f_s = C_cbd^2f_c \qquad (6)$$

$$M_s = C_sbd^2F_s = AjdF_s \qquad (7)$$

$$M_c = C_cbd^2F_c \qquad (8)$$

$$M = Rbd^2 \qquad (9)$$

When the economic steel ratio is used, $M_s = M_c$; also,

$$k_e = \frac{n}{n+r_e} \qquad (10)$$

$$p_e = \frac{k_e}{2r_e} \qquad (11)$$

Formulas 7 and 8 furnish the fundamental equations for designing and investigating rectangular reinforced-concrete beams. Formula 7, expresses the resistance of the beam for steel and is to be used when the steel ratio is below its critical value, while formula 8 gives expression to the moment of resistance of concrete and governs the design in cases when the amount of steel is above the economic ratio. C_s and C_c can be determined by formulas 4 and 5, and for finding the economic steel ratio formula 11 is available. For $n = 12$ and $n = 15$, C_s and C_c can be taken directly from the accompanying table; also, the economic ratio of steel may be ascertained from this table, as will be explained presently.

Reinforced-Concrete Tables and Their Application.—For $n = 12$ or 15 the accompanying table of properties of reinforced-

PROPERTIES OF REINFORCED-CONCRETE BEAMS

$n = 12$

p	k	j	C_s	C_c	$r = \frac{f_s}{f_c}$
.001	.1434	.9522	.0009522	.0683	71.69
.002	.1964	.9345	.001869	.0918	49.10
.003	.2347	.9218	.002765	.1082	39.12
.004	.2655	.9115	.003646	.1210	33.19
.005	.2916	.9028	.004514	.1316	29.16
.006	.3142	.8953	.005372	.1407	26.19
.007	.3344	.8885	.006220	.1486	23.89
.008	.3526	.8825	.007060	.1556	22.04
.009	.3691	.8770	.007893	.1619	20.51
.010	.3844	.8719	.008719	.1676	19.22
.011	.3985	.8672	.009539	.1728	18.11
.012	.4116	.8628	.010353	.1776	17.15
.013	.4239	.8587	.01116	.1820	16.31
.014	.4355	.8548	.01197	.1861	15.55
.015	.4464	.8512	.01277	.1900	14.88
.016	.4567	.8478	.01356	.1936	14.27
.017	.4665	.8445	.01436	.1970	13.72
.018	.4758	.8414	.01514	.2002	13.22
.019	.4847	.8384	.01593	.2032	12.76
.020	.4932	.8356	.01671	.2061	12.33

$n = 15$

p	k	j	C_s	C_c	$r = \frac{f_s}{f_c}$
.001	.1589	.9470	.0009470	.0752	79.43
.002	.2168	.9277	.001855	.1006	54.20
.003	.2584	.9139	.002742	.1181	43.06
.004	.2916	.9028	.003611	.1316	36.45
.005	.3195	.8935	.004468	.1427	31.95
.006	.3437	.8854	.005313	.1522	28.64
.007	.3651	.8783	.006148	.1603	26.08
.008	.3844	.8719	.006975	.1676	24.02
.009	.4019	.8660	.007794	.1740	22.33
.010	.4179	.8607	.008607	.1798	20.89
.011	.4327	.8558	.009413	.1851	19.67
.012	.4464	.8512	.010214	.1900	18.60
.013	.4592	.8469	.01101	.1945	17.66
.014	.4712	.8429	.01180	.1986	16.83
.015	.4825	.8392	.01259	.2025	16.08
.016	.4932	.8356	.01337	.2061	15.41
.017	.5033	.8322	.01415	.2094	14.80
.018	.5129	.8290	.01492	.2126	14.25
.019	.5220	.8260	.01569	.2156	13.74
.020	.5307	.8231	.01646	.2184	13.27

concrete beams gives for $p=\frac{A}{bd}$, varying by .001, the values of k, j, C_s, C_c, and r, which may be used in the preceding formulas. The economic percentage of steel for any given working stresses, F_s and F_c, may be determined by computing $r_e=\frac{F_s}{F_c}$ and finding in the table a value of p that corresponds, or nearly corresponds, to r_e in the column headed $r=\frac{f_s}{f_c}$.

EXAMPLE.—Find the economic ratio of steel for $n=15$, $F_s=16{,}000$, and $F_c=500$.

SOLUTION.— $$r_e=\frac{F_s}{F_c}=\frac{16{,}000}{500}=32$$

On referring to the table for $n=15$, it is found that the nearest corresponding value of r is 31.95, for which p is .005, which is the economic ratio of steel.

To Design a Beam.—The following practical examples will serve to show the way in which the table may be used in designing a beam:

EXAMPLE 1.—Let the following values be given: $n=15$, $F_s=12{,}500$, $F_c=600$, $M=500{,}000$ in.-lb., $d=22$ in., and $p=.006$. Required: (*a*) the value of b and (*b*) that of A.

SOLUTION.—(*a*) By the preceding method, find from the table the economic steel ratio for the given n, F_s, and F_c, which is .01. As this is greater than the given $p=.006$; formula 7 must be employed. Substituting given values and noting in the table that for $p=.006$, $C_s=.00531$, it is obtained $500{,}000=.00531\times b\times 22^2\times 12{,}500$. Whence, $b=15.6$ in.

(*b*) $A=pbd=.006\times 15.6\times 22=2.06$ sq. in.

NOTE.—If, in the preceding example, the given steel ratio p were greater than the economic steel ratio, formula 8 would have to be used. If the economic steel ratio were used, either formula 7 or 8 would give the same result.

EXAMPLE 2.—Let the dimensions of the beam be fixed, as $b=18$ in. and $d=27$ in. Also, let $M=800{,}000$ in.-lb., $F_s=15{,}000$, $F_c=550$, and $n=12$. Required, A.

SOLUTION.—Solving formulas 7 and 8 for C_s and C_c, respectively, and substituting known values,

$$C_s = \frac{800,000}{18 \times 27^2 \times 15,000} = .00406$$

and
$$C_c = \frac{800,000}{18 \times 27^2 \times 550} = .111$$

On referring to the table for $n = 12$, it is found that for $C_s = .00406$, $p = .0045$; also, that for $C_c = .111$, $p = .0032$. The former value of p being the greater, it must be used; therefore, $A = pbd = .0045 \times 18 \times 27 = 2.2$ sq. in.

To review a beam.—To review a beam means to investigate one that has already been built. In this case, b, d, p, and n will be known, and it will be required either to determine M for given F_s and F_c, or to find f_s and f_c for a given M.

EXAMPLE 1.—Let $b = 15$ in., $d = 30$ in., and $p = .008$. Find M for $n = 15$, $F_s = 13,500$, and $F_c = 500$.

SOLUTION.—By the method already given, it is found that the economic steel ratio is .0066. As this is less than the given value of p, formula 8 must be employed. From the table for $n = 15$ and $p = .008$, $C_c = .168$; therefore, substituting this value in formula 8, $M = .168 \times 15 \times 30^2 \times 500 = 1,134,000$ in.-lb.

EXAMPLE 2.—Let $b = 18$ in., $d = 30$ in., $p = .012$, $n = 12$, and $M = 2,000,000$ in.-lb. Find f_s and f_c.

SOLUTION.—In the table for $n = 12$, it is found that for $p = .012$, $C_s = .0104$ and $C_c = .178$. Solving formula 6 for f_s and f_c and substituting known values,

$$f_s = \frac{2,000,000}{.0104 \times 18 \times 30^2} = 11,870$$

$$f_c = \frac{2,000,000}{.178 \times 18 \times 30^2} = 690$$

Values of R for Special Constants.—For the values $n = 12$ and $n = 15$ and certain unit stresses, F_s and F_c, the calculations in the design of reinforced-concrete beams may be effected by formula 9, in which R has the value given in the following tables. The economic steel ratio for each set of units of these tables is printed in Italic. The application of this table will best be seen from the examples that follow:

EXAMPLE 1.—Let $M = 2,000,000$ in.-lb., $F_s = 16,000$, $F_c = 600$, $n = 12$, and $b = 20$ in. Find: (*a*) d and (*b*) A.

VALUES OF *R* FOR SPECIAL CONSTANTS—$n=12$

p	$F_s=16{,}000$ $F_c=500$	$F_s=16{,}000$ $F_c=550$	$F_s=16{,}000$ $F_c=600$	$F_s=16{,}000$ $F_c=650$	$F_s=16{,}000$ $F_c=700$	$F_s=16{,}000$ $F_c=750$	$F_s=16{,}000$ $F_c=800$
.001	15.24	15.24	15.24	15.24	15.24	15.24	15.24
.002	29.91	29.91	29.91	29.91	29.91	29.91	29.91
.003	44.24	44.24	44.24	44.24	44.24	44.24	44.24
.004	58.34	58.34	58.34	58.34	58.34	58.34	58.34
.00426	*61.98*						
.005	65.81	72.22	72.22	72.22	72.22	72.22	72.22
.00502		*72.49*					
.00582			*83.47*				
.006	70.33	77.37	84.40	85.94	85.94	85.94	85.94
.00666				*94.88*			
.007	74.28	81.71	89.14	96.57	99.52	99.52	99.52
.00753					*108.66*		
.008	77.78	85.56	93.34	101.12	108.90	112.96	112.96
.00844						*118.80*	
.009	80.93	89.02	97.12	105.21	113.30	121.40	126.28
.00938							*131.35*
.010	83.78	92.16	100.54	108.92	117.30	125.67	134.03
.011	86.39	95.03	103.67	112.31	120.95	129.59	138.22
.012	88.79	97.67	106.55	115.43	121.31	133.18	142.06
.013	91.01	100.11	109.21	118.31	127.41	136.51	145.61
.014	93.07	102.38	111.69	120.99	130.30	139.61	148.91
.015	95.00	104.50	114.00	123.50	133.00	142.50	152.00
.016	96.80	106.48	116.16	125.84	135.52	145.20	154.88
.017	98.50	108.35	118.19	128.04	137.89	147.74	157.59
.018	100.09	110.10	120.11	130.12	140.13	150.14	160.15
.019	101.60	111.76	121.92	132.08	142.24	152.40	162.56
.020	103.03	113.33	123.64	133.94	144.24	154.55	164.85

VALUES OF R FOR SPECIAL CONSTANTS—$n=15$

p	$F_s=16,000$ $F_c=500$	$F_s=16,000$ $F_c=550$	$F_s=16,000$ $F_c=600$	$F_s=16,000$ $F_c=650$	$F_s=16,000$ $F_c=700$	$F_s=16,000$ $F_c=750$	$F_s=16,000$ $F_c=800$
.001	15.15	15.15	15.15	15.15	15.15	15.15	15.15
.002	29.69	29.69	29.69	29.69	29.69	29.69	29.69
.003	43.87	43.87	43.87	43.87	43.87	43.87	43.87
.004	57.78	57.78	57.78	57.78	57.78	57.78	57.78
.00499	*71.30*						
.005	71.37	71.48	71.48	71.48	71.48	71.48	71.48
.00585		*82.95*					
.006	76.08	83.69	85.00	85.00	85.00	85.00	85.00
.00675			*95.04*				
.007	80.17	88.19	96.21	98.37	98.37	98.37	98.37
.00769				*107.53*			
.008	83.78	92.16	100.54	108.92	111.60	111.60	111.60
.00867					*120.37*		
.009	87.01	95.71	104.41	113.11	121.81	124.71	124.71
.00968						*133.51*	
.010	89.92	98.91	107.90	116.90	125.89	134.88	137.71
.01071							*146.94*
.011	92.57	101.83	111.08	120.34	129.60	138.85	148.11
.012	95.00	104.50	114.00	123.50	133.00	142.50	152.00
.013	97.23	106.96	116.68	126.40	136.13	145.85	155.57
.014	99.31	109.24	119.17	129.10	139.03	148.96	158.89
.015	101.23	111.36	121.48	131.60	141.73	151.85	161.97
.016	103.03	113.33	123.64	133.94	144.24	154.55	164.85
.017	104.72	115.19	125.66	136.13	146.60	157.07	167.55
.018	106.30	116.93	127.56	138.19	148.82	159.45	170.08
.019	107.79	118.57	129.35	140.13	150.91	161.69	172.47
.020	109.20	120.12	131.04	141.96	152.88	163.80	174.72

SOLUTION.—(*a*) As p is not specified, the economic ratio of steel will be used. This is given in Italic in the table for $n=12$ as .00582. The corresponding value of R is 83.47. Then, substituting in formula 9, and solving for d,

$$d=\sqrt{\frac{2{,}000{,}000}{83.47\times 20}}=34.6 \text{ in.}$$

(*b*) $A=pbd=.00582\times 20\times 34.6=4.03$ sq. in.

EXAMPLE 2.—Let $M=800{,}000$ in.-lb., $b=18$, $d=27$, $F_s=16{,}000$, $F_c=500$, and $n=12$. Find A.

SOLUTION.—Substituting given values in formula 9 and solving for R,

$$R=\frac{800{,}000}{18\times 27^2}=60.97$$

From the table the cor responding value of p is .0042. Then, $A=.0042\times 18\times 27=2.04$ sq. in.

EXAMPLE 3.—Find the safe value of M when $b=14$, $d=30$, $p=.006$, $n=15$, $F_s=16{,}000$, and $F_c=700$.

SOLUTION.—From the table for the given constants, $R=85.00$. Therefore, $M=85\times 14\times 30^2=1{,}071{,}000$ in.-lb.

Web Stresses.—Two general methods are used for preventing failure of a beam by diagonal tension. These are: (1) by bending up diagonally part of the horizontal reinforcement, and (2) by the use of special shear members, or *stirrups*.

The following formulas may be employed for the purpose of designing stirrups:

For rectangular beams reinforced at the bottom,

$$v=\frac{V}{bjd} \qquad (1)$$

For vertical stirrups,

$$P=\frac{Vc}{jd} \qquad (2)$$

For stirrups inclined at 45°,

$$P=.7\frac{Vc}{jd} \qquad (3)$$

In these formulas V is the total external vertical shear, in pounds; v, the unit shear, in pounds per square inch; P, the total stress in one stirrup, in pounds; and c, the horizontal spacing of stirrups, in inches. The other letters have the same meaning as previously given.

For T beams,

$$v=\frac{V}{b_1jd}, \qquad (4)$$

in which b_1 is the width of the stem.

If the neutral axis is in the flange, j can be found as in rectangular beams; if it is in the stem, the formulas for rectangular beams will not give the correct value of j, and in place of jd the approximate value of $d-\frac{t}{2}$ may be used, t being the thickness of the flange.

The value of v, the unit shear in concrete, should not exceed 40 lb. per sq. in., when no reinforcement is used. When web reinforcement is used, it is generally assumed that the concrete itself can take one-third of the shear. In this case, the allowable unit shear in the concrete is usually taken from 60 to 120 lb. per sq. in.

Bond Between Steel and Concrete.—In a reinforced-concrete beam the stress from the load is transmitted to the steel reinforcement by means of the adhesion, or bond, between the concrete and the steel. The amount of stress H that is transmitted to the horizontal reinforcement at the bottom at any section can be found approximately by the formula,

$$H=\frac{V}{jd},$$

in which V is the external shear at the section under consideration and $jd=D$, as before. Let f_b denote the unit bond induced at the same section and O the sum of the perimeters of the horizontal reinforcement, then O will also be the total bond area for one unit of length; therefore,

$$f_b=\frac{H}{O}=\frac{V}{jdO}$$

This value should not exceed the allowable unit adhesion between the steel and concrete. It is usually taken at about 80 lb. per sq. in.

FORMULAS FOR DOUBLE-REINFORCED BEAMS

Double reinforced-concrete beams are not economical, but sometimes they cannot be avoided. To determine the quantity of steel required in a given section to carry a certain bending moment M, first calculate the area A_e required at the bottom when the economic ratio of steel is used, and no steel is used at the top. On referring to the accompanying illustration, let xx represent the neutral axis for this arrangement, and M_e, the bending moment that the beam could resist if only this amount of steel were used. Then the steel to be added at the bottom above A_e is

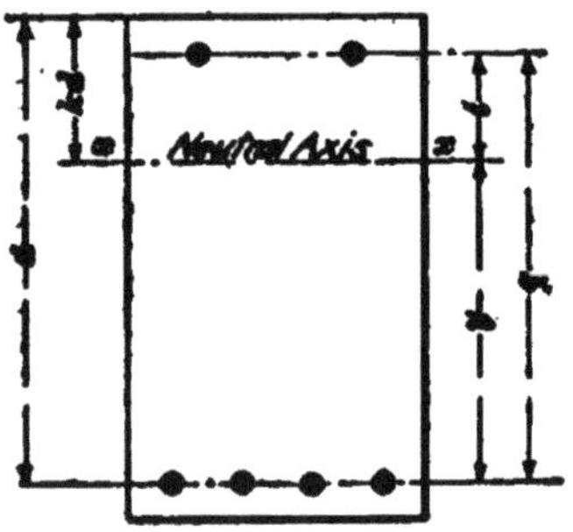

$$A_y = \frac{M - M_e}{F_s q}; \qquad (1)$$

or putting $M - M_e = M_x$, $A_y = \frac{M_x}{F_s q}$, and the area of steel to be used at the top is,

$$A_t = \frac{y}{t} A_y \qquad (2)$$

EXAMPLE.—In a certain beam b is limited to 10 in. and d to 18 in. $M = 724{,}800$ in.-lb., the beam is to be double reinforced, and designed for $n = 15$, $F_s = 16{,}000$ and $F_c = 500$.

SOLUTION.—From the table of values of R for special constants it is found that, for the constants given, $p_e = .00499$ and $R = 71.3$. Then, $M_e = Rbd^2 = 71.3 \times 10 \times 18^2 = 231{,}012$ in.-lb. Then, $M_x = 724{,}800 - 231{,}012 = 493{,}788$ in.-lb. If the compressive steel is placed, say 2 in. from the top of the beam, then $q = d - 2 = 18 - 2 = 16$. Substituting in formula 1,

$$A_y = \frac{493{,}788}{16{,}000 \times 16} = 1.93 \text{ sq. in.}$$

The total area of steel at the bottom is, therefore, $A = p_e bd + A_y = .00499 \times 10 \times 18 + 1.93 = 2.83$ sq. in. The area of steel at the top is found by formula 2. As the compressive steel is 2 in. from the top of the beam, $t = kd - 2$, and taking the value $k = .32$ from the table of properties of reinforced-concrete beams for $p = .00499$ and $n = 15$, $t = .32 \times 18 - 2 = 3.76$, and $y = d - kd = 18 - 5.76 = 12.24$. Then,

$$A_t = \frac{12.24}{3.76} \times 1.93 = 6.28 \text{ sq. in.}$$

FORMULAS FOR T-SHAPED BEAMS

When a slab and the beam supporting it are so constructed as to form a monolith, the slab may be considered as a part of the beam. Conservative practice requires that the width of the slab that may be considered as acting with the beam should not exceed one-fourth the span of the beam; it should also not exceed four times the thickness of the slab.

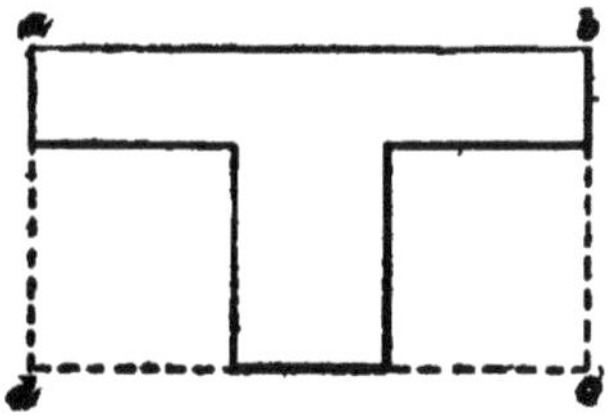

When the neutral axis does not fall below the bottom of the slab, the beam may be designed as a rectangular beam, having a section *abcd*, as in the accompanying illustration.

When the neutral axis falls below the bottom of the slab, the following approximate formula may be used:

$$M = AF_s\left(d - \frac{t}{2}\right)$$

In this formula, t is the thickness of the slab, and the other letters have the same significance as before. From it the area of steel required may be determined. To insure that the concrete is not overstressed, the maximum allowable unit stress should not exceed

$$F_c = \frac{2M}{tb\left(d - \frac{t}{2}\right)}$$

In these two formulas, the compressional area of the stem is neglected. They should, therefore, not be used when the stem forms a considerable part of the section, which will happen when the beam is large and the slab is shallow. In the latter case, it is well to neglect the T effect and consider that the beam carries the entire load.

FORMULAS FOR COLUMNS

Let, in addition to previous notation, a be the cross-sectional area of the column, a_s the cross-sectional area of the steel, and a_c the cross-sectional area of the concrete. Let, further, s_s and s_c denote the unit stresses in steel and concrete, respectively, and W the total load on column centrally loaded. Then

$$s_s = ns_c \quad (1)$$

$$W = s_c\,(a_s n + a_c) \quad (2)$$

As an example, let it be required to find W for a column 18 in. square and reinforced with eight rods $\frac{3}{4}$ in. square, using $s_c = 450$ and $n = 15$. Applying formula 1, $s_s = 450 \times 15 = 6{,}750$. To apply formula 2, substitute for a_s, $8 \times \frac{3}{4} \times \frac{3}{4} = 4.5$, and for a_c, $18 \times 18 - 4.5 = 319.5$. Then, $W = 450(4.5 \times 15 + 319.5) = 174{,}150$ lb.

FOUNDATIONS

SUBFOUNDATIONS

The *subfoundation* of a structure is that part of the natural surface of the earth on which the structure rests. The *foundation* is the lower part of the structure, which connects it with the subfoundation.

Materials for Subfoundations.—The materials usually regarded as suitable for subfoundations are solid rock, loose rock, earth, and sand.

The supporting power of a rock subfoundation may be considered as approximately equal to the resistance to crushing of the material of which the rock is composed, modified by a suitable factor of safety. The accompanying table is based on a factor of safety of 10.

Loose rock in any of its forms may make a satisfactory subfoundation, but it requ res very careful examination and, if possible, should be avoided.

SUPPORTING POWER OF ROCKS

Kind of Rock	Safe Foundation Load, Tons per Square Foot		
	From	To	Average
Granite	72	144	108
Limestone..	43	130	87
Sandstone..	30	108	69
Shale.......	3	100	52

The strength of earth subfoundations is largely affected by the quantity of water they contain; and the extent to which they may be exposed to water in the subfoundations is an important element to be considered in determining their sustaining capacity. The following table gives approximate values. The engineer, however, must in each case be guided largely by judgment based on experience and actual facts.

SAFE LOADS ON EARTH SUBFOUNDATIONS

Kinds of Material	Load in Tons per Square Foot	
	From	To
Hard pan and other indurated clays...	2	2½
Ordinary clays and clay soils, not submerged in water................	1½	2
Clay, soft and plastic...............	½	1
Ordinary soils, comparatively dry.....	1	1½
Ordinary soils, wet..................	½	1
Swamp and bog material.............	¼	½

Sand and gravel are capable of carrying very great loads; but as they are easily eroded by flowing water great care must be taken to protect them from direct contact with currents of water. Clean dry sand can bear a load of from 2 to 4 T. per sq. ft.

Depth of Subfoundation Below Surface of Ground.—Foundations in earth should be carried to such a depth below the ground surface that frost will not reach them. Nearly all moist earth expands, or heaves, with freezing, and repeated freezing is likely to soften and disintegrate it. It may also be subjected to other disturbances near the surface. The depth of foundations may be dictated by conditions other than frost. Often a good material cannot be found except at greater depths than are necessary to provide against frost.

The penetration of frost varies with the latitude. In the American Gulf States, ice seldom forms; while in the Lake region, the ground sometimes freezes to a depth of 5 or even 6 ft. Ordinarily, in the northern parts of the United States, subfoundations 4 ft. below the ground surface may be considered safe from injury by frost.

Required Area of Subfoundation.—In the case of foundations for ordinary structures where the weight is uniformly distributed over the whole of the subfoundation, the required area is equal to the total load coming on the subfoundation divided by the safe load per unit area. If the loads are irregular and the subfoundation is compressible, great care must be taken to secure an even distribution of the loads; otherwise, there is danger of uneven settlement, which may cause cracks.

Intensity of Pressure and Rule of the Middle Third.—Let AB, in the accompanying illustration, which represents the width of a rectangular subfoundation of a length equal to unity,

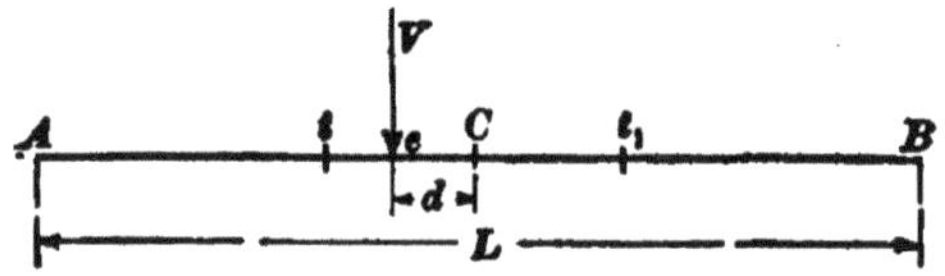

be divided into three equal parts, At, tt_1, and t_1B, and be bisected at C. If the point of intersection with AB of the resultant of all the forces acting on the structure, which point is called the *center of pressure*, is at C, the intensity of pressure is uniform throughout AB and is equal to $\frac{V}{L}$, V being the vertical component of the resultant pressure. When the center of pressure is at a point e, at a distance d from C, then the intensity is not uniform, being maximum at A and equal to

$$P_a=\frac{V}{L}+\frac{6Vd}{L^2}$$

and minimum at *B* and equal to

$$P_b=\frac{V}{L}-\frac{6Vd}{L^2}$$

When the center of pressure is at *t*, $d=\frac{L}{6}$ and $P_b=0$, while $P_a=2\times\frac{V}{L}$; that is, twice the average intensity. If the center of pressure falls between *t* and *A*, P_b becomes negative, which means that the foundation at *B* is then subjected to an uplifting force; in order, therefore, that this should not occur, the foundation must be so designed that the center of pressure will fall within tt_1, the middle third of the line *AB*. This principle is known as the *rule of the middle third.*

SPREAD FOUNDATIONS

Spread foundations are used in order to enlarge the base of a structure until it covers an area of subfoundation that can safely carry the weight of the structure. This is ordinarily accomplished by means of offsets called *footings*, as shown in Fig. 1.

Masonry Foundations.—In masonry construction the footings may be treated as cantilevers uniformly loaded. The force acting on *mn*, for instance, is the upward pressure on the part *ab* of the subfoundation. This pressure is assumed to be uniformly distributed, its intensity being equal to the total load on the structure divided by the area of the subfoundation. Likewise, the force acting on *pq* is part of the upward pressure, or reaction, on *nn*. The intensity in this case is the total load of the structure divided by the area at *nn*.

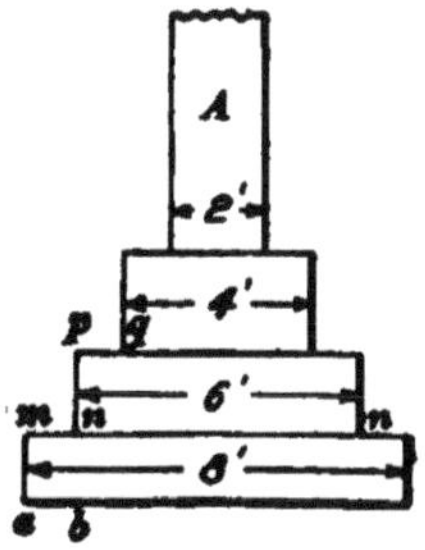

Fig. 1

EXAMPLE.—Fig. 1 shows a wall *A* 2 ft. thick carrying a load of 12 T. per lin. ft. of wall, including its own weight. The foundation of concrete is designed to have each footing project 1 ft. beyond the one above. What should be the thickness of each course, assuming the maximum allowable fiber stress of concrete in tension to be 25 lb. per sq. in.?

SOLUTION.—Since the load is 24,000 lb. per lin. ft., the intensity on the bottom course is $24{,}000 \div 8 = 3{,}000$ lb. per sq. ft.; on the second course, $24{,}000 \div 6 = 4{,}000$ lb. per sq. ft.; and on the third, $24{,}000 \div 4 = 6{,}000$ lb. per sq. ft. The respective bending moments are, therefore, $\frac{3{,}000 \times 12}{2} = 18{,}000$ in.-lb.; $\frac{4{,}000 \times 12}{2} = 24{,}000$ in.-lb., and $\frac{6{,}000 \times 12}{2} = 36{,}000$ in.-lb. Then, apply the formula $M = \frac{bd^2f}{6}$. Here, $b = 12$ in. and $f = 25$ lb. per sq. in. Solving for d, there results for the bottom course,

$$d = \sqrt{\frac{18{,}000}{50}} = 19 \text{ in., nearly;}$$

for the second course, $d = \sqrt{\frac{24{,}000}{50}} = 21.9$ in.;

and for the third course, $d = \sqrt{\frac{36{,}000}{50}} = 26.8$ in.

Steel Foundations.—In a steel spread foundation, Fig. 2,

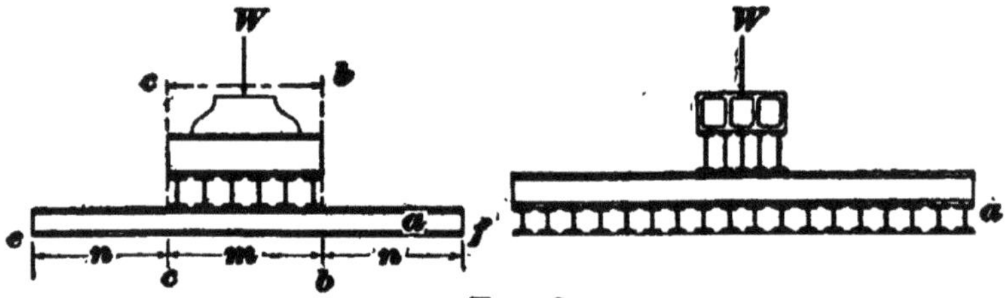

FIG. 2

the bending moment is considered to be maximum at the center of the beam, and its amount is equal to

$$M = \frac{Wn}{4}$$

EXAMPLE.—The total load carried by the bottom course of steel I beams, Fig. 2, is 360,000 lb.; the length of the beams is

10 ft.; and the width of the course next above it is 3 ft. (*a*) What is the maximum bending moment? (*b*) What size I beam may be used, assuming an extreme fiber stress of 15,000 lb. per sq. in.?

SOLUTION.—(*a*) The projection at each end of the bottom course is

$$\frac{10-3}{2}=3\tfrac{1}{2}\text{ ft., or 42 in.}$$

There are eighteen I beams in the course; therefore, the load on each is 360,000 ÷ 18 = 20,000. Substituting these values in the formula, gives

$$M=\frac{20{,}000\times 42}{4}=210{,}000\text{ in.-lb.}$$

(*b*) Referring to a steel manufacturer's handbook, it is found that the moment of inertia of an 8-in. I beam weighing 18 lb. per ft. of length is 56.9. The resisting moment of the beam is therefore

$$\frac{15{,}000\times 56.9}{4}=213{,}375\text{ in.-lb.;}$$

therefore, an 8-in. I beam may be used.

SUPPORTING POWER OF PILES

Assume that R is the resistance or bearing capacity of a pile; s, the set of pile, or distance, in inches, that the pile is driven during last blow; w, the weight of pile hammer; and h, the fall, in feet, of hammer during last blow. Then, for drop-hammer pile drivers,

$$R=\frac{2wh}{s+1} \qquad (1)$$

For steam-hammer pile drivers,

$$R=\frac{2wh}{s+.1} \qquad (2)$$

Formula 1 is called the *Engineering News formula*, because it was first published by that engineering journal. It has been very extensively adopted, as experience has proved that it is as reliable as any formula can justifiably claim to be. The uncertainties of pile driving are so great that it is useless to attempt to use a more accurate formula.

EXAMPLE.—A pile was driven with an ordinary hammer weighing 2,400 lb. The sinking under the last five blows was 22 in. The fall of the hammer during the last blows averaged 28 ft. What was the safe bearing power of the pile?

SOLUTION.—Here the value of s may be taken as the average of the total sinking during the last five blows, or $22 \div 5 = 4.4$ in. Then, $w = 2{,}400$ lb.; $h = 28$; and $s = 4.4$. Substituting these values in formula 1,

$$R = \frac{2 \times 2{,}400 \times 28}{4.4 + 1} = 24{,}889 \text{ lb.}$$

RETAINING WALLS

STABILITY OF RETAINING WALLS

VERTICAL BACK

A *retaining wall* is a wall that sustains the pressure of earth filling or backing deposited behind it after it has been built.

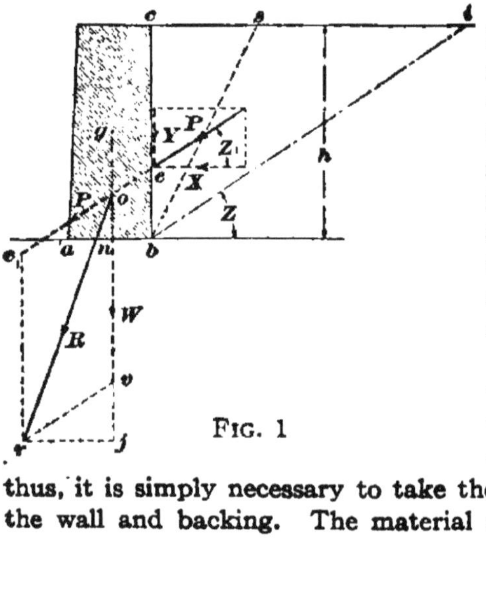

FIG. 1

Analysis of Forces. In Fig. 1 is shown a retaining wall with a vertical face and a vertical back; *ib* is the natural slope of the back filling, which on an average is 1½ horizontal to 1 vertical. The top *ci* is level with the top of the wall.

In making calculations, only 1 ft. of the length of wall and of the backing is taken; thus, it is simply necessary to take the area of the section of the wall and backing. The material composing the backing is supposed to be perfectly dry and to possess no cohesive power, which is practically true of pure sand.

It is generally assumed that the maximum pressure on a retaining wall is caused by a wedge-shaped prism of earth *bsc* included between the wall and the line *bs*, which bisects the angle *cbi*. This line is called the ***line of maximum pressure***, and the prism whose cross-section is *cbs* is called the ***prism of maximum pressure***.

The earth pressure P on the wall is the resultant of two forces X and Y, Fig. 1. The pressure X is obtained by determining the weight of the prism of maximum pressure and resolving it into two components, one perpendicular to *cb* and one parallel to *bs*. The former is the force X. For a wall with a vertical back;

$$X = \tfrac{1}{2}wh^2\tan^2(45° - \tfrac{1}{2}Z)$$

in which w is the weight per cubic foot of back filling; h, the height of the wall; and Z, the angle of repose of the back filling, which for $1\frac{1}{2}$ horizontal to 1 vertical is 33° 41′.

The force Y is the friction between the wall and the filling, due to the pressure X; and if f denotes the coefficient of friction between the material of the wall and that of the filling,

$$Y = fX$$

As is well known, f is the tangent of the angle of friction between the material of the wall and that of the back filling. This angle is shown as Z_1 in the illustration. For dry earth, it is generally taken as equal to Z. In this case, P would be parallel to *bi* and f would be .67.

The point of application e of P is assumed to be such that $be = \frac{1}{3}bc = \frac{1}{3}h$.

Pressure on Base of Wall.—When X, Y, and the position of e have been determined, the magnitude and exact position of P are most conveniently determined graphically. The total pressure R, Fig. 1, acting on the base of the wall is then the resultant of the pressure P and the weight W of the wall. Its magnitude and line of action are determined by the parallelogram oe_1rv, in which $oe_1 = P$ and $ov = W$, the point o being the intersection of the line of action of P with a vertical through the center of gravity g of the wall.

If both the wall and the foundation were absolutely incompressible and incapable of fracture or crushing, the wall would be safe from overturning if the point *n* where the line of action of *R* meets the base came anywhere inside the base of the wall; and, theoretically, the pressure *P* could be increased until *n* coincided with *a*—that is, until the line of action of the resultant pressure *R* passed through the toe *a*. But practical considerations require that, under ordinary conditions, the point *n* should fall within the middle third of the base of the wall. It must be stated, however, that the distance *a n* may safely be reduced somewhat from one-third to even one-fifth the width of the base, if the foundation is perfectly rigid and the masonry of the best. This will give a maximum intensity of pressure on the foundation at *a* 3⅕ times the intensity there would be if the center of pressure were at the center of the base.

Stability Against Sliding.—The total pressure *R* on the base may be resolved into a vertical component *oj* ($= W+Y$) and a horizontal thrust *jr* ($= X$) the latter tending to produce sliding on the base. This thrust must not exceed the product of the normal pressure *oj* and the coefficient of friction between the wall and its foundation; otherwise expressed, the angle *roj* must not exceed the angle of friction between the wall and its foundation unless some external means, such as earth placed in front of the wall at the base, is employed to strengthen the wall against sliding.

Ordinarily, the friction of the back filling is disregarded in determining the resistance to sliding. The neglect of this factor of stability against sliding is warranted in the majority of cases, because, the thickness of wall required for stability against overturning gives ample weight to resist sliding, and the added help of the filling in front of the foundation, required on account of frost and other surface influences, is generally sufficient to make up for the neglected friction of the filling. It is, however, sometimes advisable to take it into account, for though latent when there is no motion of the wall, the instant that the wall begins to move, or is about to do so, whether by overturning or by sliding, the filling begins to slide—or is ready to do so—down the back of the wall, and brings the friction into action.

BATTERED BACK

For a wall with a battered back, Fig. 2, the line of maximum pressure is the one bisecting the angle *ibt* formed by the vertical *bt* and the slope of repose. The prism of maximum pressure is one whose cross-section is *cbs*. The point of application *e* of the force *P* is such that $be = \frac{1}{3}bc$; *X* is perpendicular to *bc*, and its magnitude is determined as follows: Calculate the weight of the prism of maximum pressure for a unit length and lay it off to any convenient scale on a vertical line drawn through *e*. Let this weight be represented by *el*. Through *l* draw *lx*, parallel to *bs* and through *e* draw *ex* perpendicular to *bc*; *ex* gives the magnitude and posi-

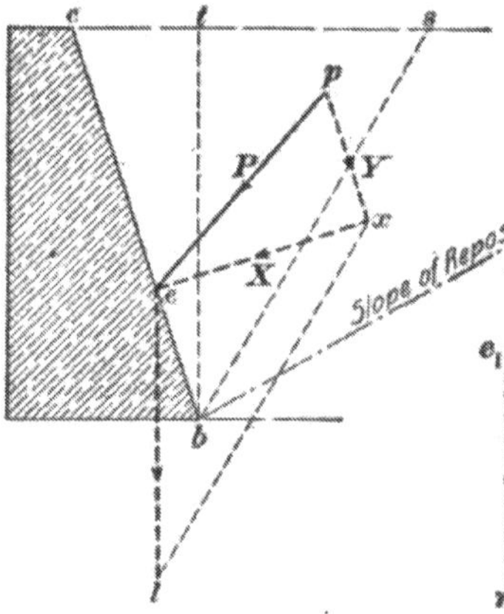

FIG. 2

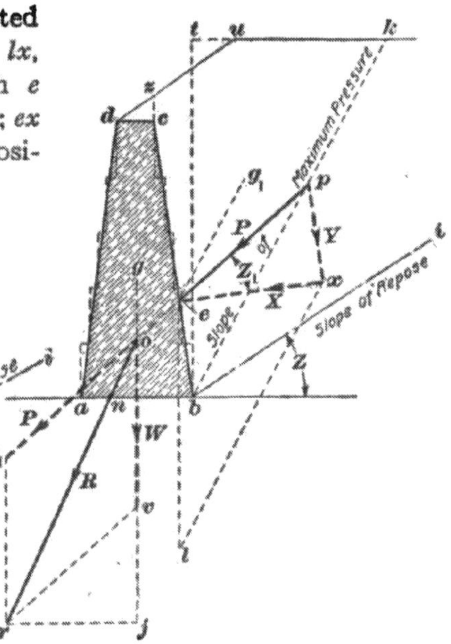

FIG. 3

tion of *X*. Then, as before, on a line at right angles to *ex*, lay off $xp = fX = Y$; *ep* then determines the position and magnitude of *P*, and *R* may now be found as in the case of a wall with a vertical back.

SURCHARGED WALL

With a surcharged wall, Fig. 3, the line of maximum pressure is determined as before, and the maximum pressure is considered as being caused by the earth lying between the broken line *bcs* and the line of maximum pressure *bk*. The general method of procedure is the same as previously described, except that in this case, *be* is no longer equal to $\frac{1}{3}$ *bc*, and the point of application *e* of the pressure *P* is located by determining the center of gravity g_1 of the area *zukbcz* and drawing the line g_1e parallel to the line of maximum pressure. The intersection *e* of this line with the back of the wall is the required point of application of *P*. Fig. 3 shows all the remaining steps that should be taken in the analysis of the retaining wall, which are the same as those already described.

SUPERIMPOSED LOADS

In case of loads resting on the top of the back filling, they must be added to the weight of the prism of maximum pressure, or of the body of earth causing the maximum pressure. The method of procedure is the same as the one already given for a surcharged wall, the modification being only in the manner of locating the center of gravity g_1, which, in this case, is the center of gravity of the system of bodies consisting of the earth filling and of the loads.

EMPIRICAL RULES

All the theories of the equilibrium and stability of retaining walls are based on assumptions that have not been conclusively proved. For this reason, empirical rules based on observation and experience are extensively employed in practice. Of these rules, those by John C. Trautwine are most widely used. They are given with slight modification in the following paragraph.

Rules for Vertical Walls.—For a vertical wall resting on a foundation of masonry suitably enlarged for a proper distribution of the load on the soil, with the top of the fill leveled off at the top of the wall, the ratio of the thickness to the height of the wall should be .35 for a wall of cut stone, or of first-class large-ranged rubble, in mortar, or of concrete; .4 for a wall of good common rubble or brick, in mortar; and .5 for a wall of dry rubble.

For a wall with a battered or stepped back, Trautwine recommends using the same average thickness as for a vertical wall, increasing the base by the same amount that the top width is decreased.

A wall with a battered face may be made to give the same stability with a materially smaller volume and average thickness than would be required if a vertical wall were used.

Rules for Surcharged Walls.—When the surcharge runs over the top of the wall, as in Fig. 3, there is a slight increase in

SURCHARGED VERTICAL WALLS—RATIO OF THICKNESS TO HEIGHT

Ratio of Surcharge to Height of Wall	Toe of Slope at Back of Wall			Toe of Slope at Front of Wall		
	Cut Stone	Rubble, or Brick in Mortar	Dry Rubble	Cut Stone	Rubble, or Brick in Mortar	Dry Rubble
.0	.35	.40	.50	.35	.40	.50
.1	.42	.47	.57	.42	.47	.57
.2	.46	.51	.61	.46	.51	.61
.3	.49	.54	.64	.49	.55	.66
.4	.51	.56	.66	.53	.60	.72
.5	.52	.57	.67	.58	.65	.79
.6	.54	.59	.69	.62	.70	.85
.7	.55	.60	.70	.65	.74	.91
.8	.56	.61	.71	.67	.77	.96
.9	.57	.62	.72	.69	.80	1.00
1.0	.58	.63	.73	.71	.82	1.04
2.0	.62	.67	.77	.81	.96	1.26
3.0	.63	.68	.78	.85	1.02	1.35
5.0	.64	.69	.79	.88	1.07	1.44
25.0	.68	.73	.83	.92	1.11	1.50

the weight resisting overturning by the addition of the triangle of earth *dcz*, as well as the larger increase in the weight of the wedge of backing pressing against the back of the wall. For a height of surcharge less than about a quarter of the height of the wall, the additional weight of the filling resting on the top of the wall will offset the extra weight of the overturning wedge; but, as the height of the surcharge increases, the overturning pressure increases rapidly, while the increased resistance due to the earth resting on the top of the wall changes only slightly with the increase in thickness of the wall. The preceding table shows the proper ratios of thickness to height for vertical walls with various amounts of surcharge. After ascertaining the thickness of the vertical wall required for restraining a surcharge bank, the form of the wall may be altered to give a battered face or back, or both, in the same way as if the top of the backing were level with the top of the wall.

HYDROSTATICS

DEFINITIONS AND GENERAL PRINCIPLES

Hydrostatics treats of the equilibrium of liquids, and of their pressures on the walls of vessels containing them and on submerged surfaces.

Liquid Bodies.—A liquid is a body whose molecules change their relative positions easily, being, however, held in such a state of aggregation that, although the body can freely change its shape, it retains a definite and invariable volume, provided the pressure and temperature are not changed. Water and alcohol are examples of liquid bodies.

A *perfect liquid* is a liquid without internal friction; that is one whose particles can move on one another with absolute freedom. On account of this characteristic property, a perfect liquid offers no resistance to a change of form.

A *viscous liquid* is a liquid that offers resistance to rapid change of form on account of internal friction, or *viscosity*. Tar, molasses, and glycerine are examples of viscous liquids.

All liquids are more or less viscous. For the purposes of hydrostatics, however, water, which is the liquid mainly dealt with, may be treated as a perfect liquid, its viscosity at ordinary temperatures being too small to be taken into account.

Compressibility.—All liquids offer great resistance to change in volume; that is, they can be compressed but little. Under the pressure of 1 atmosphere (about 14.7 lb. per sq. in.), water is compressed about $\frac{1}{20000}$ of its original volume. For engineering purposes it may be assumed that water is incompressible.

Pascal's Law.—*The pressure per unit of area exerted anywhere on a mass of liquid is transmitted undiminished in all directions; and any surface in contact with the liquid will be subjected to this pressure in a direction at right angles to the surface.*

PRESSURE OF LIQUIDS ON SURFACES

General Principles.—The pressure of a liquid on any surface immersed in it is equal to the weight of a column of the liquid whose base is the surface pressed and whose height is the perpendicular depth of the center of gravity of the surface below the level of the liquid. The pressure thus exerted is not dependent on the shape or size of the vessel containing the liquid, nor on the form of the surface, whether it be flat or curved; nor on the position of the surface, whether it be vertical, horizontal, or inclined. The pressure is normal to the immersed surface.

Let, in the accompanying illustrations, the depth of water in each dam be 12 ft. Consider a portion of the embankment or wall 1 ft. long. Then in Fig. 1 the area of the immersed surface is 12 sq. ft.; the distance of the center of gravity of the surface from the level of the water is 6 ft., and assuming the weight of water as 62.5 lb. per cu. ft., the total pressure on the surface AB is $12\times6\times62.5=4{,}500$ lb. In Figs. 2 and 3 the walls, being inclined, expose a greater surface to pressure, say 18 ft. from A to B. Then the total pressure is $18\times1\times6\times62.5=6{,}750$ lb. These pressures may be considered as forces acting, in each case, normally to the surface AB. The point of application C of the resultant pressure on the wall, called the *center of pressure*, is not at the center of gravity of the submerged area, but at one-third of the distance AB from the bottom; so that in each case $CB=\frac{1}{3}AB$.

In Fig. 1 the resultant pressure is horizontal, producing an overturning moment about the outer toe, and also tending to slide the wall along its base. In Figs. 2 and 3 the resultant pressure may be resolved into two components, one horizontal and the other vertical. The horizontal component in both cases is the same as the total pressure in Fig. 1, whereas in Fig. 2, the vertical component tends to counteract the effect of the horizontal component, and in Fig. 3, it tends to lift the wall.

Pressure on the Upper Surface of a Liquid.—If the surface of a liquid is subjected to an *external pressure*, this pressure is

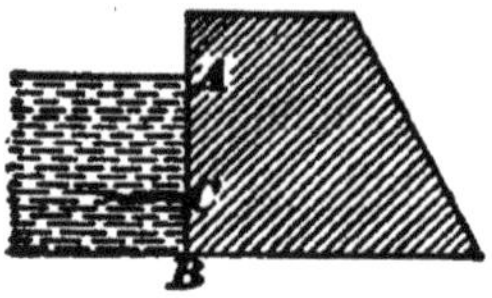

FIG. 1

FIG. 2

FIG. 3

transmitted undiminished to all parts of the enclosing vessel, and must be added to the pressure due to the weight of the liquid.

The *atmospheric pressure* is the external pressure due to the weight of the air, and may be taken to have an average value of 14.7 lb. per sq. in.

Buoyant Effort.—When a solid body is immersed in a liquid, a buoyant effort equal to the weight of the liquid displaced acts upwards and opposes the action of gravity. The weight of a body, as shown by a scale, is decreased by an amount equal to the buoyant effort, that is, by an amount equal to the weight of liquid displaced. This principle is called the *principle of Archimedes*, from the name of its discoverer.

SPECIFIC GRAVITY

The *specific gravity* of a body is the ratio between its weight and the weight of a like volume of distilled water at a temperature of 39.2° F. The weight of 1 cu. ft. of water at 39.2° F., which is the temperature of its maximum density, is 62.425 lb. For nearly all engineering purposes 62.5 lb. is used as an approximate value.

Since a column of water 1 sq. in. in cross-section and 1 ft. high is $\frac{1}{144}$ cu. ft., its weight is $62.5 \div 144 = .434$ lb.

The accompanying table gives the specific gravities and weights per cubic inch of a great variety of substances

SPECIFIC GRAVITIES OF VARIOUS SUBSTANCES

Name of Substance	Specific Gravity	Weight per Cubic Inch Pounds
Acid, acetic	1.062	.0384
Acid, muriatic	1.200	.0434
Acid, nitric	1.217	.0440
Acid, phosphoric	1.558	.0563
Acid, sulphuric	1.841	.0665
Alcohol, commercial	.833	.0301
Alcohol, pure	.792	.0286
Alder	.800	.0289
Aluminum	2.660	.0960
Antimony	6.712	.2420
Apple tree	.793	.0287
Asbestos, starry	3.073	.1110
Ash, the trunk	.845	.0305
Atmospheric air	.0012	
Bay tree	.822	.0297
Beech	.852	.0308
Beer, lager	1.034	.0374
Beeswax	.965	.0349
Bismuth	9.746	.3520
Borax	1.714	.0619
Box, Brazilian red	1.031	.0372
Box, Dutch	1.328	.0480
Box, French	.960	.0347
Brass, common	8.500	.3070
Brick	2.000	.0723
Bronze, gun-metal	8.500	.3070
Butter	.942	.0340
Cedar, American	.561	.0203
Cedar, Palestine	.613	.0221
Cedar, wild	.596	.0215
Chalk	2.784	.1006
Champagne	.997	.0360
Charcoal	.441	.0159
Cherry tree	.672	.0243
Cider	1.018	.0368
Clay	1.900	.0686
Coal, anthracite	1.640 1.436	.0592 .0519
Coal, bituminous	1.350	.0488
Coal, Maryland	1.355	.0490
Coal, Newcastle	1.270	.0459
Coal, Scotch	1.300	.0470

TABLE—(*Continued*)

Name of Substance	Specific Gravity	Weight per Cubic Inch Pounds
Common soil	1.984	.0717
Copper, pure	8.788	.3170
Copper, wire and rolled	8.878	.3210
Coral, red	2.700	.0975
Cork	.250	.0090
Earth, loose	1.360	.0491
Ebony, American	1.220	.0441
Egg	1.090	.0394
Elder tree	.695	.0251
Elm	.560	.0202
Emery	4.000	.1450
Ether, sulphuric	.739	.0267
Fat	.923	.0333
Filbert tree	.600	.0217
Fir, female	.498	.0180
Fir, male	.550	.0199
Flint, black	2.582	.0933
Flint, white	2.594	.0937
Gold, hammered	19.361	.6990
Gold, pure cast	19.258	.6960
Gold, 22 carats fine	17.486	.6320
Glass, bottle	2.732	.0987
Glass, flint	3.500	.1260
Glass, green	2.642	.0954
Glass, white	2.900	.1050
Granite, Patapsco	2.640	.0954
Granite, Quincy	2.652	.0958
Granite, Scotch	2.625	.0948
Granite, Susquehanna	2.704	.0977
Grindstone	2.143	.0774
Gum arabic	1.452	.0525
Gunpowder, loose	.900	.0325
Gunpowder, shaken	1.000	.0361
Gypsum, opaque	2.168	.0783
Hazel	.600	.0217
Honey	1.450	.0524
Human blood	1.054	.0381
India rubber	.933	.0337
Iron, cast	7.207	.2600
Iron, hammered	7.789	.2810
Iron, pure	7.768	.2810
Iron, wrought and rolled	7.780	.2810

TABLE—(*Continued*)

Name of Substance	Specific Gravity	Weight per Cubic Inch Pounds
Ivory	1.822	.0659
Lard	.947	.0342
Lead, hammered	11.388	.4110
Lead, pure	11.330	.4090
Lemon tree	.703	.0254
Lignum vitæ	1.330	.0481
Limestone	2.700	.0980
Linden tree	.604	.0218
Logwood	.913	.0330
Mahogany, Honduras	.560	.0202
Maple	.790	.0285
Marble, African	2.708	.0978
Marble, common	2.686	.0970
Marble, Egyptian	2.668	.0964
Marble, Parian	2.838	.1025
Marble, white Italian	2.708	.0978
Mercury, at +32° F	13.619	.4920
Mercury, at 60° F	13.580	.4910
Mercury, at 212° F	13.375	.4830
Mercury, solid, at −40° F	15.632	.5650
Mica	2.800	.1012
Milk	1.032	.0373
Mulberry	.897	.0324
Niter	1.900	.0686
Oak	.950	.0343
Oil, linseed	.940	.0340
Oil, olive	.915	.0331
Oil, turpentine	.870	.0314
Oil, whale	.932	.0337
Orange tree	.705	.0255
Pear tree	.661	.0239
Pearl, Oriental	2.650	.0957
Phosphorus	1.770	.0639
Pine, southern	.720	.0260
Pine, white	.400	.0144
Poplar	.383	.0138
Poplar, white Spanish	.529	.0191
Plaster of Paris	1.872 2.473	.0676 .0893
Platinum, hammered	20.337	.7350
Platinum, rolled	22.009	.8190
Platinum, wire	21.042	.7600

TABLE—(*Continued*)

Name of Substance	Specific Gravity	Weight per Cubic Inch Pounds
Proof spirit	.925	.0334
Quartz	2.660	.0961
Quicklime	1.500	.0542
Rotten stone	1.981	.0716
Salt, common	2.130	.0769
Saltpeter	2.090	.0755
Sand	2.650	.0957
Sassafras	.482	.0174
Shale	2.600	.0939
Silver, hammered	10.511	.3800
Silver, pure	10.474	.3780
Slate	2.800	.1012
Spermaceti	.943	.0341
Spruce	.500	.0181
Spruce, old	.460	.0166
Steel, cast	7.919	.2860
Steel, common soft	7.833	.2830
Steel, hardened and tempered	7.818	.2820
Stone, Bristol	2.510	.0907
Stone, common	2.520	.0910
Stone, mill	2.484	.0897
Stone, paving	2.416	.0873
Sugar	1.605	.0580
Sulphur, native	2.033	.0734
Talc, black	2.900	.0105
Tallow, calf	.934	.0337
Tallow, sheep	.924	.0334
Tallow, ox	.923	.0333
Tin, English	7.021	.2630
Tin, from Böhmen	7.312	.2640
Vinegar	1.080	.0390
Walnut	.610	.0220
Water, distilled (62.425 lb. per cu. ft.)	1.000	.0361
Water, sea	1.030	.0372
Wine	.992	.0358
Zinc, rolled	7.101	.2600

HYDRAULICS

GENERAL PRINCIPLES

Hydraulics treats of liquids in motion, particularly of the flow of water through orifices, pipes, and channels.

The quantity of water, in cubic feet, flowing through a channel or a pipe in 1 sec. is called the *discharge* of the channel or the pipe in cubic feet per second and is denoted by Q. It is equal to the mean, or average, velocity of flow through the given section multiplied by its area, or

$$Q = vA,$$

in which v is the mean velocity, in feet per second, and A the area, in square feet. If the area of the channel or pipe varies, the mean velocities vary inversely as the corresponding cross-sections; or,

$$\frac{v_a}{v_b} = \frac{A_b}{A_a},$$

A_a, v_a and A_b, v_b denoting, respectively, areas and corresponding velocities at two different cross-sections.

Hydrostatic Head and Pressure Head.—When water contained in any vessel or pipe discharges freely into the atmosphere, the velocity of discharge v, in feet per second, if frictional and other resistances are neglected, is equal to

$$v = \sqrt{2gh}$$

in which h is the vertical distance in feet of the point of discharge from the level of the water, and $g = 32.16$. This velocity is produced by the pressure due to the weight of a column of water of the height h, the latter being called the *static* or *hydrostatic head.*

The water in the pipe or vessel may be subjected to an external pressure, thus giving an intensity of pressure greater than that due to the static head, or owing to losses during the flow, it may have an intensity of pressure which is smaller than that due to the static head. Let p be the intensity of pressure in pounds per square inch, and v' the velocity due to this pressure; then,

$$v' = \sqrt{2g\frac{p}{w}}$$

in which w is the weight of a column of water 1 sq. in. in cross-section and 12 in. high, usually taken as .434 lb. The term $\frac{p}{w}$ represents the head necessary to produce the pressure p and is called the *pressure head.*

The pressure head in a water pipe can be measured by the height to which the water will rise in a tube inserted in the pipe. Such a tube or gauge is called a *piezometer* or *piezometric tube.*

Velocity Head.—When water in a pipe or a channel is flowing to a level h ft. lower than the starting point, if frictional and other resistances are not considered, the velocity attained during the flow is $v = \sqrt{2gh}$, which is the same as the velocity attained by a body falling through a height h. Solving for h,

$$h = \frac{v^2}{2g}$$

The expression $\frac{v^2}{2g}$ is called the *velocity head.*

Loss of Head.—Owing to frictional and other resistances, a loss of energy occurs in flowing water, thus reducing the theoretical velocity of the flow, and, consequently, the discharge. This loss is usually expressed as a fractional factor of the theoretical velocity head $\frac{v^2}{2g}$, the factor being called the *coefficient of hydraulic resistance.*

Flow of Water Through a Standard Orifice.—When the water in flowing through a hole touches the opening on the inside edges only, the hole is a *standard orifice.* The theoretical discharge is

$$Q = Av$$

in which A is the area of the orifice, and $v = \sqrt{2gh}$, h being the head or the distance of the center of the orifice from the level surface of the water. The actual discharge is reduced on account of frictional resistances and contraction of the jet. The friction reduces the velocity to 98% of the theoretical velocity and the contraction reduces the cross-section of the issuing jet to 62% of the area of the orifice. The actual discharge is, therefore, $Q_a = .98 \times .62\ Q = .61\ A\ \sqrt{2gh}$.

WEIRS

A *weir* is a dam or obstruction placed across a stream for the purpose of diverting the water and causing it to flow through a channel of known dimensions, which channel may be a notch or opening in the obstruction itself. The notch is usually rectangular in form.

There are two general types of weirs, namely, those with end contractions, as in Fig. 1 (*a*), and those without end contractions, as in Fig. 1 (*b*).

Crest of the Weir.—The edge of the notch over which the water flows, as shown in cross-section at *a*, Fig. 1 (*c*) and (*d*), is

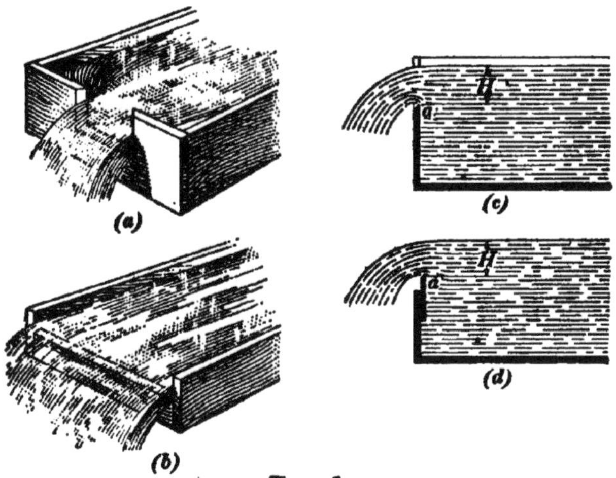

FIG. 1

called the *crest* of the weir. In all weirs, the inner edge of the crest is made sharp, so that, in passing over it, the water touches along a line. The same statement applies to the inner edge of both the top and the ends of the notch in weirs with end contractions. For very accurate work, the edges of the notch should be made with a thin plate of metal having a sharp inner edge, as shown in Fig. 1 (*d*); but for ordinary work the edges of the board in which the notch is cut may be chamfered off to an angle of about 30°, as shown at (*c*). The top edge of the notch must be straight and set perfectly level, and the sides must be set carefully at right angles to the top. Means for admitting air under the falling sheet of water must be made; otherwise, there will be formed a partial vacuum that tends to increase the discharge. The sides of a weir without end contractions should be smooth and straight and should project a slight distance beyond the crest.

Standard Dimensions for Weirs.—The distance from the crest of the weir to the bottom of the feeding canal or reservoir should be at least three times the head H, Fig. 1 (*c*) and (*d*); and, with a weir having end contractions, the distance from the vertical edges to the sides of the canal should also be at least three times the head. The water must approach the weir quietly and with little velocity; theoretically, it should have no velocity. It is often necessary to place one or more sets of baffle boards or planks across the stream at right angles to the flow, and at varying depths from the surface, to reduce the velocity of the water as it approaches the weir.

Theoretical and Actual Discharge of Weirs.—The *theoretical discharge* of a weir is

$$Q = 5.347\, bH^{\frac{3}{2}}$$

in which b is the length of the crest and H is the effective depth producing the discharge. When the velocity of approach is inappreciable, the effective depth is the distance from the crest of the weir to the surface of the water at a point up-stream beyond the curve assumed by the flowing water as it approaches the weir; but, when the velocity of approach v is considerable,

$$Q = 5.347\, b\, (H+h)^{\frac{3}{2}}$$

in which $h = \dfrac{v^2}{2g}$.

The *actual discharge* of weirs is, when the velocity of approach is not considered,

$$Q = 3.33\left(b - \frac{n}{10}H\right) H^{\frac{3}{2}}$$

and when the velocity of approach is considered

$$Q = 3.33\left(b - \frac{n}{10}H\right)\left[(H+h)^{\frac{3}{2}} - h^{\frac{3}{2}}\right]$$

In these formulas n denotes the number of end contractions; hence, for a weir with two end contractions, $n = 2$; for a weir with one end contraction, $n=1$; and for a weir with no end contractions, $n=0$. In the last case, the two preceding formulas become, respectively,

$$Q=3.33\,bH^{\frac{3}{2}}$$

and
$$Q=3.33b\left[(H+h)^{\frac{3}{2}}-h^{\frac{3}{2}}\right]$$

The velocity of approach can be determined by first finding Q from the formula $Q=3.33\,bH^{\frac{3}{2}}$, and dividing it by the area of the cross-section of the channel; the quotient will be the velocity of approach v and h will equal $.01555\,v^2$.

EXAMPLE 1.—A weir with end contractions is 5 ft. long and the measured head is .872 ft. Calculate the discharge on the assumption that the velocity of approach is negligible.

SOLUTION.—Substituting the given values in the proper formula, $Q=3.33\times(5-\frac{2}{10}\times.872)\times.872^{\frac{3}{2}}=13.085$ cu. ft. per sec.

The preceding formulas are known as Francis's formulas and are recommended for heads from 5 to 19 in. For lower heads, the formula of Fteley and Stearns, which follows, is recommended:

$$Q=3.31b(H+1.5h)^{\frac{3}{2}}+.007b$$

For higher heads, Bazin's formula is recommended:

$$Q=\left(.405+\frac{.00984}{H}\right)\left[1+.55\left(\frac{H}{H+p}\right)^2\right]bH\sqrt{2gH}$$

The last two formulas are applicable only to weirs with no end contractions. In these formulas, p is the distance from the bottom of the channel to the crest; the other letters have the same significance as before.

Triangular Weir.—The form of weir shown in Fig. 2 may be used for small flows where the head lies between the limits of .02 and 1 ft. For a right-angled weir with sharp inner edges,

$$Q=2.54H^{\frac{5}{2}}$$

FIG. 2

EXAMPLE.—Calculate the discharge of a triangular weir whose effective head is 9 in.

SOLUTION.—Substituting the given values in the formula,

$$Q=2.54\times.75^{\frac{5}{2}}=1.24 \text{ cu. ft. per sec.}$$

FLOW OF WATER IN CHANNELS

A *channel* is the bed of a long body of water flowing under the action of gravity. An artificial channel whose bed is formed by the natural soil is called a *canal*, and when the bed is artificial, like a flume or a sewer pipe, it is called a *conduit*. A *ditch* is a small canal.

The *slope* s of a channel is the ratio of the fall h to the length l in which the fall occurs; or

$$s=\frac{h}{l}$$

The *wetted perimeter* of a cross-section of a channel is the part of the boundary in contact with the water. The *hydraulic radius* of a channel is the ratio of the area of the cross-section of the water in the channel to the wetted perimeter.

Chezy's Formula.—The fundamental formula for the velocity of flow in a channel is $v=c\sqrt{rs}$,
in which s is the slope of the channel; r, the hydraulic radius; and c, a variable coefficient whose value is given by Kutter's formula, which is,

$$c=\frac{23+\frac{1}{n}+\frac{.00155}{s}}{.5521+\left(23+\frac{.00155}{s}\right)\frac{n}{\sqrt{r}}}$$

In this formula n is the *coefficient of roughness*, whose values are as follows:

Character of Channel	*Value of n*
Clean well-planed timber	.009
Clean, smooth, glazed iron and stoneware pipes	.010
Masonry smoothly plastered with cement, and for very clean smooth cast-iron pipe	.011
Unplaned timber, ordinary cast-iron pipe, and selected pipe sewers, well laid and thoroughly flushed	.012
Rough iron pipes and ordinary sewer pipes laid under the usual conditions	.013
Dressed masonry and well-laid brickwork	.015
Good rubble masonry and ordinary rough or fouled brickwork	.017

Character of Channel	*Value of n*
Coarse rubble masonry and firm compact gravel.	.020
Well-made earth canals in good alinement......	.0225
Rivers and canals in moderately good order and perfectly free from stones and weeds.........	.025
Rivers and canals in rather bad condition and somewhat obstructed by stones and weeds...	.030
Rivers and canals in bad condition, overgrown with vegetation and strewn with stones and other detritus, according to condition.	.035 to .050

EXAMPLE.—Find the discharge of a rough-plank sluice 24 in. wide, when the depth of the water in the sluice is 15 in. and the fall 3 in. in 100 ft.

SOLUTION.—The slope $s=.25\div 100=.0025$; the wetted perimeter $p=2+(2\times 1.25)=4.5$ ft.; and the area A of the water cross-section $=2\times 1.25=2.5$ sq. ft. The hydraulic radius is, therefore, $r=2.5\div 4.5=.5556$. The value of n for unplaned timber is .012; therefore,

$$c=\frac{23+\frac{1}{.012}+\frac{.00155}{.0025}}{.5521+\left(23+\frac{.00155}{.0025}\right)\times\frac{.012}{\sqrt{.5556}}}=114.7$$

Substituting the values found in Chezy's formula,

$$v=114.7\sqrt{.5556\times .0025}=4.27 \text{ ft. per sec.}$$

Therefore, the discharge is

$$Q=Av=2.5\times 4.27=10.675 \text{ cu. ft. per sec.}$$

Discharge of Large Streams.—The discharge of a large body of water, when it is impracticable to construct a weir, is determined by measuring, on one hand, the mean velocity v at a cross-section of flowing water by means of floats or by the use of special instruments, and, on the other hand, by ascertaining the area A of that cross-section. Then, the discharge

$$Q=Av$$

The *current meter* affords the most convenient and accurate method of measuring velocities of a stream. One form of this instrument is shown in Fig. 1. The number of revolutions of the buckets b depending on the velocity of the flow is recorded electrically on the dials m and n. The relation between this

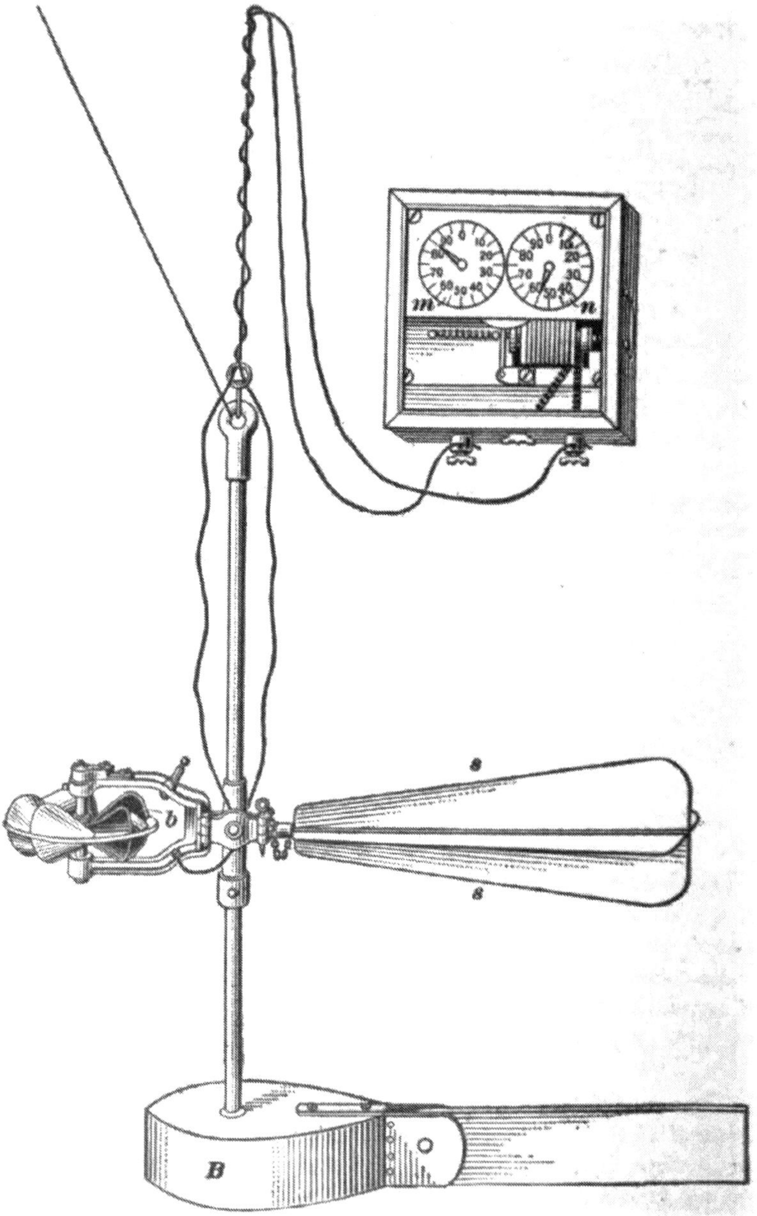

FIG. 1

number and the velocity of current, called the rating of the instrument, is usually effected by drawing the meter at a given speed through still water. The part *s* is a rudder and the part *B* a ballast for use in very deep water. The approximate mean velocity of flow at a cross-section of a stream may be determined by measuring the velocity of the depth at .6 below the surface at the deepest part of the cross-section. When accurate results are required, measurements should be taken at different parts of the section as well as at different depths of the same section and the average calculated. The ordinary method of procedure is as follows:

A range at right angles to the stream is selected (see Fig. 2) and divided into any desired number of parts. Soundings are taken along the points of division, and at the same points the

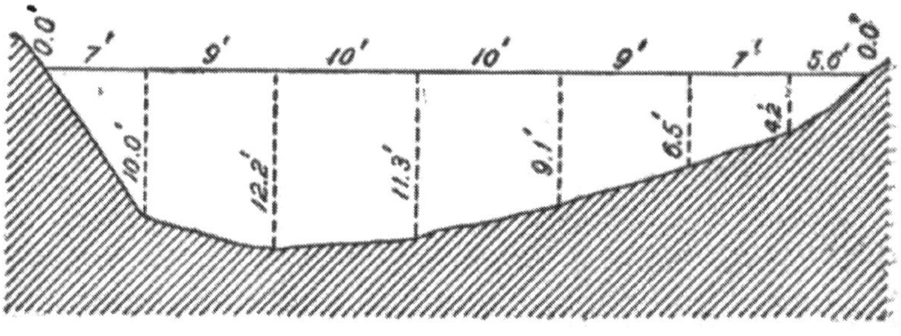

FIG. 2

mean velocities are determined by moving the meter vertically at a uniform rate from the surface of the water to the bottom and back to the surface. The mean velocity of a division multiplied by the corresponding area gives the partial discharge of that division. The sum of the partial discharges is the total discharge of the river.

Determination of Discharge by Floats.—The discharge of streams is best determined by means of *rod floats*, which are wooden or hollow tin cylinders weighted at the lower end. They should be placed as near the bottom of the stream as possible. A suitable portion of the stream between two cross-sections at right angles to it is selected. The sections are divided into a suitable number of parts, soundings are taken at each division point, and the float is timed between the corresponding division points. The partial areas of the two cross-sections are determined, and the mean of the areas of the corresponding division multiplied by the corresponding velocity will give the partial discharge of that division. The sum of the partial discharges is the total discharge of the river.

The mean velocity as observed by a rod float is to be taken as the actual mean velocity only when the float is made to pass close to the bottom. When the float is immersed only to a depth i, the actual mean velocity is,

$$V_a = V_m\left[1.0 - 0.116\left(\frac{d-i}{d} - .1\right)\right]$$

in which V_m is the measured mean velocity and d the depth of water at which the measurement was taken.

When a surface float is used, the actual mean velocity may be obtained approximately by multiplying the measured mean velocity by .8.

FLOW OF WATER IN PIPES

In determining the flow of water in pipes, the discharge in cubic feet per second is $Q = .7854d^2v$, in which d is the diameter of the pipe in feet and v the actual velocity, in feet per second. The theoretical velocity is $v_t = \sqrt{2gh}$, h being the *static head*. This head h, which is available before the flow begins, sustains losses during the flow due to skin friction between the water and the pipe, to resistances at entrance, to bends and elbows, and to other causes, resulting in a reduction of the theoretical velocity. The actual velocity is,

$$v = \sqrt{\frac{2gh}{1 + f \times \frac{l}{d} + c}}$$

in which l is the length and d the diameter of the pipe, both in feet; f, the coefficient of resistance for friction; and c, the sum of all coefficients for losses due to entrance, bends, valves, etc. For a pipe whose length is more than 1,000 times its diameter, called a *long pipe*, the value of $1+c$ is very small in comparison

VALUES OF THE COEFFICIENT OF FRICTION f FOR SMOOTH CAST- OR WROUGHT-IRON PIPES

Diameter Inches	Velocity, in Feet per Second									
	1	2	3	4	5	6	8	10	12	15
1/2	.0440	.0345	.0301	.0289	.0282	.0276	.0265	.0258	.0252	.0248
3/4	.0408	.0332	.0298	.0284	.0277	.0271	.0261	.0254	.0248	.0244
1	.0380	.0324	.0294	.0281	.0274	.0268	.0258	.0251	.0246	.0241
1 1/4	.0357	.0316	.0289	.0277	.0270	.0264	.0255	.0248	.0243	.0239
1 1/2	.0336	.0303	.0283	.0273	.0266	.0260	.0251	.0244	.0240	.0236
2	.0316	.0292	.0277	.0268	.0262	.0256	.0247	.0240	.0236	.0233
3	.0297	.0280	.0268	.0260	.0254	.0249	.0240	.0234	.0233	.0229
4	.0285	.0271	.0260	.0252	.0247	.0242	.0235	.0229	.0224	.0221
6	.0274	.0259	.0249	.0243	.0238	.0233	.0225	.0220	.0216	.0213
8	.0264	.0250	.0240	.0234	.0229	.0225	.0218	.0212	.0209	.0206
10	.0257	.0244	.0234	.0227	.0223	.0219	.0213	.0208	.0205	.0202
12	.0250	.0237	.0228	.0221	.0217	.0214	.0208	.0203	.0200	.0197
14	.0244	.0230	.0222	.0216	.0212	.0208	.0201	.0197	.0195	.0192
16	.0235	.0224	.0215	.0210	.0206	.0203	.0196	.0193	.0191	.0189
18	.0228	.0217	.0209	.0204	.0200	.0198	.0193	.0189	.0187	.0185
20	.0222	.0212	.0204	.0199	.0196	.0193	.0188	.0185	.0183	.0180
24	.0210	.0200	.0193	.0189	.0186	.0184	.0180	.0177	.0176	.0173
30	.0197	.0188	.0181	.0176	.0174	.0172	.0169	.0166	.0165	.0163
36	.0185	.0177	.0170	.0166	.0164	.0162	.0159	.0157	.0156	.0154
42	.0168	.0163	.0159	.0157	.0155	.0154	.0151	.0149	.0148	.0147
48	.0158	.0154	.0150	.0148	.0146	.0145	.0143	.0141	.0140	.0139
54	.0149	.0146	.0143	.0140	.0138	.0136	.0134	.0132	.0132	.0131
60	.0141	.0138	.0136	.0133	.0132	.0130	.0128	.0126	.0125	.0125
72	.0134	.0131	.0128	.0127	.0125	.0124	.0122	.0120	.0120	.0119

with the loss due to friction and is therefore neglected. The formula used for long pipes, is $v = \sqrt{\frac{2ghd}{fl}}$

The value of f depends not only on the roughness of the pipe, but also on its diameter and the velocity of flow. Its values for a smooth pipe are given in the preceding table. For rough pipes, these values should be multiplied by 2.

COEFFICIENTS FOR ANGULAR BENDS
(a = angle of bend in degrees)

a	10°	20°	40°	60°	80°	90°	100°	110°	120°	130°	140°	150°
k_s	.017	.046	.139	.364	.74	.984	1.26	1.56	1.86	2.16	2.43	2.81

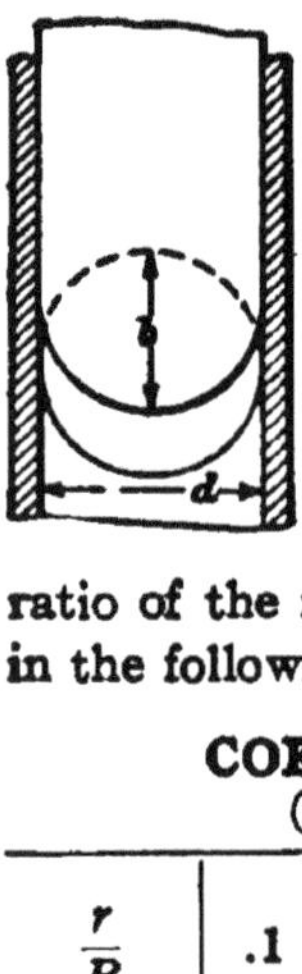

When the pipe is shorter than 1,000 diameters and the first of the preceding formulas is used, the component parts forming the value of c must be ascertained and the results added and substituted in the formula. The coefficient k_s for angular bends can be taken from the accompanying table giving its value for different angles. The coefficient for circular bends is $c_e = \frac{a}{180} k_c$, in which a is the angle of the bend and k_c is a constant depending on the ratio of the radius of the pipe to that of the bend and is given in the following table for circular bends.

COEFFICIENTS FOR CIRCULAR BENDS
(r = radius of pipe. R = radius of bend)

$\frac{r}{R}$	.1	.2	.3	.4	.5	.6	.7	.8	.9	1.0
k_c	.131	.138	.158	.206	.294	.440	.661	.977	1.408	1.978

Valves.—With reference to the accompanying illustration, the coefficients of resistance j for different ratios of b to d are as follows:

COEFFICIENTS FOR VALVES

$\frac{b}{d}$	0	$\frac{1}{8}$	$\frac{1}{4}$	$\frac{3}{8}$	$\frac{1}{2}$	$\frac{5}{8}$	$\frac{3}{4}$	$\frac{7}{8}$
j	.0	.07	.26	.81	2.1	5.5	17	98

Sudden Change of Section.—When the area of a cross-section of a pipe is suddenly enlarged or contracted, in the latter case, by inserting a smaller pipe or by an obstruction, the coefficient of hydraulic resistance is

$$\left(\frac{A}{a}-1\right)^2$$

in which A is the area of the larger cross-section, and a that of the smaller one.

HYDRAULIC TABLES FOR LONG PIPES

The following table is compiled for pipes whose lengths are more than 1,000 times their diameter. The data given are: (1) the velocity of flow in feet per second; (2) the corresponding slope $\frac{h}{l}$, that is, the available head per unit length of pipe; and (3) the discharges, in cubic feet per second, both for a clean and an extremely foul pipe, thus giving the extreme limits between which the discharge may vary

HYDRAULIC TABLE FOR CAST-IRON PIPES

4-In. Pipe				6-In. Pipe			
v Ft. per Sec.	$s=\frac{h}{l}$	Discharge, Cu. Ft. per Sec.		*v* Ft. per Sec.	$s=\frac{h}{l}$	Discharge, Cu. Ft. per Sec.	
		Clean Pipe	Foul Pipe			Clean Pipe	Foul Pipe
.2	.0000563	.01745	.01234	.2	.0000360	.03927	.02777
.4	.0002215	.03491	.02468	.4	.0001419	.07854	.05554
.6	.0004916	.05236	.03702	.6	.0003143	.11781	.08330
.8	.0008621	.06981	.04936	.8	.0005516	.15708	.11107
1.0	.0013283	.08727	.06170	1.0	.0008508	.19635	.13884
1.2	.0018913	.10472	.07405	1.2	.0012071	.23562	.16661
1.4	.0025487	.12217	.08639	1.4	.0016236	.27489	.19437
1.6	.0032955	.13962	.09873	1.6	.0020951	.31416	.22214
1.8	.0041346	.15708	.11107	1.8	.0026274	.35343	.24991
2.0	.0050596	.17453	.12341	2.0	.0032238	.39270	.27768
2.2	.0060679	.19198	.13575	2.2	.0038647	.43197	.30544
2.4	.0071461	.20944	.14809	2.4	.0045564	.47124	.33321
2.6	.0083111	.22689	.16043	2.6	.0053137	.51051	.36098
2.8	.0095660	.24434	.17278	2.8	.0061238	.54978	.38875
3.0	.010914	.26179	.18511	3.0	.0069738	.58905	.41651
3.2	.012341	.27925	.19745	3.2	.0078837	.62832	.44428
3.4	.013846	.29670	.20980	3.4	.0088569	.66759	.47205
3.6	.015426	.31415	.22214	3.6	.0098810	.70685	.49982
3.8	.017106	.33160	.23448	3.8	.010956	.74612	.52758
4.0	.018836	.34906	.24682	4.0	.012080	.78540	.55535
4.2	.020684	.36651	.25916	4.2	.013263	.82467	.58312
4.4	.022601	.38396	.27150	4.4	.014490	.86393	.61089
4.6	.024604	.40142	.28384	4.6	.015771	.90321	.63866
4.8	.026672	.41887	.29618	4.8	.017093	.94248	.66642
5.0	.028824	.43632	.30852	5.0	.018470	.98175	.69420
6.0	.040634	.52359	.37023	6.0	.026059	1.1781	.83303
7.0	.054394	.61086	.43194	7.0	.034861	1.3745	.97187
8.0	.070089	.69812	.49364	8.0	.044736	1.5708	1.1107
9.0	.087345	.78538	.55534	9.0	.055913	1.7671	1.2495
10.0	.10672	.87265	.61705	10.0	.068283	1.9635	1.3884
11.0	.12777	.95991	.67875	11.0	.081945	2.1598	1.5272
12.0	.15045	1.0472	.74046	12.0	.096714	2.3562	1.6661
13.0	.175620	1.1344	.80216	13.0	.11277	2.5525	1.8049
14.0	.20258	1.2217	.86387	14.0	.12994	2.7489	1.9438
15.0	.23213	1.3090	.92557	15.0	.14874	2.9452	2.0826

TABLE—(*Continued*)

8 In. Pipe				10-In. Pipe			
v Ft. per Sec.	$s=\frac{h}{l}$	Discharge, Cu. Ft. per Sec.		v Ft. per Sec.	$s=\frac{h}{l}$	Discharge, Cu. Ft. per Sec.	
		Clean Pipe	Foul Pipe			Clean Pipe	Foul Pipe
.2	.0000260	.06981	.04937	.2	.0000202	.10908	.07713
.4	.0001027	.13963	.09873	.4	.0000798	.21817	.15427
.6	.0002274	.20944	.14810	.6	.0001770	.32725	.23140
.8	.0003988	.27925	.19746	.8	.0003109	.43633	.30853
1.0	.0006147	.34907	.24683	1.0	.0004798	.54542	.38566
1.2	.0008758	.41888	.29619	1.2	.0006824	.65450	.46280
1.4	.0011775	.48870	.34556	1.4	.0009186	.76359	.53993
1.6	.0015212	.55851	.39492	1.6	.0011883	.87267	.61706
1.8	.0019041	.62832	.44428	1.8	.0014870	.98175	.69419
2.0	.0023283	.69814	.49365	2.0	.0018179	1.0908	.77133
2.2	.0027902	.76794	.54301	2.2	.0021825	1.1999	.84846
2.4	.0032937	.83776	.59238	2.4	.0025737	1.3090	.92559
2.6	.0038403	.90757	.64174	2.6	.0029940	1.4181	1.0027
2.8	.0044173	.97739	.69111	2.8	.0034446	1.5272	1.0799
3.0	.0050372	1.0472	.74047	3.0	.0039223	1.6363	1.1570
3.2	.0057026	1.1170	.78984	3.2	.0044322	1.7453	1.2341
3.4	.0064055	1.1868	.83921	3.4	.0049690	1.8544	1.3113
3.6	.0071568	1.2566	.88857	3.6	.0055394	1.9635	1.3884
3.8	.0079338	1.3264	.93793	3.8	.0061477	2.0726	1.4655
4.0	.0087462	1.3963	.98730	4.0	.0067820	2.1817	1.5427
4.2	.0096015	1.4661	1.0367	4.2	.0074509	2.2908	1.6198
4.4	.010488	1.5359	1.0860	4.4	.0081484	2.3998	1.6969
4.6	.011414	1.6057	1.1354	4.6	.0088746	2.5089	1.7741
4.8	.012369	1.6755	1.1848	4.8	.0096285	2.6180	1.8512
5.0	.013363	1.7453	1.2341	5.0	.010410	2.7271	1.9283
6.0	.018873	2.0944	1.4809	6.0	.014722	3.2725	2.3140
7.0	.025231	2.4435	1.7278	7.0	.019746	3.8179	2.6997
8.0	.032477	2.7925	1.9746	8.0	.025409	4.3633	3.0853
9.0	.040650	3.1416	2.2214	9.0	.031734	4.9087	3.4710
10.0	.049440	3.4907	2.4683	10.0	.038805	5.4542	3.8566
11.0	.059370	3.8397	2.7151	11.0	.046593	5.9996	4.2423
12.0	.070118	4.1888	2.9619	12.0	.055020	6.5450	4.6280
13.0	.081818	4.5378	3.2087	13.0	.064255	7.0904	5.0136
14.0	.094343	4.8870	3.4556	14.0	.074158	7.6359	5.3993
15.0	.10799	5.2360	3.7024	15.0	.084709	8.1813	5.7850

TABLE—(*Continued*)

12-In. Pipe				14-In. Pipe			
v Ft. per Sec.	$s=\frac{h}{l}$	Discharge, Cu. Ft. per Sec.		v Ft. per Sec.	$s=\frac{h}{l}$	Discharge, Cu. Ft. per Sec.	
		Clean Pipe	Foul Pipe			Clean Pipe	Foul Pipe
.2	.0000165	.15708	.11107	.2	.0000137	.21380	.15118
.4	.0000649	.31416	.22214	.4	.0000541	.42760	.30236
.6	.0001437	.47124	.33321	.6	.0001199	.64140	.45353
.8	.0002519	.62832	.44428	.8	.0002105	.85520	.60471
1.0	.0003881	.78540	.55535	1.0	.0003246	1.0690	.75589
1.2	.0005525	.94248	.66642	1.2	.0004613	1.2828	.90707
1.4	.0007435	1.0996	.77750	1.4	.0006195	1.4966	1.0583
1.6	.0009600	1.2566	.88857	1.6	.0008010	1.7104	1.2094
1.8	.0012069	1.4137	.99963	1.8	.0010051	1.9242	1.3606
2.0	.0014751	1.5708	1.1107	2.0	.0012281	2.1380	1.5118
2.2	.0017728	1.7279	1.2218	2.2	.0014738	2.3518	1.6630
2.4	.0020910	1.8850	1.3328	2.4	.0017393	2.5656	1.8141
2.6	.0024319	2.0420	1.4439	2.6	.0020251	2.7794	1.9653
2.8	.0027986	2.1991	1.5550	2.8	.0023319	2.9932	2.1165
3.0	.0031902	2.3562	1.6661	3.0	.0026577	3.2070	2.2677
3.2	.0036075	2.5133	1.7771	3.2	.0030089	3.4208	2.4189
3.4	.0040457	2.6704	1.8882	3.4	.0033799	3.6346	2.5700
3.6	.0045093	2.8274	1.9993	3.6	.0037683	3.8484	2.7212
3.8	.0049951	2.9845	2.1103	3.8	.0041775	4.0622	2.8724
4.0	.0055025	3.1416	2.2214	4.0	.0046054	4.2760	3.0236
4.2	.0060445	3.2987	2.3325	4.2	.0050587	4.4898	3.1748
4.4	.0066097	3.4557	2.4435	4.4	.0055312	4.7036	3.3259
4.6	.0071981	3.6129	2.5546	4.6	.0060231	4.9175	3.4771
4.8	.0078089	3.7699	2.6657	4.8	.0065336	5.1312	3.6283
5.0	.0084421	3.9270	2.7768	5.0	.0070629	5.3450	3.7795
6.0	.011955	4.7124	3.3321	6.0	.0099784	6.4140	4.5353
7.0	.016059	5.4978	3.8875	7.0	.013373	7.4831	5.2913
8.0	.020696	6.2832	4.4428	8.0	.017160	8.5520	6.0471
9.0	.025791	7.0685	4.9982	9.0	.021502	9.6210	6.8030
10.0	.031591	7.8540	5.5535	10.0	.026279	10.690	7.5589
11.0	.037925	8.6393	6.1089	11.0	.031604	11.759	8.3148
12.0	.044775	9.4248	6.6642	12.0	.037381	12.828	9.0707
13.0	.052234	10.210	7.2195	13.0	.043645	13.897	9.8265
14.0	.060214	10.996	7.7750	14.0	.050358	14.966	10.583
15.0	.068913	11.781	8.3303	15.0	.057689	16.035	11.338

TABLE—(*Continued*)

16-In. Pipe				18-In. Pipe			
v Ft. per Sec.	$s=\frac{h}{l}$	Discharge, Cu. Ft. per Sec.		v Ft. per Sec.	$s=\frac{h}{l}$	Discharge, Cu. Ft. per Sec.	
		Clean Pipe	Foul Pipe			Clean Pipe	Foul Pipe
.2	.0000115	.27926	.19746	.2	.0000010	.35343	.24991
.4	.0000456	.55852	.39493	.4	.0000393	.70685	.49982
.6	.0001013	.83778	.59239	.6	.0000872	1.0603	.74972
.8	.0001782	1.1170	.78986	.8	.0001531	1.4137	.99963
1.0	.0002743	1.3963	.98732	1.0	.0002363	1.7671	1.2495
1.2	.0003902	1.6756	1.1848	1.2	.0003367	2.1206	1.4994
1.4	.0005257	1.9548	1.3823	1.4	.0004542	2.4740	1.7494
1.6	.0006794	2.2341	1.5797	1.6	.0005869	2.8274	1.9993
1.8	.0008508	2.5133	1.7772	1.8	.0007375	3.1808	2.2492
2.0	.0010429	2.7926	1.9746	2.0	.0009005	3.5343	2.4991
2.2	.0012512	3.0718	2.1721	2.2	.0010816	3.8877	2.7490
2.4	.0014776	3.3511	2.3696	2.4	.0012776	4.2411	2.9989
2.6	.0017215	3.6304	2.5670	2.6	.0014882	4.5945	3.2488
2.8	.0019819	3.9097	2.7645	2.8	.0017130	4.9480	3.4987
3.0	.0022583	4.1889	2.9620	3.0	.0019515	5.3014	3.7486
3.2	.0025528	4.4682	3.1594	3.2	.0022087	5.6548	3.9985
3.4	.0028617	4.7474	3.3569	3.4	.0024802	6.0083	4.2484
3.6	.0031915	5.0267	3.5543	3.6	.0027671	6.3617	4.4983
3.8	.0035426	5.3059	3.7518	3.8	.0030711	6.7151	4.7482
4.0	.0039104	5.5852	3.9493	4.0	.0033897	7.0685	4.9982
4.2	.0042968	5.8645	4.1468	4.2	.0037225	7.4220	5.2481
4.4	.0046999	6.1437	4.3442	4.4	.0040694	7.7754	5.4979
4.6	.0051173	6.4230	4.5417	4.6	.0044303	8.1289	5.7479
4.8	.0055530	6.7022	4.7391	4.8	.0048047	8.4822	5.9978
5.0	.0060051	6.9815	4.9366	5.0	.0051928	8.8357	6.2477
6.0	.0085129	8.3778	5.9239	6.0	.0073881	10.603	7.4972
7.0	.011427	9.7742	6.9113	7.0	.0099341	12.370	8.7468
8.0	.014657	11.170	7.8986	8.0	.012816	14.137	9.9963
9.0	.018474	12.567	8.8859	9.0	.016052	15.904	11.246
10.0	.022528	13.963	9.8732	10.0	.019610	17.671	12.495
11.0	.027117	15.359	10.861	11.0	.023603	19.438	13.745
12.0	.032104	16.756	11.848	12.0	.027940	21.206	14.994
13.0	.037480	18.152	12.835	13.0	.032615	22.973	16.244
14.0	.043240	19.548	13.823	14.0	.037624	24.740	17.494
15.0	.049480	20.945	14.810	15.0	.043050	26.507	18.743

TABLE—(*Continued*)

20-In. Pipe				24-In. Pipe			
v Ft. per Sec.	$s=\frac{h}{l}$	Discharge, Cu. Ft. per Sec. Clean Pipe	Discharge, Cu. Ft. per Sec. Foul Pipe	v Ft. per Sec.	$s=\frac{h}{l}$	Discharge, Cu. Ft. per Sec. Clean Pipe	Discharge, Cu. Ft. per Sec. Foul Pipe
.2	.0000087	.43633	.30853	.2	.0000069	.62832	.44428
.4	.0000343	.87267	.61706	.4	.0000271	1.2566	.88857
.6	.0000762	1.3090	.92559	.6	.0000601	1.8850	1.3328
.8	.0001340	1.7453	1.2341	.8	.0001057	2.5133	1.7771
1.0	.0002071	2.1817	1.5427	1.0	.0001633	3.1416	2.2214
1.2	.0002955	2.6180	1.8512	1.2	.0002328	3.7699	2.6657
1.4	.0003986	3.0544	2.1597	1.4	.0003133	4.3983	3.1100
1.6	.0005149	3.4907	2.4683	1.6	.0004060	5.0266	3.5543
1.8	.0006456	3.9270	2.7768	1.8	.0005098	5.6548	3.9985
2.0	.0007896	4.3633	3.0853	2.0	.0006231	6.2832	4.4428
2.2	.0009481	4.7997	3.3938	2.2	.0007491	6.9115	4.8871
2.4	.0011187	5.2360	3.7024	2.4	.0008848	7.5398	5.3314
2.6	.0013015	5.6723	4.0109	2.6	.0010310	8.1681	5.7756
2.8	.0014985	6.1087	4.3195	2.8	.0011872	8.7965	6.2200
3.0	.0017093	6.5450	4.6280	3.0	.0013517	9.4248	6.6642
3.2	.0019343	6.9814	4.9365	3.2	.0015315	10.053	7.1085
3.4	.0021718	7.4177	5.2451	3.4	.0017218	10.681	7.5528
3.6	.0024239	7.8540	5.5535	3.6	.0019222	11.310	7.9971
3.8	.0026913	8.2903	5.8621	3.8	.0021327	11.938	8.4413
4.0	.0029731	8.7267	6.1706	4.0	.0023532	12.566	8.8857
4.2	.0032680	9.1630	6.4792	4.2	.0025862	13.195	9.3300
4.4	.0035740	9.5993	6.7877	4.4	.0028308	13.823	9.7742
4.6	.0038945	10.036	7.0963	4.6	.0030842	14.451	10.219
4.8	.0042253	10.472	7.4047	4.8	.0033492	15.080	10.663
5.0	.0045709	10.908	7.7133	5.0	.0036225	15.708	11.107
6.0	.0064746	13.090	9.2559	6.0	.0051492	18.850	13.328
7.0	.0087030	15.272	10.799	7.0	.0069021	21.991	15.550
8.0	.011224	17.453	12.341	8.0	.0089551	25.133	17.771
9.0	.014084	19.635	13.884	9.0	.011208	28.274	19.993
10.0	.017239	21.817	15.427	10.0	.013775	31.416	22.214
11.0	.020746	23.998	16.969	11.0	.016611	34.557	24.435
12.0	.024555	26.180	18.512	12.0	.019701	37.699	26.657
13.0	.028660	28.362	20.054	13.0	.022964	40.840	28.878
14.0	.033057	30.544	21.597	14.0	.026450	43.983	31.100
15.0	.037863	32.725	23.140	15.0	.030294	47.124	33.321

TABLE—(*Continued*)

30-In. Pipe				36-In. Pipe			
v Ft. per Sec.	$s=\frac{h}{l}$	Discharge, Cu. Ft. per Sec.		v Ft. per Sec.	$s=\frac{h}{l}$	Discharge, Cu. Ft. per Sec.	
		Clean Pipe	Foul Pipe			Clean Pipe	Foul Pipe
.2	.0000051	.98175	.69419	.2	.0000040	1.4137	.99963
.4	.0000203	1.9635	1.3884	.4	.0000158	2.8274	1.9993
.6	.0000451	2.9452	2.0826	.6	.0000351	4.2411	2.9989
.8	.0000793	3.9270	2.7768	.8	.0000620	5.6548	3.9985
1.0	.0001224	4.9087	3.4710	1.0	.0000958	7.0685	4.9982
1.2	.0001748	5.8905	4.1651	1.2	.0001364	8.4822	5.9978
1.4	.0002360	6.8723	4.8594	1.4	.0001841	9.8960	6.9975
1.6	.0003057	7.8540	5.5535	1.6	.0002388	11.310	7.9971
1.8	.0003828	8.8357	6.2477	1.8	.0002995	12.723	8.9966
2.0	.0004677	9.8175	6.9419	2.0	.0003665	14.137	9.9963
2.2	.0005611	10.799	7.6361	2.2	.0004394	15.551	10.996
2.4	.0006620	11.781	8.3303	2.4	.0005182	16.964	11.996
2.6	.0007710	12.763	9.0244	2.6	.0006033	18.378	12.995
2.8	.0008879	13.745	9.7187	2.8	.0006944	19.792	13.995
3.0	.0010119	14.726	10.413	3.0	.0007910	21.206	14.994
3.2	.0011450	15.708	11.107	3.2	.0008958	22.619	15.994
3.4	.0012861	16.690	11.801	3.4	.0010065	24.033	16.994
3.6	.0014346	17.671	12.495	3.6	.0011236	25.447	17.993
3.8	.0015912	18.653	13.190	3.8	.0012467	26.860	18.993
4.0	.0017552	19.635	13.884	4.0	.0013764	28.274	19.993
4.2	.0019307	20.617	14.578	4.2	.0015139	29.688	20.992
4.4	.0021141	21.598	15.272	4.4	.0016574	31.101	21.992
4.6	.0023055	22.580	15.967	4.6	.0018072	32.515	22.992
4.8	.0025046	23.562	16.661	4.8	.0019630	33.929	23.991
5.0	.0027114	24.544	17.355	5.0	.0021248	35.343	24.991
6.0	.0038507	29.452	20.826	6.0	.0030224	42.411	29.989
7.0	.0052048	34.361	24.297	7.0	.0040731	49.480	34.987
8.0	.0067183	39.270	27.768	8.0	.0052802	56.548	39.985
9.0	.0084223	44.178	31.238	9.0	.0066324	63.617	44.983
10.0	.010323	49.087	34.710	10.0	.0081261	70.685	49.982
11.0	.012446	53.996	38.180	11.0	.0097947	77.754	54.979
12.0	.014758	58.905	41.651	12.0	.011612	84.822	59.978
13.0	.017257	63.813	45.122	13.0	.013575	91.890	64.976
14.0	.019941	68.723	48.594	14.0	.015683	98.960	69.975
15.0	.0122808	73.631	52.064	15.0	.017934	106.03	74.972

TABLE—(*Continued*)

42-In. Pipe				48-In. Pipe			
v Ft. per Sec.	$s=\frac{h}{l}$	Discharge, Cu. Ft. per Sec.		v Ft. per Sec.	$s=\frac{h}{l}$	Discharge, Cu. Ft. per Sec.	
		Clean Pipe	Foul Pipe			Clean Pipe	Foul Pipe
.2	.0000031	1.9242	1.3606	.2	.0000025	2.5133	1.7771
.4	.0000123	3.8485	2.7213	.4	.0000010	5.0266	3.5543
.6	.0000272	5.7727	4.0819	.6	.0000223	7.5398	5.3314
.8	.0000480	7.6970	5.4425	.8	.0000395	10.053	7.1085
1.0	.0000745	9.6212	6.8031	1.0	.0000614	12.566	8.8857
1.2	.0001066	11.545	8.1638	1.2	.0000880	15.080	10.663
1.4	.0001442	13.470	9.5245	1.4	.0001192	17.593	12.440
1.6	.0001872	15.394	10.885	1.6	.0001544	20.106	14.217
1.8	.0002357	17.318	12.246	1.8	.0001944	22.619	15.994
2.0	.0002896	19.242	13.606	2.0	.0002388	25.133	17.771
2.2	.0003487	21.167	14.967	2.2	.0002878	27.646	19.548
2.4	.0004127	23.091	16.328	2.4	.0003410	30.159	21.326
2.6	.0004816	25.015	17.688	2.6	.0003986	32.672	23.103
2.8	.0005562	26.940	19.049	2.8	.0004601	35.186	24.880
3.0	.0006364	28.864	20.409	3.0	.0005261	37.699	26.657
3.2	.0007214	30.788	21.770	3.2	.0005966	40.212	28.434
3.4	.0008108	32.712	23.131	3.4	.0006713	42.726	30.211
3.6	.0009061	34.636	24.491	3.6	.0007506	45.239	31.988
3.8	.0010070	36.560	25.852	3.8	.0008335	47.752	33.765
4.0	.0011130	38.485	27.213	4.0	.0009204	50.266	35.543
4.2	.0012247	40.409	28.573	4.2	.0010127	52.779	37.320
4.4	.0013415	42.333	29.934	4.4	.0011091	55.292	39.097
4.6	.0014644	44.258	31.295	4.6	.0012090	57.806	40.874
4.8	.0015914	46.182	32.655	4.8	.0013137	60.318	42.651
5.0	.0017235	48.106	34.016	5.0	.0014226	62.832	44.428
6.0	.0024562	57.727	40.819	6.0	.0020317	75.398	53.314
7.0	.0033128	67.349	47.622	7.0	.0027502	87.965	62.200
8.0	.0042928	76.970	54.425	8.0	.0035621	100.53	71.085
9.0	.0054114	86.590	61.228	9.0	.0044705	113.10	79.971
10.0	.0066364	96.212	68.031	10.0	.0054881	125.66	88.857
11.0	.0079976	105.83	74.834	11.0	.0066217	138.23	97.742
12.0	.0094796	115.45	81.638	12.0	.0078581	150.80	106.63
13.0	.011088	125.07	88.440	13.0	.0092092	163.36	115.51
14.0	.012816	134.70	95.245	14.0	.010665	175.93	124.40
15.0	.014652	144.32	102.05	15.0	.012191	188.50	133.28

TABLE—(*Continued*)

54-In. Pipe				60-In. Pipe			
v Ft. per Sec.	$s=\frac{h}{l}$	Discharge, Cu. Ft. per Sec.		v Ft. per Sec.	$s=\frac{h}{l}$	Discharge, Cu. Ft. per Sec.	
		Clean Pipe	Foul Pipe			Clean Pipe	Foul Pipe
.2	.0000021	3.1808	2.2492	.2	.0000018	3.9270	2.7768
.4	.0000084	6.3617	4.4983	.4	.0000071	7.8540	5.5535
.6	.0000188	9.5425	6.7475	.6	.0000160	11.781	8.3303
.8	.0000332	12.723	8.9966	.8	.0000283	15.708	11.107
1.0	.0000515	15.904	11.246	1.0	.0000439	19.635	13.884
1.2	.0000738	19.085	13.495	1.2	.0000629	23.562	16.661
1.4	.0001000	22.266	15.744	1.4	.0000853	27.489	19.437
1.6	.0001302	25.447	17.993	1.6	.0001108	31.416	22.214
1.8	.0001639	28.627	20.242	1.8	.0001398	35.343	24.991
2.0	.0002012	31.808	22.492	2.0	.0001721	39.270	27.768
2.2	.0002425	34.989	24.741	2.2	.0002074	43.197	30.544
2.4	.0002872	38.170	26.990	2.4	.0002456	47.124	33.321
2.6	.0003356	41.350	29.239	2.6	.0002871	51.051	36.098
2.8	.0003876	44.532	31.488	2.8	.0003320	54.978	38.875
3.0	.0004440	47.712	33.737	3.0	.0003795	58.905	41.651
3.2	.0005031	50.893	35.987	3.2	.0004308	62.832	44.428
3.4	.0005652	54.074	38.236	3.4	.0004853	66.759	47.205
3.6	.0006313	57.255	40.485	3.6	.0005420	70.685	49.982
3.8	.0007009	60.435	42.734	3.8	.0006008	74.612	52.758
4.0	.0007739	63.617	44.983	4.0	.0006627	78.540	55.535
4.2	.0008508	66.797	47.232	4.2	.0007290	82.467	58.312
4.4	.0009311	69.978	49.481	4.4	.0007982	86.393	61.089
4.6	.0010147	73.159	51.731	4.6	.0008698	90.321	63.866
4.8	.0011017	76.340	53.980	4.8	.0009449	94.248	66.642
5.0	.0011920	79.521	56.229	5.0	.0010230	98.175	69.419
6.0	.0016915	95.425	67.475	6.0	.0014507	117.81	83.303
7.0	.0022889	111.33	78.721	7.0	.0019625	137.45	97.187
8.0	.0029629	127.23	89.966	8.0	.0025472	157.08	111.07
9.0	.0037164	143.14	101.21	9.0	.0032037	176.71	124.95
10.0	.0045744	159.04	112.46	10.0	.0039303	196.35	138.84
11.0	.0055265	174.94	123.70	11.0	.0047330	215.98	152.72
12.0	.0065670	190.85	134.95	12.0	.0056058	235.62	166.61
13.0	.0076956	206.75	146.19	13.0	.0065686	255.25	180.49
14.0	.0089117	222.66	157.44	14.0	.0076059	274.89	194.37
15.0	.010215	238.56	168.69	15.0	.0087173	294.52	208.26

TABLE—(*Continued*)

72-In. Pipe				72-In. Pipe			
v Ft. per Sec.	$s=\frac{h}{l}$	Discharge, Cu. Ft. per Sec. Clean Pipe	Foul Pipe	v Ft. per Sec.	$s=\frac{h}{l}$	Discharge, Cu. Ft. per Sec. Clean Pipe	Foul Pipe
.2	.0000014	5.6548	3.9985	3.8	.0004771	107.44	75.971
.4	.0000056	11.310	7.9971	4.0	.0005274	113.10	79.971
.6	.0000126	16.964	11.996				
.8	.0000223	22.619	15.994	4.2	.0005796	118.75	83.969
1.0	.0000346	28.274	19.993	4.4	.0006341	124.41	87.967
				4.6	.0006909	130.06	91.967
1.2	.0000496	33.929	23.991	4.8	.0007498	135.72	95.964
1.4	.0000672	39.584	27.990	5.0	.0008110	141.37	99.963
1.6	.0000876	45.239	31.988				
1.8	.0001105	50.893	35.987	6.0	.0011567	169.64	119.96
2.0	.0001360	56.548	39.985	7.0	.0015643	197.92	139.95
				8.0	.0020166	226.19	159.94
2.2	.0001642	62.203	43.984	9.0	.0025354	254.47	179.93
2.4	.0001945	67.858	47.982	10.0	.0031094	282.74	199.93
2.6	.0002272	73.512	51.980				
2.8	.0002621	79.168	55.980	11.0	.0037561	311.01	219.92
3.0	.0002994	84.822	59.978	12.0	.0044626	339.29	239.91
				13.0	.0052285	367.56	259.90
3.2	.0003402	90.477	63.976	14.0	.0060540	395.84	279.90
3.4	.0003837	96.132	67.975	15.0	.0069379	424.11	299.89
3.6	.0004292	101.79	71.973				

PIPE SYSTEMS FOR WATER SUPPLY

COMPUTATION OF A PIPE SYSTEM

In the accompanying illustration is shown a typical town, lying on both sides of a stream and divided into eight sections by dotted lines. The elevations, referred to the adopted datum, are shown by figures in parentheses. The lengths of the lines and the amount to be delivered at each point such as *B*, *C*, etc., are also shown. It is required to find the proper size of pipes to serve such a town, assuming the population to be 50,000, and the water consumption to be 19 cu. ft. per sec.

The branch *B* has an elevation of 550 ft. at the point *B*; therefore, the piezometric elevation at O_1 must be greater than 550, so that water may flow from O_1 toward *B*. A 24-in. pipe will first be tried for the main *A*. On referring to the hydraulic tables for cast-iron pipes, it is found that, for a diameter of 24 in. and a discharge of 19 cu. ft. per sec., the value of *s*, or $\frac{h}{l}$, is .0052; and, since $l = 10{,}000$, this gives $h = 10{,}000 \times .0052 = 52$ ft. as the required head between *R* and O_1. As this is greater

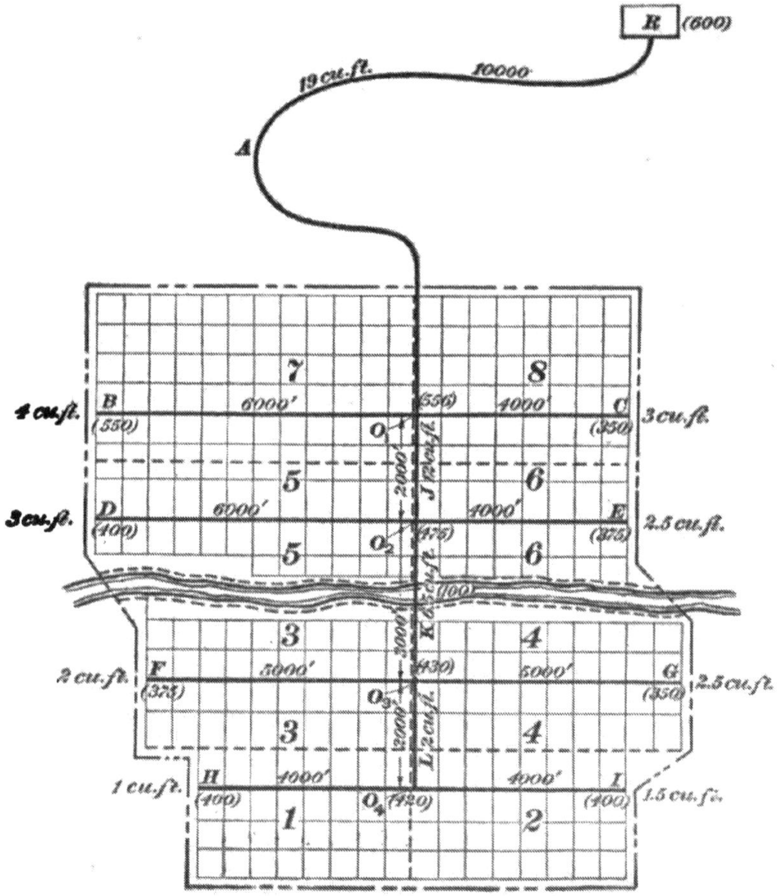

than the actual difference in elevation (600−556) between *R* and O_1, the assumed diameter is too small. Trying a 30-in. pipe, the value of $\frac{h}{l}$ is found to be .0017; therefore, $h = 10{,}000$

$\times .0017 = 17$ ft. This makes the piezometric elevation at O_1 $600-17=583$ ft. As this is greater than 556, the elevation of O_1, and also greater than 550, the elevation of B, the 30-in. pipe may be used for the main A. The heads for pipes B and C are, respectively, $583-550=33$ and $583-350=233$. The corresponding values of $\frac{h}{l}$ are $33 \div 6{,}000 = .0055$, and $233 \div 4{,}000 = .0583$. Knowing these values and the discharges, the diameters can be taken from the table. They are 14 in. for pipe B and 8 in. for pipe C.

In carrying the main to the next branch point O_2, the possibilities of choice of size are greater. But since the point H, 11,000 ft. away, is at an elevation of 400, it is desirable to reduce the head as little as may be, and it will be assumed that an effective head of 50 ft. will give necessary pressures without making the pipes too large. The effective head in J being 50 ft. in 2,000, the value of $\frac{h}{l}$ is $50 \div 2{,}000 = .025$; and from the table, the pipe necessary to carry 12 cu. ft. per sec. with this value of $\frac{h}{l}$ is found to be between 14 and 16 in. Using the 14-in. pipe, the value of $\frac{h}{l}$ is .033; $h = 2{,}000 \times .033 = 66$ ft., and, therefore, the piezometric elevation at O_2 is $583-66=517$ ft.

Proceeding as for the branches B and C, the value of $\frac{h}{l}$ for E is found to be .0355, which, by the table, requires an 8-in. pipe; for D, $\frac{h}{l} = .0195$, which, by the table, requires a 10-in. pipe.

Still bearing in mind the elevation of 400 at H, an effective head of 50 ft. will be assumed between O_2 and O_3, so that the piezometric elevation at the junction O_3 will be $517-50 = 467$. The pipe K, then, will have a value of $\frac{h}{l}$ of $50 \div 3{,}000 = .017$; and it is found by the table, that for a delivery of 6.5 cu. ft. per sec., a 14-in. pipe is a little too large; it may, however, be used. The table gives, for that pipe, $\frac{h}{l} = .012$, and therefore, $h = 3{,}000 \times .012 = 36$ ft. The piezometric elevation at the junction O_3 is, then, $517 - 36 = 481$. Proceeding as before, it is found that each of the branches *F* and *G* requires an 8-in. pipe.

Assuming an effective head of 30 ft. for *L*, the value of $\frac{h}{l}$ is $30 \div 2{,}000 = .015$, and the pipe *L* is found to be between an 8- and a 10-in. pipe. For the 10-in. pipe, and the delivery of 2 cu. ft. per sec., the value of $\frac{h}{l}$ is .0057; therefore, $h = .0057 \times 2{,}000 = 11.4$ and the piezometric elevation at O_4 is $481.0 - 11.4 = 469.6$. The branches *I* and *H* are found to require diameters of 8 and 6 in., respectively.

HYDRAULIC GRADE LINE

The *hydraulic grade line*, or *hydraulic gradient*, is a line drawn through a series of points to which water would rise in piezometer tubes attached to a pipe through which water flows. With a straight smooth pipe of uniform cross-section, the hydraulic grade line is a straight line extending from the reservoir to the end of the pipe.

In the accompanying illustration is shown a horizontal pipe leading from a reservoir to a stop-valve *S*. When the valve is open so that water from the pipe discharges freely into the atmosphere, the hydraulic grade line is the line *adfg*. The distance of the point *a* below the surface of the water in the reservoir represents the head absorbed in overcoming the resistances of entrance to the pipe, and in producing the velocity with which the water flows. In the same way, the difference in the height to which the water rises in any two piezometer tubes represents the head absorbed in overcoming the resistance to flow in the pipe between the points at which the tubes are inserted.

The flow of water through the pipe *P* would be the same whether the pipe were horizontal, as shown in the illustration, or whether it were laid along the grade line *adfg*. The flow would also be the same if the reservoir were deepened and the pipe laid along the line *a'd'f'*. The pressures in the pipe, however, would vary greatly with the different positions. If the pipe were laid along the line *adfg*, there would be little or no pressure in any part of it. In the horizontal position, however,

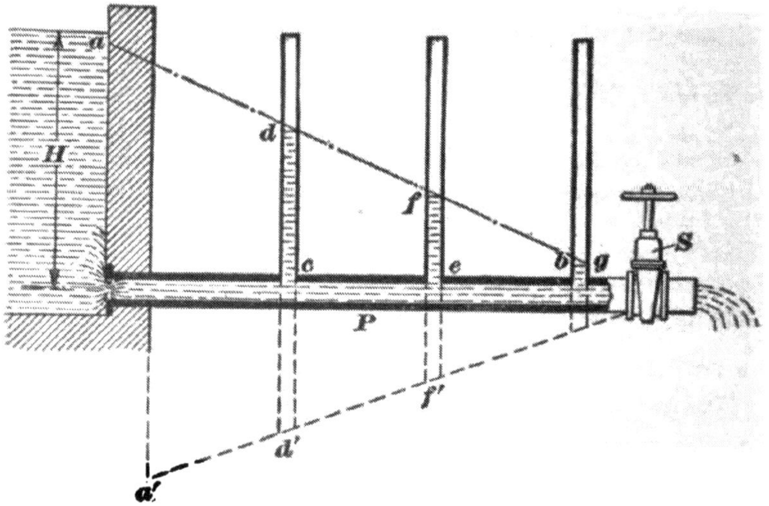

and still more in the position *a'd'f'*, there would be pressure at all points, the pressure for any point in the pipe being equivalent to the head represented by the vertical distance from that point to the hydraulic grade line.

Position of Hydraulic Grade Line.—In laying a line of pipe to connect two points lying at different levels, it is of the utmost importance to ascertain the position of the hydraulic grade line. In order that the pipe may flow full, no part of it should rise above the hydraulic grade line.

The Siphon.—The part of a pipe that rises above the hydraulic gradient is called a *siphon*. If the siphon is kept filled, the flow through it will take place in accordance with the laws given for pipes laid below the hydraulic gradient, and the same formulas apply.

The total head producing the flow in a siphon is the vertical distance from the discharge end of the pipe to the level of the water in the reservoir, but the pressure in all parts of the pipe that rise above the line will be less than the atmospheric pressure. Air always tends to collect in the highest point of a siphon, and means must be provided for its removal, in order to keep up the flow. This is effected by means of an air pump or air valve. Such means of removing the air should be provided for whenever circumstances make it unavoidable to place part of a pipe above the hydraulic gradient.

CAST-IRON PIPES

The thickness of a cast-iron pipe may be computed by the following formula:

$$t=\frac{(p+p')d}{6,600}+.25$$

in which t is the thickness of pipe, in inches; p, the static pressure, due to the head above the pipe, in pounds per square inch; d, the diameter of pipe, in inches; and p', the allowance for water hammer (shocks caused by opening of valves).

The following are values of p' for different diameters:

Diameter of Pipe *Inches*	*Value of p'* *Pounds per Square Inch*
3 to 10	120
12	110
16	100
20	90
24	85
30	80
36	75
40 to 60	70

EXAMPLE.—Determine the thickness of a cast-iron pipe 14 in. in diameter to withstand a pressure of 130 lb. per sq. in.

SOLUTION.—Here, $d=14$ and $p=130$. The value of p' corresponding to a diameter of 14 in. is a mean between the values corresponding to the diameters 12 and 16, or 105. Substituting these values in the formula,

$$t=\frac{(130+105)\times 14}{6,600}+.25=.75 \text{ in.}$$

Weight of a Cast-Iron Pipe Line.—To ascertain by a rapid approximation the weight, in tons (2,000 lb.), of a cast-iron pipe line, the following formula may be used:

$$T=28mt(d+t)$$

in which T is the weight, in tons, and m the length, in miles. In estimating, about 5% may be added to cover breakage, specials, and contingencies.

EXAMPLE.—What is the weight of 17 mi. of pipe 16 in. in diameter and .7 in. thick?

SOLUTION.—Substituting given values in the formula, $T = 28\times17\times.7\times(16+.7) = 5{,}564$ T. Adding 5%, the required weight is $5{,}564+5{,}564\times.05 = 5{,}842$ T.

The following table gives the nominal diameter, thickness, weight per foot and per length of 12 feet with standard sockets, for four different pressures.

RIVETED STEEL PIPE

Thickness of Riveted Steel Pipe.—The thickness of a riveted steel pipe may be computed by the following formula:

$$t = \frac{pd}{20{,}000} + .3$$

in which t is the thickness, in inches; d, the diameter of pipe, in inches; p, the pressure, in pounds per square inch, due to static head.

EXAMPLE.—Determine the thickness of a riveted steel pipe 36 in. in diameter, to withstand a pressure of 125 lb. per sq. in.

SOLUTION.—Here, $p = 125$ and $d = 36$. Substituting these values in the formula,

$$t = \frac{125\times36}{20{,}000} + .3 = .53 \text{ in.}$$

Flow in Riveted Pipes.—On account of their special construction, riveted steel pipes offer greater resistance to flow than do cast-iron pipes. Sufficient data are not available from which a satisfactory value for f can be found. The formula most generally used for the velocity in riveted pipes is Chezy's formula supplemented by Kutter's formula with a value for n varying between .013 and .015.

WOODEN-STAVE PIPES

Wooden-stave pipes are composed of wooden staves held together with round steel rods called bands. They are well adapted for carrying water for long distances and in quantities that necessitate large diameters. Their cheapness in first cost, in transportation, and in laying will lead to their use in cases where iron and steel are precluded on account of their cost. Other advantages of wooden pipes are that they are free from tuberculation, and have a tendency to wear even smoother than when first made. On this account, the flow may be computed by using for the coefficient f the values applying to smooth iron pipe; it may be safely assumed that this value will hold, even when the pipes become old, provided, however, that the velocity of flow in the pipe is at least 2 ft. per sec., so that no fungous growths can form.

Formulas for Stave Pipes.—The following formulas may be used in the design of wooden-stave pipes:

$$d=\frac{D+2t}{80}$$

and

$$s=\frac{65(D+2t)^2}{16(pD+200t)},$$

in both of which d is the diameter of bands, in inches; D, the inside diameter of pipe, in inches; t, the thickness of pipe, in inches; p, the water pressure in pipe, in pounds per square inch; and s, the distance between bands, in inches.

DIMENSIONS OF PIPE STAVES

(Recommended by A. L. Adams)

Nominal Diameter of Pipe Inches	Stock Sizes for Staves Inches	Thickness of Finished Staves Inches
22	2 ×6	1 3/8
24	2 ×6	1 3/8
27	2 ×6	1 7/16
30	2 ×6	1 1/2
36	2 ×6	1 9/16
42	2 ×6	1 5/8
48	2 ×6	1 11/16
54	2 1/2×8	2 3/8
60	3 ×8	2 1/2
66	3 ×8	2 9/16
72	3 ×8	2 5/8

STANDARD THICKNESSES AND WEIGHTS OF CAST-IRON PIPE

Nominal Inside Diam. Inches	Class A 100-Ft. Head 43 Lb. Pressure			Class B 200-Ft. Head 86 Lb. Pressure			Class C 300-Ft. Head 130 Lb. Pressure			Class D 400-Ft. Head 173 Lb. Pressure			Nominal Inside Diam. Inches
	Thickness Inches	Weight, in Pounds per Foot	Weight, in Pounds per Length	Thickness Inches	Weight, in Pounds per Foot	Weight, in Pounds per Length	Thickness Inches	Weight, in Pounds per Foot	Weight, in Pounds per Length	Thickness Inches	Weight, in Pounds per Foot	Weight, in Pounds per Length	
4	.42	20.0	240	.45	21.7	260	.48	23.3	280	.52	25.0	300	4
6	.44	30.8	370	.48	33.3	400	.51	35.8	430	.55	38.3	460	6
8	.46	42.9	515	.51	47.5	570	.56	52.1	625	.60	55.8	670	8
10	.50	57.1	685	.57	63.8	765	.62	70.8	850	.68	76.7	920	10
12	.54	72.5	870	.62	82.1	985	.68	91.7	1,100	.75	100.0	1,200	12
14	.57	89.6	1,075	.66	102.5	1,230	.74	116.7	1,400	.82	129.2	1,550	14
16	.60	108.3	1,300	.70	125.0	1,500	.80	143.8	1,725	.89	158.3	1,900	16
18	.64	129.2	1,550	.75	150.0	1,800	.87	175.0	2,100	.96	191.7	2,300	18
20	.67	150.0	1,800	.80	175.0	2,100	.92	208.3	2,500	1.03	229.2	2,750	20
24	.76	204.2	2,450	.89	233.3	2,800	1.04	279.2	3,350	1.16	306.7	3,680	24
30	.88	291.7	3,500	1.03	333.3	4,000	1.20	400.0	4,800	1.37	450.0	5,400	30
36	.99	391.7	4,700	1.15	454.2	5,450	1.36	545.8	6,550	1.58	625.0	7,500	36
42	1.10	512.5	6,150	1.28	591.7	7,100	1.54	716.7	8,600	1.78	825.0	9,900	42
48	1.26	666.7	8,000	1.42	750.0	9,000	1.71	908.3	10,900	1.96	1,050.0	12,600	48
54	1.35	800.0	9,600	1.55	933.3	11,200	1.90	1,141.7	13,700	2.23	1,341.7	16,100	54
60	1.39	916.7	11,000	1.67	1,104.2	13,250	2.00	1,341.7	16,100	2.38	1,583.3	19,000	60

NOTE.—The above weights are for 12-ft. lengths and standard sockets; proportionate allowance to be made for any variation therefrom.

It is not advisable to use bands less than $\frac{3}{8}$ in. in diameter, as they are likely to cut into the wood. By reducing the distance between the bands, stave pipes can be made to stand very heavy pressures, but above about 85 lb. per sq. in., the cost of construction is equal to or greater than for steel pipe.

POWER REQUIRED FOR PUMPING

In calculating the power required in pumping water through a height h, the work performed in overcoming the resistances to flow must be taken into account. Let Q be the discharge in cubic feet per second; f, the coefficient of resistance due to friction; c, the sum of the coefficients of resistance due to entrance, bends, valves, etc.; d, the diameter and l the length of the pipe, both in feet. Then, the work performed by the pump in a second, in foot-pounds, is

$$U = 62.5Q\left[h + .0252\left(f \times \frac{l}{d} + c\right)\frac{Q^2}{d^4}\right]$$

and the number of horsepower is

$$\text{H. P.} = .1136Q\left[h + .0252\left(f \times \frac{l}{d} + c\right)\frac{Q^2}{d^4}\right]$$

EXAMPLE 1.—It is desired to raise 15 cu. ft. of water per sec. by pumping to a reservoir 300 ft. above and 2 mi. distant from the pumping well. What horsepower will be necessary to do this work through a main 24 in. in diameter, having four bends, assuming a value of f as .018, a value of .5 for the coefficient of resistance at entrance, and the value of the coefficient for each bend as .9.

SOLUTION.—Here, $Q = 15$, $h = 300$, $l = 5{,}280 \times 2 = 10{,}560$, $d = \frac{24}{12} = 2$, $f = .018$, $c = .5 + .9 \times 4 = 4.1$. Substituting these values in the formula

$$\text{H. P.} = .1136 \times 15 \times \left[300 + .0252 \times \left(.018 \times \frac{10{,}560}{2} + 4.1\right) \times \frac{15^2}{2^4}\right] = 571 \text{ H. P.}$$

Cost of Pumping Water.—The cost of pumping water is approximately as follows: In a general way, the cost of water may be estimated at a certain amount per million or per thousand gallons. It is found by experience that, in the best and largest plants, where the engines are of the most economical form, and where the plant is specially designed, it costs at the rate of about 5c. for each million gallons lifted 1 ft. For smaller plants, the costs of lifting 1,000,000 gal. 1 ft. are about as follows:

Capacity of Plant Gallons per Day	*Cost of Lifting Cents*
10,000,000 or more	5
1,000,000	10
100,000	15

Intermediate quantities may be estimated at intermediate proportionate amounts.

SEWERAGE

SEWERAGE SYSTEMS

A *storm-water system* is a sewerage system that carries storm water only; a *separate system* is one that carries house sewage only; and a system that carries both storm water and house refuse is called a *combined system*.

A storm-water system should be adequate for the prompt removal of the rainfall from the surface during violent storms, including also such animal and vegetable refuse from the streets as will necessarily be removed with the storm water. If this is accomplished, and the drains are located at sufficient depth, efficient drainage will be provided for the subsoil.

The separate system should be able to carry off promptly from houses all sink, laundry, and closet wastes, without offensive odors, and without interruption. It should keep itself clean, that is, free from deposits; it should not pollute the soil through which the pipes pass; and it should have an outlet that is without objection.

The separate system is less costly than the combined system, is more strictly sanitary, and is especially adapted for towns and villages that are built on porous soil that allows the storm water to be readily discharged at convenient outlets.

Capacity Required for Storm-Water Sewer.—Various formulas are used for the capacities of storm-water sewers. Of these the formula proposed by Buerkli, a German authority, is probably the most reliable. It is as follows:

$$E = fre,$$

in which $e = \sqrt[4]{SA^3}$ and is tabulated in the accompanying table. In these formulas, E is the total flow in cubic feet per second from a sewer district containing A acres; S, the average surface slope (presumably toward and along the drain), in feet per thousand feet through drainage district; f, a coefficient relating to "the proportion of rainfall that will reach the sewer"; r, the coefficient representing rate of rainfall, in inches per hour, "during period of greatest intensity of rain."

VALUES OF e, OR $\sqrt[4]{SA^3}$

Acres = A	$S = 2.5$	$S = 5$	$S = 10$	$S = 15$	$S = 20$	$S = 25$	$S = 50$
40	20.00	23.78	28.28	31.30	33.64	35.57	42.29
60	27.10	32.24	38.34	42.43	45.59	48.21	57.33
80	33.64	40.00	47.57	52.64	56.57	59.81	71.13
100	39.76	47.29	56.23	62.23	66.87	70.71	84.09
120	45.59	54.22	64.47	71.35	76.67	81.07	96.41
160	56.57	67.27	80.00	88.53	95.14	100.60	119.63
200	66.87	79.53	94.57	104.66	112.47	118.92	141.42
300	90.64	107.79	128.19	141.86	152.44	161.19	191.68
400	112.47	133.74	159.05	176.02	189.15	200.00	237.84
500	132.96	158.09	188.02	208.09	223.61	236.44	281.17
600	152.44	181.28	215.58	238.58	256.37	271.08	322.37
800	189.15	224.92	267.50	296.03	318.11	336.36	400.00
1,000	223.61	265.90	316.23	349.96	376.06	397.64	472.87
1,200	256.37	304.84	362.57	401.24	431.17	455.90	542.16
1,500	303.08	360.39	428.62	474.34	509.71	538.96	640.93
2,000	376.06	447.21	531.83	588.57	632.46	668.74	795.27
2,500	444.57	528.68	628.72	695.79	747.67	790.57	940.15

To the coefficient f in the Buerkli formula are given values ranging from .31 in rural districts and suburbs to .75 in cities well built up, with a mean value of .62. By *mean value* is here meant that value which best represents the most usual conditions.

The quantity *r*, though commonly stated as the rate of rainfall during the greatest downpour, has been shown to be scarcely more than an arbitrary coefficient. In climates where the intensity of rainfall varies greatly with the duration of the storm, it is necessary, in using *r*, to fix on a definite length of time as representing the duration of a typical storm, and this is equivalent to arbitrarily fixing the value of *r*. In using the Buerkli formula, the European practice is to give *r* values ranging from 1.75 to 2.5 in. per hr., but recent American practice gives *r* values of from 2 to 3.5, and even higher, for sewers designed to carry all the storm water. In St. Louis, Mo., a value of .75 for *f* and values for *r* varying from 3.02, for a district containing 100 A. to 3.51, for a district containing 2,000 A., were used. Observations taken in Rochester, N. Y., of rainstorms lasting less than 1 hr. indicate that, for the conditions in that city, storms lasting 51 minutes give the greatest flow. For storms of this duration, the value of *r* will be about 2. A value of 2.75 is about the mean of American practice.

Capacity Required for a Separate System.—The design of the sewers of a separate system is based on the quantity of sewage delivered, and provision must also be made for carrying subsoil and ground water. No rule can be given for determining the amount of subsoil water that may be added to the flow. For 8-in. pipes, it runs from 5,000 gal. per mi. per da. to 25,000 gal. or more. It can only be determined approximately from previous experience in similar cases and from what knowledge can be secured as to the subsurface conditions.

Sewage Discharge and Water Supply.—The available records of sewer gaugings for American cities are not sufficient to indicate accurately the quantity of sewage per capita that must be provided for. Records of water supply, however, are abundant, and, since the sewer gaugings that have been made indicate that the quantity of sewage from a given district is somewhat less than the quantity of water consumed by its inhabitants, the statistics of water supply are useful and are the main factor in estimating the sewage discharge.

In using the records of a public water supply for this purpose, it must be remembered that often there are factories that have a private water-supply, and these may often discharge a considerable volume of sewage, which should be provided for. The provision necessary for subsoil water has already been referred to. That the amount of actual sewage will generally be less than the water supply will be evident when it is considered that all the water used for sprinkling, and some of that used for cleaning, either soaks into the ground or evaporates. In manufacturing districts, also, considerable quantities of water are used that do not reach the sewers.

The common practice among American engineers is to proportion the sewers of the separate system so that, when running half full, they will discharge a quantity of sewage equal to the maximum hourly water consumption, this maximum being taken equal to 1.5 times the average. The remaining capacity is reserved for extreme variations in flow and for ventilation. The conditions of flow are then as follows:

Average daily flow, 100%; sewer one-third full.

Average maximum daily flow, 150%; sewer one-half full.

Total capacity of sewer, 300%; sewer full.

The average daily flow is assumed to be such as may reasonably be expected when the territory is fairly well developed and the buildings all connected with the sewers.

EXAMPLE.—What capacity should the main sewer of a city of 25,000 population have, the water consumption being 85 gal. per head each day, assuming the sewer to be flowing half full?

SOLUTION.—The total water consumption is $25{,}000\times 85 = 2{,}125{,}000$ gal. per da. The discharge from the sewer is $2{,}125{,}000 \times 1.50 = 3{,}187{,}500$ gal. per da. Reducing this quantity to cubic feet per second, the capacity of the sewer is found to be

$$\frac{3{,}187{,}500}{7.48\times 24\times 60\times 60} = 4.9 \text{ cu. ft. per sec.}$$

SEWER COMPUTATIONS

Sewer computations are made by Chezy's and Kutter's formulas, given under the heading Hydraulics. For sewer work, two values of n in Kutter's formula are used: .013 for vitrified pipe and .015 for concrete and brick sewers. For both of these values, the accompanying tables give velocities and discharges for sewers of various sizes laid on different grades.

VELOCITY AND DISCHARGE FOR CIRCULAR PIPE SEWERS FLOWING FULL

(*n* = .013; *Q in cubic feet per second; v in feet per second*)

Diam. Inches	Grade 1 in 10		Grade 1 in 20		Grade 1 in 30		Grade 1 in 40		Grade 1 in 50		Grade 1 in 60	
	v	*Q*	*v*	*Q*	*v*	*Q*	*v*	*Q*	*v*	*Q*	*v*	*Q*
6	7.99	1.57	5.65	1.11	4.61	.905	3.99	.784	3.57	.701	3.26	.639
8			7.10	2.48	5.79	2.02	5.02	1.75	4.48	1.57	4.09	1.43
9			7.78	3.44	6.35	2.81	5.50	2.43	4.92	2.17	4.49	1.98
10			8.44	4.60	6.89	3.76	5.97	3.25	5.39	2.94	4.87	2.66
12					7.93	6.23	6.86	5.39	6.14	4.82	5.60	4.40
15							8.12	9.96	7.26	8.91	6.63	8.13
18									8.31	14.7	7.59	13.4
20									8.98	19.6	8.19	17.9
21											8.49	20.4

Diam. Inches	Grade 1 in 70		Grade 1 in 80		Grade 1 in 100		Grade 1 in 150		Grade 1 in 200		Grade 1 in 300	
6	3.01	.592	2.82	.553	2.52	.495	2.06	.403	1.78	.349	1.45	.284
8	3.79	1.32	3.54	1.24	3.17	1.11	2.58	.902	2.23	.780	1.82	.636
9	4.15	1.84	3.89	1.72	3.47	1.54	2.83	1.25	2.45	1.08	2.00	.883
10	4.51	2.46	4.22	2.30	3.77	2.06	3.07	1.68	2.66	1.45	2.17	1.18
12	5.18	4.07	4.85	3.81	4.34	3.41	3.54	2.78	3.06	2.40	2.49	1.96
15	6.13	7.53	5.74	7.04	5.13	6.30	4.19	5.14	3.62	4.44	2.95	3.62
18	7.02	12.4	6.57	11.6	5.87	10.4	4.79	8.48	4.15	7.33	3.38	5.97
20	7.58	16.5	7.09	15.5	6.34	13.8	5.18	11.3	4.48	9.77	3.65	7.97
21	7.86	18.9	7.35	17.7	6.57	15.8	5.36	12.9	4.64	11.2	3.78	9.10
24	8.65	27.2	8.09	25.4	7.24	22.7	5.91	18.6	5.11	16.1	4.17	13.1
30					8.48	41.6	6.92	34.0	5.99	29.4	4.89	24.0
36							7.86	55.6	6.80	48.1	5.55	39.2

Diam. Inches	Grade 1 in 400		Grade 1 in 600		Grade 1 in 1000		Grade 1 in 1500		Grade 1 in 2000		Grade 1 in 3000	
6	1.25	.246	1.02	.199								
8	1.57	.549	1.28	.446	.982	.343						
9	1.73	.762	1.40	.620	1.08	.476	.871	.385				
10	1.87	1.02	1.52	.831	1.17	.638	.946	.516				
12	2.16	1.69	1.75	1.38	1.35	1.06	1.09	.856	.936	.735		
15	2.55	3.13	2.08	2.55	1.60	1.96	1.29	1.59	1.11	1.36	.893	1.10
18	2.92	5.16	2.38	4.20	1.83	3.24	1.48	2.62	1.28	2.25	1.03	1.81
20	3.16	6.89	2.57	5.61	1.98	4.32	1.60	3.50	1.38	3.01	1.11	2.42
21	3.27	7.87	2.66	6.41	2.05	4.94	1.66	4.00	1.43	3.44	1.15	2.77
24	3.60	11.3	2.93	9.22	2.26	7.10	1.83	5.76	1.58	4.96	1.27	4.00
30	4.23	20.7	3.44	16.9	2.65	13.0	2.15	10.6	1.86	9.11	1.50	7.36
36	4.80	33.9	3.91	27.7	3.02	21.3	2.45	17.3	2.11	14.9	1.71	12.1

VELOCITY AND DISCHARGE FOR CIRCULAR BRICK SEWERS FLOWING FULL

($n = .015$; Q in cubic feet per second; v in feet per second)

Diam. Inches	Grade 1 in 75		Grade 1 in 100		Grade 1 in 200		Grade 1 in 400		Grade 1 in 1000		Grade 1 in 1500		Grade 1 in 3000	
	v	Q	v	Q	v	Q	v	Q	v	Q	v	Q	v	Q
24	7.03	22.1	6.09	19.1	4.30	13.5	3.03	9.52	1.90	5.97	1.54	4.84	1.07	3.36
30			7.16	35.2	5.06	24.8	3.57	17.5	2.24	11.0	1.82	8.92	1.26	6.20
36					5.76	40.7	4.07	28.7	2.56	18.1	2.08	14.7	1.45	10.2
42					6.43	61.8	4.54	43.7	2.85	27.4	2.32	22.3	1.62	15.6
48					7.06	88.7	4.98	62.6	3.13	39.4	2.55	32.0	1.78	22.4
54					7.65	122	5.40	85.9	3.40	54.1	2.77	44.0	1.94	30.8
60							5.81	114	3.66	71.8	2.98	58.5	2.09	41.0
66							6.20	147	3.90	92.8	3.18	75.5	2.23	53.0
72							6.57	186	4.14	117	3.37	95.4	2.37	67.0
78							6.93	230	4.37	145	3.56	118	2.50	83.0
84							7.28	280	4.59	177	3.74	144	2.63	101
90							7.62	337	4.81	212	3.92	173	2.76	122
96							7.95	400	5.02	252	4.09	206	2.88	145
102									5.23	296	4.26	242	3.00	170
108									5.42	345	4.42	281	3.12	198
114									5.62	398	4.59	325	3.23	229
120									5.81	456	4.74	372	3.34	262

VELOCITY AND DISCHARGE FOR EGG-SHAPED SEWERS (NEW FORM) FLOWING FULL

($n = .015$; Q *in cubic feet per second;* v *in feet per second*)

Size Inches	Grade .0100		Grade .0070		Grade .0040		Grade .0020		Grade .0010		Grade .0005		Grade .0003		Grade .0001	
	v	Q	v	Q	v	Q	v	Q	v	Q	v	Q	v	Q	v	Q
24×36	6.69	29.8	5.59	25.0	4.22	18.8	2.98	13.3	2.09	9.33	1.46	6.51	1.11	4.97		
30×45	7.86	54.8	6.57	45.8	4.96	34.6	3.50	24.4	2.46	17.1	1.72	12.0	1.32	9.16		
36×54	8.94	89.7	7.48	75.0	5.65	56.7	3.98	40.0	2.80	28.1	1.96	19.7	1.50	15.1		
42×63			8.33	114.0	6.29	85.9	4.44	60.6	3.12	42.7	2.19	29.9	1.68	22.9		
48×72					6.90	123.0	4.87	86.8	3.43	61.2	2.41	42.9	1.85	33.0	1.03	18.3
54×81							5.28	119.0	3.72	84.0	2.61	59.0	2.01	45.3	1.12	25.3
60×90							5.67	158.0	4.00	111.0	2.81	78.3	2.16	60.2	1.21	33.7
66×99							6.04	204.0	4.26	144.0	3.00	101.0	2.31	77.9	1.30	43.8
72×108							6.40	257.0	4.52	181.0	3.18	128.0	2.45	98.4	1.38	55.5
78×117							6.75	318.0	4.77	225.0	3.36	158.0	2.59	122.0	1.46	69.0
84×126									5.01	273.0	3.53	193.0	2.72	149.0	1.54	84.3
90×135									5.24	329.0	3.69	232.0	2.85	179.0	1.62	102.0

Egg-shaped sewers have a larger hydraulic radius than circular sewers when the flow is shallow; consequently, they reduce the likelihood of deposits. The general form of the cross-section of an egg-shaped sewer is shown in the accompanying illustration. The part above the line OO' is a semicircle, the part below the line OO' is formed by the three arcs DE, EG and GA, the arcs DE and GA having equal radii. It will be noticed that three different radii are used in constructing the figure; namely, $CA = CB = CD = r$ for the upper semicircle, $OD = OE = O'G = O'A = r_1$ for the two side arcs DE and GA, and $C'E = C'F = C'G = r_0$ for the lower arc EFG, commonly called the ***invert.***

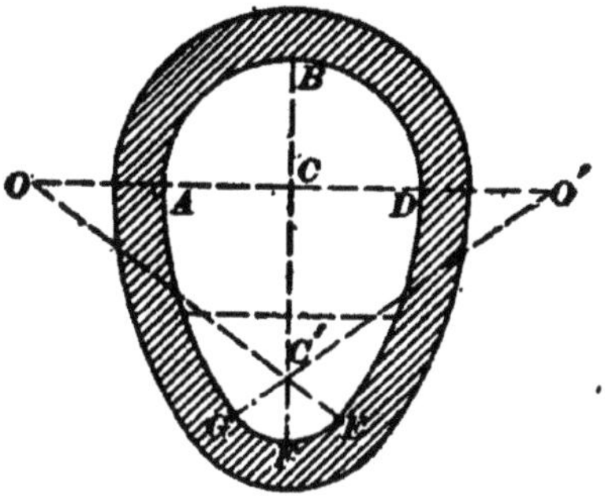

The proportions of the different dimensions, as well as other useful data, are given in the accompanying table, and the discharge and velocities for various grades in the preceding table.

ELEMENTS OF CROSS-SECTION OF EGG-SHAPED SEWERS

	Element	Value for New Form
1.	Horizontal diameter	$2r$
2.	Vertical diameter	$3r$
3.	Radius of bottom arc	$\frac{1}{4}r$
4.	Radius of side arcs	$2\frac{2}{3}r$
5.	Distance between centers	$1\frac{3}{4}r$
6.	Distance $OC = O'C$	$1\frac{2}{3}r$
7.	Wetted perimeter, full	$7.8409r$
8.	Wetted perimeter, $\frac{2}{3}$ full	$4.6994r$
9.	Wetted perimeter, $\frac{1}{3}$ full	$2.6651r$
10.	Area of flow, full	$4.4602r^2$
11.	Area of flow, $\frac{2}{3}$ full	$2.8894r^2$
12.	Area of flow, $\frac{1}{3}$ full	$1.0171r^2$
13.	Hydraulic radius, full	$.5688r$
14.	Hydraulic radius, $\frac{2}{3}$ full	$.6148r$
15.	Hydraulic radius, $\frac{1}{3}$ full	$.3817r$
16.	Angle COC'	46° 23′ 50″
17.	Angle $EC'G$	87° 12′ 22″

DIMENSIONS OF SEWER PIPES

The standard lengths of sewer pipes are 2, 2½, and 3 ft. The latter is the most desirable, because it reduces the number of joints in the pipe line. In diameter, they are made 4, 5, 6, 8, 9, 10, 12, 18, 21, and 24 in. Special sizes, such as 20, 24, 27, 30, and 36 in., are also carried by some factories.

Thickness and Strength.—The practice of factories is to make pipe of two thicknesses, one known as *standard pipe* and the other known as *double-strength pipe*. The accompanying table shows the thickness that well-made pipe should have by the custom of the best factories.

THICKNESS OF SEWER PIPE

Kind of Pipe	Diameter, in Inches										
	6	8	9	10	12	15	18	21	24	30	36
Standard.........	⅝	¾	13/16	⅞	1	1⅛	1¼	1⅜	1⅝	2	2¼
Double-strength..					1	1¼	1½	1¾	2	2½	3

DEPTHS OF SOCKETS FOR STANDARD AND FOR DEEP-AND-WIDE SOCKET

Kind of Socket	Diameter, in Inches										
	6	8	9	10	12	15	18	21	24	30	36
Standard.........	1⅝	1¾	1¾	1⅞	2	2¼	2¼	2⅜	2½	3	3½
Deep-and-wide...	2¼	2½	2½	2½	3	3¼	3½	3¾	4	4½	5

Tests indicate that standard pipe as made can carry a uniform load of about 2,000 lb. per lin. ft. of pipe, and double-strength pipe, about 4,000 lb. The load that sewer pipes must carry is the weight of the earth in the trench above them, with the additional weight of a wagon wheel or a steam-roller wheel, either of which may add 1 T. loading to the pipe. A 12-in. pipe in an 8-ft. trench will have a mass of $1\times8\times1=8$ cu. ft. of earth, or about 1,000 lb. with 2,000 lb. pressure on the top resting on it. Only a fraction of this loading, however, is transmitted to the pipe, the rest being supported by the sides of the trench. A factor of safety of 3 should be employed. It is safe practice to use double-strength pipe when the pipe is in a trench less than 6 ft. deep, and heavy surface loads may be expected. Under other conditions, standard pipe may be used, though double-strength pipe is always safer.

Depth of Socket.—There are two types of socket, the *standard* and the *deep-and-wide*, or *deep*, socket. The depths of socket, in inches, are shown in the preceding table. The advantage of the deep-and-wide socket lies in the fact that the jointing material can be rammed into the sockets to a greater depth, and there is therefore less leakage through the joints.

BRICK AND CONCRETE SEWERS

Brick Sewers.—Sewers of a larger diameter than 24 in. are generally built of brick or concrete and can be made in any desired form.

For ordinary conditions, the following empirical formula will generally be found satisfactory for indicating the number of rings required:

$$R=.4+\frac{D(H-D)}{25}$$

in which R is the number of 4-in. rings or courses; D, the internal diameter of a circular sewer, or horizontal diameter of an egg-shaped sewer; and H, the total depth of the trench—all in feet.

Any fraction greater than .25 in the value of R should be considered as 1.

Concrete Sewers.—The concrete used for sewers should be of first-class quality, carefully proportioned to have as small a percentage of voids as possible. The concrete must be strong, to take up the tensile stresses in the arch; and impervious, to keep ground water out of the sewers. A mixture of 1-2-4 may be used for the arch, and a mixture of 1-2½-5 for the bottom. The mixing must be very thorough, and the tamping into place carefully done. For sewer work, the mixture should be so wet that a spade can be readily thrust down into the mass to work the mixture into homogeneity.

The thickness of circular concrete sewers built in firm and stable ground and at a depth not exceeding 12 ft. may be taken to be approximately as follows:

Diameter *Feet*	*Thickness of Sewer* *Inches*
3	4
6	6
9	8
12	10

This thickness must be varied, however, with the character of the soil and the depth of cutting. In wet, running soils, the lower half of the sewer may be from two to four times these thicknesses, with extra thickness at the sides. In trenches 30 ft. deep, the thickness of the arch may be twice the thickness given.

ROADS AND PAVEMENTS

HIGHWAYS

GRADES, CROSS-SECTION, AND CURVES

In order that a road may be satisfactory for travel, it must be dry and solid, and have easy grades, easy curves, and a smooth surface. These conditions refer to the use of the road, but there are other conditions that are essential to economic construction and maintenance; namely, (1) that the length of the road shall be a minimum; (2) that its surface shall be so placed with reference to the natural surface of the ground that the amount of excavation and embankment shall be a minimum; and (3) that it shall be so located as to be free from landslides, washouts, and snowdrifts. These different conditions often conflict with one another, and there is generally a great deal of difficulty in reconciling them. The question of cost frequently becomes the controlling factor, but it is not always wise to cut down initial cost to the lowest amount possible; such apparent economy may result in the construction of a road requiring for its maintenance much trouble and expense, which might have been avoided by a small extra cost in the original construction. A better plan, and one that should always be followed, is to arrange the road so that future improvements can be made.

Minimum and Maximum Grades.—In order that efficient drainage may be provided for the roadway, the minimum grade should generally not be flatter than 1%, and should never be materially flatter than one-half of 1%, except on first-class pavements. In general, the maximum grade should not be steeper than 9% for earth roads, 6½% for gravel roads, and 3% for macadam roads, in any case where it is possible to keep within these limits; and, preferably, should never be steeper than about 3 to 5% for any kind of road.

As a result of investigations, it has been deduced that, dependent on the amount of traffic and the cost of construction and maintenance of the road, the most advantageous gradients vary for mountainous country between 5 and 3%; for hilly country, between 3 and 2½%; and for gently rolling country, between 2½ and 1%.

Form of Cross-Section.—One of the best forms for highways is a parabolic arc, as is shown in the accompanying illustration. Its construction is as follows:

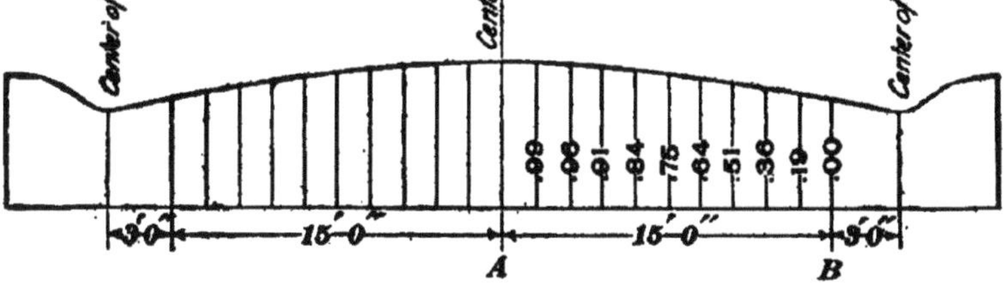

Divide the width *AB* between the edge of the gutter and the center of the wheelway into ten equal parts, and at the points of division erect perpendiculars, the lengths of which measured from the line joining the edges of the gutters are determined by multiplying the rise at the center by the number given on each perpendicular in the figure. The rise at the center should be as follows: For earth roads, $\frac{1}{40}$ of the width; for gravel roads, $\frac{1}{50}$ of the width; and for broken-stone roads, $\frac{1}{60}$ of the width.

EXAMPLE.—Find the ordinates for an earth road 30 ft. wide.

SOLUTION.—The center height must be $\frac{30}{40}=.75$ ft. The distance between the center of the road and the edge of the gutter is 15 ft.; the points of division are, therefore, 1.5 ft. apart. The ordinates are as follows (see the illustration):

At the center..............................	.75 ft.
At $1\frac{1}{2}$ ft. from the center	$.75\times.99=.74$ ft.
At 3 ft. from the center	$.75\times.96=.72$ ft.
At $4\frac{1}{2}$ ft. from the center...........	$.75\times.91=.68$ ft.
At 6 ft. from the center	$.75\times.84=.63$ ft.
At $7\frac{1}{2}$ ft. from the center...........	$.75\times.75=.56$ ft.
At 9 ft. from the center............	$.75\times.64=.48$ ft.
At $10\frac{1}{2}$ ft. from the center..........	$.75\times.51=.38$ ft.
At 12 ft. from the center...........	$.75\times.36=.27$ ft.
At $13\frac{1}{2}$ ft. from the center..........	$.75\times.19=.14$ ft.
At 15 ft. from the center....................	.00 ft.

Width of Roadway.—The width of the wheelway required to accommodate two lines of travel is 18 ft.; for a single line of travel, 8 ft. is sufficient, but suitable turnouts must be provided at frequent intervals.

Curves.—The straight parts of the roads must be joined by curves, the least permissible radius of which depends on the length of the teams using the road. As a rule, the greatest possible radius should be used, and no curve should have a radius of less than 50 ft. The curves may be either circular or parabolic. A parabolic curve is often preferred, on account of the ease with which it can be laid out.

DRAINAGE

Water is the greatest enemy of roads. Through its solvent action, it softens and dissolves the materials of which the road is constructed, and by its expansion while freezing disrupts the roadbed by lifting and displacing its component parts. Hence, the speedy and efficient removal of water is imperative for the preservation of a road.

The surface drainage, that is, the removal of the rain water from the surface of a road, is provided by gutters connected with side ditches or underground drains and by giving the road a suitable cross-section and grade. Very often it is also necessary to provide for the removal of the underground water. A wet substratum cannot give a firm subfoundation for a road, and will invariably destroy its efficiency under traffic. Sandy soils, unless saturated with water, do not present any difficulty in securing a dry and solid foundation, especially if the fall of the natural drainage is away from the line of the road, in which case gutters and side ditches for the removal of the rainwater will generally be found sufficient. The clay soils are naturally retentive of water, although they are not readily saturated; when they reach the state of saturation, they become very unstable and are incapable of supporting heavy loads; it is, therefore, necessary to provide a suitable system of subsoil drainage.

Rock requires little attention to drainage, except where the strata are interspersed with seams of clay and are inclined toward the road, in which case means must be provided for the removal of the water in order to prevent slips.

The removal of the subsoil water is effected by constructing underground drains or deep side ditches that discharge into the natural streams.

The main points to be attended to in the construction of all types of drains are:

1. *The Fall or Grade.*—This should rarely exceed 1 in. in 5 ft. Excessive inclination is likely to cause injury by washing in consequence of the high velocity of the water.

2. *The Area of the Drain.*—This should be in proportion to the amount of water to be removed. In using tile drains 3 in. should be the minimum size.

3. *The Filling.*—In filling the trenches, care must be taken that the material used does not choke or stop the waterway.

4. *The Materials.*—In order to avoid large maintenance expenses only durable materials should be employed.

5. *The Depth.*—The drains should be placed at a sufficient depth to accomplish the object sought. A deep drain will be more effective than a shallow one.

6. *The Inlet and Outlet.*—The ends of the drain should be such as to allow free passage of the water, and should be well protected.

CONSTRUCTION OF ROADS

Natural Roads.—Earth, or natural, roads consist of either clay or loam or sand and gravel. They form the larger part of the country roads of the United States, and under favorable conditions, furnish a sufficiently satisfactory wheelway for light traffic. By reason, however, of improper location, neglect, and insufficient drainage, the average country road is in a condition far from satisfactory during a large part of the year. By changing the location and providing drainage where necessary, and by prompt and systematic repairs, the condition of natural roads may be greatly improved without much additional expenditure. In the formation of natural roads, each soil requires different treatment to produce satisfactory results.

Sandy roads are in best condition when moist. Side-ditching, beyond a slight depth to carry away the surface water in long rainy spells, is not desirable, as it tends to facilitate the drying of the sand. When clay is available, a coating 6 in. thick, spread over the sand and mixed with it by harrowing, will produce a good roadway.

Sand roads should be as narrow as practicable, and the sides should be lined with as much vegetation as possible. Trees along the sides will aid in keeping the surface moist, and the falling leaves will assist in binding the sand together. The spreading of straw, hay, or sawdust over the surface will greatly improve the road.

In clay soils, the first essential is thorough drainage of the subsoil by either subsoil drains, deep side ditches, or both. The surface of the portion intended for the wheelway should be cleared of all vegetable matter, then graded and formed to a suitable cross-section by means of a road grader. If sand is available, the clay surface should be plowed, then covered with a layer of sand 6 in. thick, then harrowed and finally rolled. This will provide a good wheelway during dry weather. If sand is not available, the clay may be improved by burning it, and then spreading and rolling it well. Trees and vegetation should not be permitted along the sides of a clay road as they exclude the sun and keep the road damp and muddy.

Gravel Roads.—Natural roads may be improved by using a surface of gravel. The gravel should be of hard material capable of resisting abrasion, and in order that it may bind well together, it should consist of pebbles of various sizes from 2 in. down to the size of a pea. The binding is effected by fine dust which fills the voids that cannot be filled by the small pebbles. The fine material may consist of sand, clay, or loam to the amount of one-eighth to one-fourth of the bulk.

The thickness of the gravel covering will depend on the extent and weight of traffic. It ranges from 4 in. for very light traffic to 12 in. for the heaviest traffic. The gravel is spread on the prepared roadbed in layers 4 in. thick, and each layer is compacted by a roller of suitable weight, a heavy roller being used for small and a light roller for coarse or large gravel. A small quantity of water should be sprinkled over the gravel in advance of the rolling; and, when all the layers are compacted, a small quantity of clay or loam may be spread over the surface and rolled without water, after which the roadway may be opened to the traffic.

Oiled Roads.—Sand, clay, and gravel roads may be much improved by the application of crude petroleum oil; that having an asphaltic base is the best. The oiling lays the dust and, to a certain extent, serves as a binding material, forming a crust that wears well under traffic. The oil is applied by sprinkling while the road is dry, being mixed with the earth or gravel by harrowing and then compacted by rolling. Two applications are made. For the first one from about $\frac{3}{4}$ to $1\frac{1}{4}$ gal. per sq. yd. is required. The second application of about $\frac{1}{2}$ gal. per sq. yd. is made a few months after the first one.

Broken-Stone Roads.—A broken-stone road consists of a layer of broken rock spread on the previously prepared natural soil, and consolidated to a firm uniform surface by rolling with steam rollers. To secure satisfactory results, certain essential points must be observed. The stone must be of suitable quality, and must be placed on a suitable roadbed. The bed must be thoroughly drained, and all disintegrated or worn-out material and vegetable matter must be removed. The subgrade must be brought to a uniform surface, free from hollows, and must be thoroughly consolidated. The voids in the mass of the broken stone must be eliminated by rolling and by adding

FIG. 1

fine dust; this dust should not be mixed with the stone, but should be applied after the stones have received a slight compaction by rolling. The broken stones should not be left loose to be compacted by the traffic, but should be consolidated by rolling with a roller of suitable weight to bring each piece of stone into close and firm contact with the adjacent pieces. Two systems of construction are employed:

Macadam's system consists essentially in spreading and compacting one or more uniform layers of suitable rock, broken into pieces of nearly uniform size, directly on an earth foundation that has been previously formed to the proper grade and cross-section and thoroughly compacted by rolling. A cross-section of a macadam road is shown in Fig. 1.

Telford's system is much the same as Macadam's, except that the layer of broken stone forming the wearing surface is spread on a paved foundation. This paved foundation is formed by blocks of stone from 3 to 8 in. in depth, set close together on their broadest edges. The cross-section of a telford roadway is shown in Fig. 2. The blocks of stone are set on the earth foundation, and their sizes are graduated according to their position, as shown.

FIG. 2

Each of these systems has its place in the successful construction of roads. The choice depends entirely on the character and condition of the natural soil. If this is composed of clay, not easily drained, a telford foundation will be preferable; but, if the soil is easily drained, a foundation will not be required and the macadam system will be found the cheaper and better adapted to the conditions.

The varieties of rock most suitable for road metal are trap, syenite, granite, chert, limestone, mica-schist, and quartz. These are named in the order of their relative values. Sandstone, clayey slate, and rock of indurated clayey material are not suitable for this purpose. Sandstone has practically no binding properties; the fragments do not bind together to form a solid mass, but remain simply an accumulation of separate fragments, which soon become ground and crushed into sand by the traffic. Clayey stones have poor binding qualities, and when saturated with water become very soft and are easily crushed into mud. The broken stone is applied in layers of from 3 to 5 in. The first layer is spread uniformly over the road, sprinkled with water, and rolled with a suitable roller. Upon this a second layer, and sometimes even a third layer, depending on the depth required, is treated in the same manner. When the last course has been properly completed, a layer of stone dust, which is usually called the *binder*, is spread to a depth of ½ to ¾ in., after which the road is again sprinkled with water and rolled until consolidation is complete.

A common rule requires that the stone shall be broken small enough to pass through a 2½-in. ring. It is also a not uncommon practice to use somewhat larger pieces in the bottom courses of the roadway than at the top, the stones at the bottom being from 2 to 3 in. in greatest dimension and those at the surface not more than 2 in. This is probably a good practice, though it may be doubtful whether it is sufficiently advantageous to warrant the additional expense of separating the sizes.

The thickness of the covering of broken stone should not be less than 4 in. and a thickness greater than 12 in. is seldom required. Macadam considered 10 in. of well-compacted broken stone on a solid, well-drained earth foundation sufficient for a roadway sustaining the heaviest traffic. A thickness of from 8 to 10 in. is generally considered sufficient.

Bituminous Macadam Roads.—The introduction of extensive automobile traffic upon our highways has made the maintenance of macadam roads very difficult. The heavy wheels disturb the binding material, and the rapid air-currents produced by the cars carry the binding dust off, thus exposing the surface stones to the action of rain and frost. To prevent the rapid destruction resulting from such traffic, the method of sprinkling with oil has been extensively practiced. Oiling prevents the binding dust from flying off the surface, and under the rolling action of the traffic this dust binds again with the surface stones. This remedy is, however, of only temporary nature; repeated applications are required, and besides it is not always effective. The more recent practice of dealing with macadam roads is to protect them by covering them with bituminous materials. A macadam road so treated is called a *bituminous macadam road*. There are many methods of constructing this form of road, chief among them being the surface method, the penetration method, and the mixing method.

The *surface method* consists in applying the bituminous material to the surface of a macadam-finished road; it is especially adapted for roads that have already been built. Before applying the bituminous material, all dust and dirt must be removed from the surface. The material is then applied either cold or hot, at a temperature of from 100° to 250° F., and in quantities from $\frac{1}{3}$ to $\frac{2}{3}$ gal. per sq. yd. Means must be provided also for an even distribution of the bituminous material. After this has been done, a thin layer of sand or stone chips is spread on the surface and rolled with a heavy roller.

In the *penetration method*, the bituminous material takes the place of the stone-dust binder used in the ordinary macadam road. The macadam is built in the manner previously described, but, instead of the stone binder, hot bitumen is poured in quantities of about $1\frac{1}{3}$ to $1\frac{2}{3}$ gal. per sq. yd. Before rolling, stone chips about $\frac{1}{3}$ in. in size are spread over the surface. After rolling, another coat of bitumen, at the rate of about $1\frac{1}{3}$ gal. per sq. yd. is applied. Stone chips are then spread again and rolled until a firm and smooth surface is obtained.

When the *mixing method* is employed, the bitumen is mixed with the upper layer of broken stone before placing the latter on the road. This method is similar to the one known as bitulithic pavement and described under the heading City Pavements. The difference lies chiefly in the manner of selecting and grading the stones. This is done with great care in the bitulithic pavement, the aggregate of which consists of stones of different sizes proportioned so as to reduce the voids to a minimum.

Concrete Highways.—The destructive effect of modern traffic on the public highways has also led to extensive experiments in the construction of road surfaces in which Portland cement is used as a binder. Although still in the experimental stage, this form of construction promises a great development in the near future.

In constructing concrete pavements, a great variety of methods are employed, and many of them are patented. The types of construction most in use are: the one-course pavement, the two-course pavement, and the grouted pavement.

The *one-course pavement* consists of one layer of Portland-cement concrete about 6 to 8 in. deep laid on a properly prepared subfoundation. The cement used should be of the best quality, the aggregate should consist of hard and tough material, and the proportion of the different materials must be such as to fill all the voids.

The *two-course pavement* consists of a layer of Portland-cement concrete about 5 in. thick upon which is laid a 1½- to 2-in. wearing surface consisting of cement mortar prepared from the best Portland cement and a fine aggregate properly graded and capable of resisting abrasion. To secure proper binding between the two courses, the top course should be placed before the concrete in the base course has set. The advantage of this type of construction is that in many cases it allows the use of a cheaper grade of material for the concrete in the lower course. On the other hand, the one-course type of construction eliminates the danger of a loose-top such as is liable to occur in the two-course type of construction.

The *grouted pavement* is a two-course pavement in which the first layer is formed of broken stone instead of concrete. The broken stone is firmly compacted by rolling, and a Portland-cement grout is poured upon it until it flushes the surface. Upon this surface is then spread a thin layer of stone of about the size of peas, after which it is again rolled and grouted.

In all types of concrete pavements, care must be taken to prevent cracks that are liable to result from expansion and contraction of the concrete. This is usually done by providing expansion joints, which should be arranged transversely at intervals of about 50 ft., and longitudinally between the gutter and the roadway proper. The expansion joints are usually made about 1 in. wide and are filled with tar paper or bituminous cement.

Care must also be taken to prevent the surfaces of concrete roads from being too smooth and slippery. This is usually accomplished by roughening the finished surface with a stiff broom or a brush before the mortar has set.

CITY PAVEMENTS

GENERAL EXPLANATIONS

A good pavement should be: (1) impervious, in order not to retain water or surface liquids, but to facilitate their discharge into the side gutters; (2) such as to afford a secure foothold for horses, and not to become polished and slippery from use; (3) hard, tough, and durable, so as to resist wear and disintegration; (4) adapted to the grade; (5) suited to the traffic; (6) smooth and even, so as to offer the minimum resistance to traction; (7) comparatively noiseless; (8) such as to yield very little dust or mud; (9) easily cleaned; and (10) economical with regard to first cost and maintenance.

It is also desirable that the pavement should be of such material and construction that it can be readily taken up in places and quickly and substantially relaid, in order to give access to water, gas, and sewer pipes.

A pavement consists of two more or less distinct parts; namely, the wearing surface, and the foundation by which the wearing surface is supported. The wearing surface receives and sustains the traffic, but is not of itself capable of distributing the weight of the traffic over a sufficient area of yielding ground, which office is performed by the foundation.

Pavement Materials.—The materials commonly used for the wearing surfaces of pavements are stone, wood, asphalt, and brick. For the foundations, hydraulic-cement concrete, bituminous concrete, brick, broken stone, gravel, sand, and plank are employed.

The selection of the paving material depends on the character of the expected traffic, on the cost, and to a certain extent on the grade of the street. The maximum grade on which the different materials may be used is about as follows: Asphalt and wood, 4%; brick, 7%; stone blocks, 15%. The width of a street, too, influences the selection. For instance, it would not be advisable to place wood on a narrow street lined with high buildings, because, owing to the exclusion of light and air, the pavement would decay rapidly.

SYSTEMS OF CONSTRUCTION

Broken-Stone Pavement; Macadam.—Macadam's system of broken-stone pavement is generally found very satisfactory for roadways in suburban districts. The construction of broken-stone roads is treated under the heading Highways.

Stone Pavements.—The stone used for pavements is generally obtained from the granitic, sandstone, and limestone rocks. Among the varieties of granite, those containing a large percentage of feldspar or mica are unsuitable for paving. The feldspar rapidly decays in consequence of the action of the air and water. The micaceous stones are too easily laminated. The limestones, when used for paving, wear unevenly, and under the action of frost are quickly split and broken.

The most enduring pavements are made of granite or sandstone blocks. The best material for the foundation of such pavements is hydraulic-cement concrete from 4 to 9 in. in thickness, according to the nature of the traffic. When sufficient time has been allowed for the concrete to set and dry, a *cushion coat* of suitable material is spread over it to receive the paving blocks. For this purpose, a $\frac{3}{4}$- to 1-in. layer of fine clean and dry sand for granite blocks and somewhat deeper for sandstone blocks is very appropriate. A still better cushion coat is afforded by a $\frac{1}{2}$-in. layer of asphaltic cement.

The paving blocks should be rectangular in form and of uniform dimensions. A depth of 7 in. is generally considered suitable; in which case the width should be from 3 to 4 in. and the length from 9 to 12 in. The blocks must be rammed with a ram weighing not less than 50 lb. The joints between the blocks must be filled with an impervious material, for which the most suitable is bituminous concrete composed of asphaltic cement and gravel. In applying this filling, the joints should be first filled with gravel to a depth of about 2 in.; then the hot pitch should be poured in, filling the joints to the depth of about 1 in. above the gravel; then the gravel and pitch should be added alternately until the joints are filled to within ½ in. of the top; the remainder should then be completely filled with pitch over which fine gravel should be sprinkled. The joint thus formed is impervious to moisture; it adds considerably to the strength of the pavement and makes it less noisy.

Stone-block pavements are very durable and economical, are easily accessible for repairs and afford a good foothold for horses; on the other hand, they have considerable tractive resistance and are very noisy.

Wooden-Block Pavements.—The best, as well as the simplest, form of wooden pavements consist of rectangular or cylindrical blocks that are set on a solid foundation with the fibers vertical and have the joints thus formed filled with an impervious cement. Hydraulic-cement concrete forms the best foundation. A cushion coat composed either of dry sand, hydraulic-cement mortar, or asphaltic cement ½ in. thick is spread over the concrete in which the blocks are embedded. Rectangular blocks are generally required to be 3 in. in width, 6 in. in depth, and about 9 in. in length; cylindrical blocks, from 4 to 8 in. in diameter and 6 in. in depth. Each block should be of uniform cross-section throughout its length, with its ends truly perpendicular to its axis. After the blocks have been rammed properly, the joints must be filled with Portland-cement grout; or, a better result is obtained by filling the lower 2 or 3 in. with bituminous cement and the remainder with hydraulic-cement grout. In cylindrical-block pavements, it is advantageous to add gravel to the bituminous cement in order to fill the large spaces between the blocks.

The most suitable woods for pavement are not the hardwoods but close-grained pitchy soft woods. These wear longer than the hardwoods, and afford a better foothold for horses. Chemical treatments of paving blocks have very little effect on the wearing properties of the wood, and their use is of doubtful economic value. Blocks not treated chemically expand in the direction perpendicular to the fibers about 1 in. in 8 ft. Wood attains the full amount of expansion in from 12 to 18 mo. Provision must be made for this either by leaving the joints near the curbs temporarily open or by omitting the course near the curbs. The pavement is finished properly after the expansion has ceased.

Brick Pavements.—When constructed in a proper manner and of suitable materials, brick pavements form a smooth durable surface that is well adapted to moderate traffic. Bricks suitable for paving should not contain more than 1% of lime, and should be burned specially for the purpose. When tested on their flat sides, they should offer a resistance to crushing of not less than 8,000 lb. per sq. in. They should not absorb more than 5% of their weight of water, and should be so tough that, when struck a quick blow on the edge with a 4-lb. hammer, the edge will not spall or chip. The bricks should be of uniform size, straight, square on edges, and free from fire-cracks or checks. When broken, the fracture should appear smooth and the texture uniform, and when struck together, the pieces should have a firm, metallic ring.

Many methods of construction have been tried. The best modern practice is to use a hydraulic-cement foundation, constructed as described for granite-block pavements. On this foundation a layer of fine, clean, dry sand should be spread to a uniform depth of ½ in., as a cushion coat to receive the bricks. It is essential that the sand for the cushion coat should be perfectly free from moisture; if necessary, it should be dried by artificial heat. The cushion coat is sometimes made as deep as 2 in.

After the brick has been properly laid, it should be sprinkled with water for about 15 min., the water being applied from a hose or can fitted with a rose spray. Shortly after the sprinkling, the surface of the pavement should be inspected, and all the bricks that appear wet or damp should be removed and replaced with new bricks. The bricks are then pressed with a light hand hammer, after which they are thoroughly rammed with a 2- to 5-T. roller. When the bricks have been settled to a firm and solid bearing, the joints are filled full either with a grout composed of equal parts of hydraulic cement and fine, clean, sharp sand, or with a tar filler composed of No. 6 coal-tar distillate. After the joints have been filled, the entire surface is covered with a layer of sand ½ in. deep, which after a few days is swept up and removed.

Asphalt Pavement.—*Asphalt* is the solid form of bitumen, either in a state of purity or combined with other matter. *Bitumen* is a complex hydrocarbon considered to be the ultimate product of the decomposition of certain vegetable and animal matter. The best known sources of asphalt are those on the island of Trinidad, in the West Indies, and in the state of Bermudez, Venezuela, where it is usually found in the form of large deposits, or lakes. It is rarely found in a pure state and it is usually refined by a heating process, the product obtained being called *refined asphalt*. Many of the refined asphalts are too brittle for use. To remedy this defect, the asphalt is mixed with a softening agent called the *flux*. The resulting mixture is called *asphalt cement* or *asphaltic cement*. The agents most extensively employed for a flux are maltha and residuum oil, the latter of which is obtained by the distillation of petroleum. A concrete in which the matrix consists of asphalt cement or coal tar is called *bituminous concrete*.

It is very essential that all asphalting pavements be sustained by a solid unyielding foundation, as the asphalt is suitable for a wearing surface only. The foundation is made either of hydraulic-cement concrete or of bituminous concrete. The former is more durable and is, therefore, generally preferred. On the other hand, with hydraulic cement the bond between the foundation and the wearing surface is not very perfect. When bituminous concrete is used a layer of clean, well-screened, broken stone is spread on the prepared roadbed to the proper depth, and thoroughly consolidated by rolling, as in the construction of broken-stone roads, after which a coating of coal tar or bituminous cement is spread on it. The proportions used should be about 1 gal. of cement to each square yard of foundation. Bituminous concrete is less expensive than hydraulic-cement concrete.

In order to effect a more complete bond, an intermediate layer of bituminous concrete known as the *binder course*, is commonly placed between the concrete foundation and the asphalt wearing surface. It is composed of clean broken stone of small size mixed with bituminous paving cement. The stones should vary in size from ½ in. in smallest to 1 in. in greatest dimension, and should be thoroughly screened. The stones, which are heated to a temperature of from 230° to 300° F., should be mixed with the paving cement in the proportion of from ¾ to 1 gal. of cement to 1 cu. ft. of stone. This mixture should be spread, while hot, on the base course to such a depth as will consolidate to a thickness of about 1½ in.; it should then be rammed and rolled, before it loses its plastic condition, until thoroughly compacted. The binder course is substantially the same for both a hydraulic and a bituminous base.

The material for the wearing surface is laid on the foundation or binder course, sometimes in one coat and sometimes in two coats. When one coat is laid, the ingredients are made up by either one of the following two formulas:

	Ingredients	*Proportions Per Cent.*
I	Asphaltic cement	12 to 15
	Sand	83 to 70
	Pulverized carbonate of lime	5 to 15
II	Asphaltic cement	13 to 16
	Sand	63 to 58
	Stone dust	28 to 23
	Pulverized carbonate of lime	3 to 5

When two coats are laid, the first coat should contain from 2 to 4% more asphaltic cement. The asphaltic cement and the sand should be heated separately to a temperature of about 400° F. The proper amount of pulverized carbonate of lime, while cool, should be mixed with the hot sand. This compound should then be mixed with the asphaltic cement at the required temperature and in the right proportions. In order that the materials may be properly mixed, a special apparatus suited to the purpose should be used.

Laying Asphalt.—*Two Coats.* The first coat of asphalt is called the *cushion coat*, and the second the *surface coat*. The cushion coat should be laid directly on the binder course, or on the concrete foundation when no binder course is used, and should be of such depth as to give a thickness of ½ in. when consolidated by rolling. The materials for the surface coat, which is laid on the cushion coat, should be delivered on the pavement in carts, at a temperature of about 250° F.; when the temperature of the air is below 50°, each cart should be equipped with a suitable heating apparatus that will prevent the paving material from cooling below the proper temperature.

The material of the surface coat should be carefully spread on the cushion coat to such a depth as will give a uniform surface and a thickness of 2 in. after being consolidated; hot iron rakes should be used for this purpose. The material should first be moderately compressed by hand rollers; a small amount of hydraulic cement should then be spread lightly over it, after which it should be thoroughly compacted by continued rolling with a heavy steam roller for not less than 5 hr. for each 1,000 sq. yd. of surface.

One Coat.—When the pavement is given only one coat of asphaltic material, it is laid in much the same manner as just described for the surface coat. The material should be delivered in carts, at a temperature not below 250° nor above 310° F.; while in the carts, it should be protected with canvas covers when the temperature of the air is below 50° F. It should be spread on the foundation to such depth as will give a uniform surface and a thickness of 2½ in. after being consolidated. The material should first be moderately compressed by hand rollers, and a small amount of hydraulic cement should be spread lightly over it, the same as described for the surface coat, after which it should be thoroughly compacted by rolling with a steam roller weighing not less than 5 T., followed by a second roller weighing not less than 10 T.; the rolling should be continued for not less than 10 hr. for each 1,000 sq. yd. of surface.

Bitulithic Pavements.—A bitulithic pavement is composed of broken stone ranging in size from 2 in. to dust, mixed in the necessary proportions to reduce the voids to about 10%, and cemented together by a bituminous cement manufactured either from coal tar, from asphalt, or from a combination of both. The pavement is constructed in much the same manner as an asphalt pavement. The foundation is composed of a 4-in. layer of broken stone compacted by rolling. The interstices are filled and the surface is covered with bituminous cement. The material for the wearing surface is heated to about 250° F., spread while hot, and compacted by rolling with a 10-T. roller to a thickness of about 2 in. The surface is then covered with a liquid bituminous cement, on which, while it is in a sticky condition, there is spread a layer of sand or stone dust to a depth of about ½ in. The rolling is then repeated, after which the pavement is ready for use.

CITY STREETS

Width.—The roadway of a city street should be of such a width as to accommodate the traffic. For business streets, a width of roadway from 40 to 80 ft. is required, and for residence streets it should generally be from 24 to 36 ft. The sidewalks on business thoroughfares usually extend from the curbing to the building line, and on residence streets the width is about one-fifth to one-sixth the width of the roadway. The outer edges of the sidewalks on residence streets are commonly placed about 2 ft. from the fence line.

Height of Crown.—Let w be the width of the roadway, in feet; p, the per cent. of grade; and q, a coefficient given in the table on page 411. Then the height of crown in feet is

$$c = qw + \frac{wp(70q-1)}{800}$$

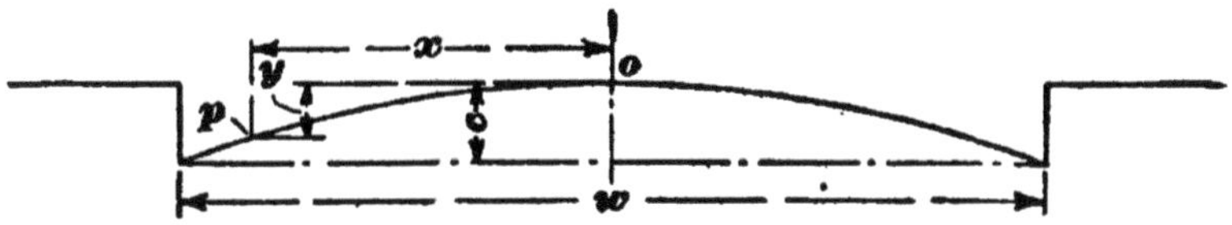

FIG. 1

When the grade is comparatively level, the height of crown is determined in the same manner as for highways, previously given. Expressed by a formula

$$c = qw$$

Form of Crown.—For laying out a *curving crown*, Fig. 1, the method given under Highways may be used, or the following formula may be employed:

$$y=\frac{4cx^2}{w^2},$$

in which x and y are, respectively, the abscissa and ordinate to any point p in the surface line of the cross-section with reference to the origin o.

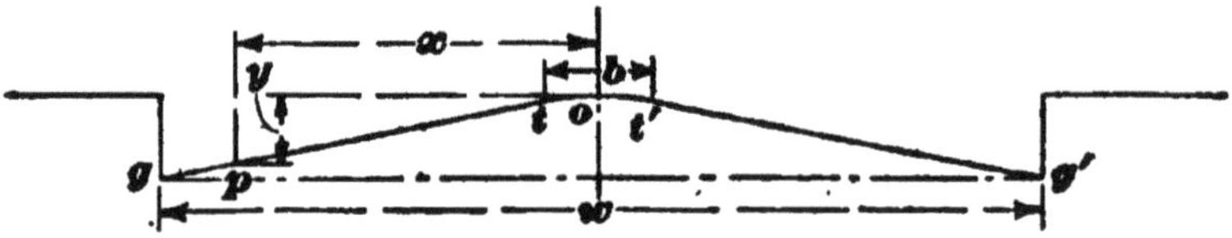

FIG. 2

For a *sloping crown*, Fig. 2, the portions tg and $t'g'$ have a uniform slope of

$$s=\frac{4c}{2w-b},$$

in which b is the width of the parabolic portion tt'. This parabolic portion may be constructed by the formula

$$y_c=\frac{sx_c^2}{b},$$

in which x_c and y_c are the coordinates of any point with reference to o as an origin. The ordinate at the tangent point t is $y_t=\frac{sb}{4}$, and the coordinates to any point p along the straight slope line $t\,g$ are related by the formula

$$y=s\left(x-\frac{b}{4}\right)$$

Grades.—In order that the surface water may be promptly and effectually removed from a roadway, the rate of grade for the street should never be less than one-fourth of 1%, that is, .25 ft. per 100 ft.; the grade should not be as flat as this except in extreme cases and with first-class pavements, such as brick or asphalt. A minimum grade of one-half of 1%, is as flat as should generally be used, and a grade as steep as 1% is very desirable. Where the grade line has the same elevation at the intersecting streets at both ends of a block, instead of making the grade level between those streets, it should be elevated in the center of the block sufficiently to cause the water to flow in each direction toward the intersecting streets. If the street is sewered, the grade may be depressed at the center of the block by locating catch basins there; generally, however, it is better to elevate the grade at the center of the block.

VALUES OF q IN FORMULA FOR HEIGHT OF CROWN

Character of Roadway	Value of q
Common earth roadways	$\frac{1}{20}$
Ordinary gravel roadways	$\frac{1}{30}$
Broken-stone roadways	$\frac{1}{40}$
Wooden-block pavement	$\frac{1}{70}$
Cobblestone pavement	$\frac{1}{30}$
Granite-block and concrete pavements	$\frac{1}{50}$
Well-laid brick pavement	$\frac{1}{100}$
First-class asphalt pavement	$\frac{1}{120}$

Lateral Slopes of Sidewalks.—For the purpose of drainage, sidewalks should have a slight lateral slope toward the curb. On business streets that are closely built up, in which the entire width between the curb and the building line is occupied by the sidewalk, this lateral slope of the sidewalk will fix the elevations on the building line. The edge of the sidewalk adjacent to the curb will be placed at the elevation of the curb, that is, at the street grade, and the edge of the sidewalk adjacent to the building line will be higher or above grade an amount equal to the width of the sidewalk in feet multiplied by the lateral slope per foot. In some cities, a lateral slope of 2½%, or 1 in 40, is given to the sidewalks; a slope of 2%, or 1 in 50, however, is generally very satisfactory for this purpose. All that portion of the street between the curb and the property line should have this uniform lateral slope, whether wholly occupied by the sidewalk or not.

MEMORANDA

MEMORANDA

MEMORANDA

Holding An Important Position

LOUISIANA & ARKANSAS RAILWAY COMPANY

Operating Department

F. W. GREEN, General Superintendent

I enrolled for the Railroad Engineering Course with the International Correspondence Schools in the spring of 1899, and have completed the same with the exception of a few drawing plates. At that time I was chief clerk to the superintendent of the K. C. S. A. Railway, Kansas City, Mo. Since then I have advanced to one position after another until I am now the general superintendent of the L. & A. Railway, directing the work of 700 men. I am always glad to recommend your Schools to those who desire to educate themselves, and I find the volumes of your Library of Technology the best and most concise upon their respective subjects that I have ever seen.

F. W. GREEN, Stamps, Ark.

NOW SERVES THE STATE

P. H. BINGHAM, 288 Madison Ave., N. Y., a graduate of the I.C.S. Civil Engineering Course, was working in a shirt factory for about $12 a week when he first began his correspondence training. He declares that this enabled him to take a position with the New York State Highway Commission, where he was afterward given charge of the drafting room. He is now in the service of the New York State Conservation Commission studying the flow of streams with relation to water supply and hydro-electric development. His salary has increased more than 150 per cent. as the result of his Course with the I.C.S.

SALARY INCREASED 100 PER CENT. EACH YEAR

GILBERT SMITH, Concho, W. Va., was a driver earning about $30 a month when he enrolled for the Surveying and Mapping Course. Four years later he secured a position as mine superintendent, having charge of 200 men, a position which he still holds. His salary has increased nearly 100 per cent. each year since enrolment.

NOW EARNS $8 INSTEAD OF $2

CHAS. W. WINCHESTER, Winchester, Wyo., states that he was working for $2 a day at common labor when he took up our Surveying and Mapping Course. With the help of this Course he has now become a surveyor and civil engineer, receiving $8 a day.

AN I.C.S. EDUCATION MADE THE DIFFERENCE

Our student, J. M. CUNNINGHAM, Carlsbad, N. Mex., was a cowboy receiving about $2 a day when he enrolled for the Surveying and Mapping Course. Having studied the Course, he bought a transit and went to work for himself. He is now county surveyor of Eddy County, receiving $10 a day and expenses.

FIVE TIMES HIS FORMER SALARY

T. G. BANKS, Freeport, Texas, was employed as a teacher when he began to study the I.C.S. Civil Engineering Course. This enabled him to take up railroad work and he has gradually advanced through the engineering department until he is now superintendent of the Houston & Brazos Valley Railway. His salary is five times what it was when he began the study of his Course and he declares that his success was made possible by his work with the Schools.

Now in Government Employ

I was working on a farm filling what was practically a laborer's position when I first enrolled with the I. C. S., for the Surveying and Mapping Course, which I completed between November, 1899, and May, 1900. I afterwards enrolled for the Civil Engineering Course. At that time I had received an ordinary high-school education, taking me part way through algebra and complete plane geometry. This was the only training I had in mathematics. I am sure that the knowledge I gained was well worth the effort and the expenditure, since it enabled me to obtain a position in the U. S. Geological Survey, with which I have been employed ever since. On the following year after I entered service I was able to pass the Civil Service examination for permanent employment and have received promotions from time to time since then. I feel that I owe my start in engineering work to the education received through the I. C. S. I attribute a considerable part of whatever success I may have obtained to this education and to the habits of perseverance and industry acquired during this course of study. I am entitled a topographical engineer and receive a salary of $2,000 a year. WM. O. TUFTS,

U. S. Geological Survey,

Washington, D. C.

CIVIL ENGINEER BECAME INSPECTOR OF IMPORTANT CONSTRUCTIONS

PETER BRADLEY, 119 Genesee St., Trenton, N. J., was no longer young when he enrolled for the Bridge Engineering Course. Through the help of our instruction, he became an inspection engineer for Stowell & Cunningham, Albany, N. Y. He has recently had charge of inspection of steel and wire for the new Manhattan and Williamsburg bridge, New York City. His salary has been increased from $85 to $200 a month.

PRESIDENT OF A CONTRACTING FIRM

One of our graduates, DAVID THOMAS, Woodbury, N. J., has established a successful engineering and contracting business since his enrolment with the I. C. S. He was earning $75 a month when he enrolled for the Bridge Engineering Course. He is now president of his own company, making a specialty of reinforced concrete construction and sewage disposal plants.

MULTIPLIED HIS INCOME SEVERAL TIMES

At the age of 23, WILLIAM S. SHARPE, 138 Arlington Ave., Arlington Heights, Ohio, subscribed for the Bridge Engineering Course. He was then working as a machinist for $10.50 a week. Later he became general superintendent for the Springfield Bridge and Iron Company, employing 125 men. For the past 4 years he has been in business for himself as a general contractor in concrete, steel, and bridge work. His present income is several times what it was when he enrolled.

NOW GENERAL MANAGER—SALARY $3,000

JOACHIN FORTIN, 131 Rue St. Pierre, Que., Canada, had taken a commercial college course when he enrolled for our Civil Engineering Course. He has advanced by the following steps: Clerk, draftsman, leveler, transitman, and now general manager of La Cie. Electrique Dorchester. He has under his direction two field engineers, five office clerks, six foremen, and from 90 to 115 laborers. His salary is $3,000 a year.

NOW HOLDS A MUNICIPAL POSITION

Before he enrolled with the I. C. S. for the Surveying and Mapping Course, ROBERT A. VESPER, Arlington, Mass., was a rodman. After graduating, he advanced from time to time until he is now superintendent of streets for the town of Arlington, at a salary of $1,500 a year.

Holds An Important Position

I was earning probably $20 a month on an average when I took up a Course in Civil Engineering with the International Correspondence Schools. At that time I had received little more than a high school education. My classmates were going to college and I was greatly distressed because I was not able to follow their example; but I gave my spare time to the study of your Course, being employed on a corps by the city engineer of Uniontown, Pa. Within a year I was made chief draftsman in his office. Later, at the age of 19, I took a position as engineer in charge of three coke plants for the H. C. Frick Company. After holding various positions, in 1906 I accepted a place with the Pennsylvania State Highway Department. I am still employed by the State, holding the position of assistant engineer at a salary of $200 a month and expenses. I have 50 men employed in my engineering department at present.

H. W. CLAYBAUGH,
Franklin, Pa.

EARNINGS INCREASED 10 TIMES

Every step in the career of G. A. Collins, Seattle, Wash., has been upwards. He was working as a chainman for $30 a month when he enrolled for the I.C.S. Railroad Engineering Course. Since then he has held numerous positions as locating engineer, bridge engineer, and chief engineer. After serving on the Washington State Railway Commission he has become irrigation engineer for the Kilbourne & Clark Company, making a specialty of irrigation work and the installation of power plants. His earnings have increased about 10 times since he enrolled with the I.C.S.

NO LONGER COMPETES WITH THE MULE

O. T. Reece, Wellington, Kans., when 48 years old, found himself working on a railroad bridge gang, competing with the mule and the steam engine. He enrolled for a Course in Railroad Engineering, and afterwards for the Civil Engineering Course. He has been appointed by the court on the Board of Commissioners of the Drainage Department, and he also enjoys a fine private practice as an engineer with a field of work constantly widening. His income has been increased more than 500 per cent.

DOUBLED HIS EARNINGS

C. J. Cook, Deposit, N. Y., had received only a high-school education and was working as signal man in a railroad tower at $40 a month, when he enrolled for the Civil Engineering Course. This enabled him to take up civil engineering and to become superintendent of construction on a state highway job. He is now consulting civil engineer, earning twice what he did at the time of enrolment.

250 PER CENT. LARGER

Rollo Keesler, Anderson, Ind., was working as a draftsman at the time he enrolled for the Civil Engineering Course. This enabled him to enter the engineering department of the Union Traction Company, where he is now office engineer in the roadway department. His salary has increased 250 per cent.

NOW SUPERINTENDENT

When F. B. Hayes, superintendent of the Pendleton City, Ore., water commission, enrolled with the I.C.S. for the Civil Engineering Course, he was employed as a clerk. Although he had received only a common-school education he was able to master his Course and to undertake the construction of a $200,000 gravity system water works. His salary, of course, has been increased, being now about double what he received at the time of enrolment.

EARNS $7 A DAY

LEONARD OERTLI, Hamilton, Mont., a graduate of our Surveying and Mapping Course, considers that the money spent on his course was the best investment he ever made. At the time of enrolment he was receiving about $25 a month. After studying his course he was twice elected county surveyor. He is now chief engineer of the Orchard Lands Company, receiving $7 a day.

HIS COURSE HELPED HIM

HOWARD P. STROUGH, Maxwell, Cal., says he is certain that the work he has done with the Schools in his Railroad Engineering Course, particularly in Hydromechanics, enabled him to become resident engineer on the main canal of the Sacramento Valley Irrigation Company.

FOUR TIMES HIS FORMER SALARY

JACOB ORR, Auburn, Ill., taught school for 13 years, never receiving more than $480.00 a year. He then took our Complete Railroad Engineering Course, working Saturdays and nights, running levels for drain tile and surveying in the mines. Six years ago he accepted the position of surveyor for the Black Diamond Coal Company, and later undertook in addition the surveying for the Cora Coal Company. He now makes surveys for these and five other mines, besides carrying on a large practice in surveying and drainage work. His salary is now about four times what it was when he enrolled.

RAPID AND GRATIFYING ADVANCEMENT

HARRY B. JOHNSON, Hayward, Wis., enrolled for the Surveying and Mapping Course when he was 29 years old, and while working 11 hours a day for $2.50 to support a family of four. At that time he hardly knew the use of the magnetic needle. Less than a year later he was doing county surveying at $5 a day and expenses for actual time employed. He is now earning $100 a month in the employ of the American Immigration Company, as surveyor and engineer.

THE I. C. S. TAUGHT HIM WHAT HE KNOWS

J. A. VERMETTE, St. Eleuthere, Quebec, was working for $45 a month, having only a commercial high school education. He says that our Railroad Engineering Course taught him everything he knows as an engineer. He has also become proficient in the English language since his enrolment, not knowing how to write anything but French until some 5 years ago. He is now timber inspector for the Transcontinental Railway Commission at a salary of $125 a month.

Earns As Much As $360 A Month

E. A. SHAFFER, Narrows, Ore., had no knowledge whatever of engineering when he enrolled with us for our Civil Engineering Course. At that time he was earning $40 a month as a flagman. Fourteen months later he accepted a position as transitman at a salary of $125 a month. Two years after enrolment he passed the examination for United States mineral surveyor with a percentage of 98. In November of the same year, 1910, he was elected county engineer, which he afterwards resigned to go into private practice. During the summer of 1911, he was in charge of engineering parties numbering at times 16 men which brought him on an average $360 a month and expenses. Mr. Shaffer says he will always be glad to answer any inquiries from prospective students and he cordially recommends the I. C. S. Courses.

WORKING ON THE BIG DITCH

CHARLES LOGASA, Cristobal, Canal Zone, Panama, enrolled for the Surveying and Mapping Course at the age of 18. On September 16, 1910, he left the position of chief draftsman in the city engineer's department, Omaha, Neb., having taken a Civil Service Examination for topographical draftsman, and entered the office of the chief of engineers at Culebra, Canal Zone, Panama. He worked in this office for 4 months, at $125 a month, and is now in the office of the Chief Engineer of the Panama Railroad at a salary of $150 a month.

500 PER CENT. INCREASE

One of our graduates, J. FRED FREEMAN, Parsons, Kans., was working as a grocery clerk when he enrolled for the Mechanical Drawing Course. Having obtained his diploma he then enrolled for the Railroad Engineering Course, entering the engineering department of the M. K. & T. Railway. Eight months later he was advanced to the position of rodman. He says that if it had not been for his I.C.S. Course he might still be in the grocery business, dissatisfied with his work, instead of holding the position as draftsman for his company with an increase in salary of 500 per cent. over what he received at the time of enrolment.

SALARY INCREASED 125 PER CENT.

D. F. HARVEY, Beaver Falls, Pa., began working for the Pittsburg & Lake Erie Railroad Company driving stakes on the engineering corps. He had reached the position of chainman when he concluded that the I. C. S. would help him to advance. He, therefore, enrolled for o r Railroad Engineering Course. This has helped him to rise to the position of supervisor of tracks for the company named above at a salary which has been increased 125 per cent. since enrolment.

OUR COURSE IS EASY TO STUDY

EARL PERRY, 1306 Ford St., Fort William, Ontario, Canada, was an axman of a surveying party when he enrolled for the Railroad Engineering Course. Although he had only a very poor public school education, and knew nothing whatever of logarithms, geometry, and trigonometry, he has mastered quite easily these subjects through his I. C. S. Instruction Papers. While working on his Course, he advanced to the position of instrument man for the J. S. Metcalf Company with an increase in his salary of more than 100 per cent.

EARNS FROM $8 TO $12 A DAY

Another successful student of the I. C. S. is J. WESLEY TURNER, Lakeland, Fla., who had only a general knowledge of surveying and was receiving but $1.50 a day when he enrolled with the I. C. S. for the Railroad Engineering Course. He now makes subdivisions of lands for realty companies, receiving from $8 to $12 a day.

Now President and Treasurer

Geo. D. Case, Painted Post, N. Y., enrolled for the Bridge Engineering Course while he was clerking in a dry-goods store for $40 a month. His previous education was confined to the district schools with one year preparatory school work. His studies enabled him to advance from time to time in the engineering line, and to pass the New York State examination as bridge designer. Although offered an appointment at $2,100 a year, he refused and obtained an interest in the Lane Bridge Company. He is now president and treasurer of this company doing a business of $250,000 a year.

HIS COURSE BROUGHT SUCCESS

WM. E. JOHNSON, 124 6th St., N. E., Washington, D. C., was employed as a minor salesman at $5 a week when he enrolled with the I.C.S. for the Civil Engineering Course. The following winter he took the Civil Service examination for topographical draftsman and in June, 1910, he was appointed to a position on the Coast Survey at $20 a week. He has steadily increased his income since that date and is now earning at the rate of $200 a month. This success he attributes to his Course with the I.C.S.

HIS COURSE MADE THE DIFFERENCE

Although GEORGE D. WOLFE was laboring in a coal mine at $35 a month, having received but little education, the Course in Civil Engineering which he took with the I.C.S. has enabled him to occupy various important positions as an engineer about the mines and in railroad and other construction work. He has sunk shafts, driven tunnels, completed heavy bridges, and located difficult railroad lines. He is now located as an engineer at Nowata, Okla., making about $150 a month plus expenses.

A YOUNG MAN'S ADVANCEMENT

GROVE V. PURCHASE, Room 105, Court House, Syracuse, N. Y., is only 24 years old at the present time. When he enrolled 4 years ago for the Civil Engineering Course, he was earning $2.50 a day as a chainman. He is now in the employ of the Superintendent of Highways of Onondaga County, having charge of the construction of two roads, 11 miles in all, to cost some $70,000. His salary has been advanced to $110 a month.

A GRADUATE'S SUCCESS

When C. JEROME NEWCOMB, 92 Union Ave., Jamaica, N. Y., enrolled for the Bridge Engineering Course, he was employed as a salesman in a wholesale metal house. Since obtaining his diploma he has received one advancement after another until he is now in full charge of the drafting room and construction work for the Conservation and Public Service Company, of New York City, a responsible and well-paid position. He is an enthusiastic friend of the I.C.S. and praises the bound volumes

500 PER CENT. INCREASE

JAMES R. PENNER, 497 West Ave., Buffalo, N. Y., enrolled for the Engineering Course while an attendant at the Rome State Hospital. As a structural draftsman for the Lackawanna Steel Company he now earns 500 per cent. more than when he took up work with the I.C.S.

What a Former Carpenter Has Accomplished

When I first enrolled with the I. C. S. for the Surveying and Mapping Course I was working as a carpenter, earning $2.50 a day. I afterwards subscribed for the Civil Engineering Course. At the present time I am city engineer for our town here. I have designed and built practically all of the reinforced concrete structures that they have erected around here, and my salary has been increased very considerably. My position as city engineer pays me $1,200 a year; and besides this I have built up an extensive business along the line of civil engineering and structural work. I believe I enjoy a good reputation among men in my line in this part of the country. My education was limited entirely to common school work. I owe all my success to the careful reading and study of your Bound Volumes, without which it would have been impossible for me to make progress. M. E. BANNON,

Marquette Bldg., Fort Madison, Iowa

HOW A BARBER ROSE

OSCAR F. WEISSGERBER, 224 N. Oak St., Owatonna, Minn.; was working as a barber when he enrolled for the Surveying and Mapping Course. At the time he was earning about $6 a week. He soon secured a position on a surveying corps at $45 a month. He is now superintendent of construction for the Fielding & Shepley Company, having at times 60 men and 30 teams at work under his direction. His salary is $5 a day.

DID NOT UNDERSTAND FRACTIONS

When G. E. LINN, Stevenson, Wash., enrolled for the Surveying and Mapping Course, his education was so limited that he did not understand fractions. He was working on a surveying corps at the time as a chainman at $40 a month. He has since filled important positions in irrigation work, the U. S. Reclamation Service, and as county engineer, receiving as high as $150 a month. He is now a successful engineer and surveyor in business for himself.

OFTEN EARNS $10 A DAY

PAUL F. MUELLER, Amarillo, Tex., was in the service of a mapping company when he enrolled for the Railroad Engineering Course. He says this has been invaluable to him, enabling him to take charge of various railroad engineering projects in Oklahoma and Texas, including a $200,000 paving job, a $40,000 sanitary sewer, and a complete purification plant and irrigation farm for his home city. His salary has never been less than $100 a month, and has often averaged $10 a day.

INCOME RISES TO $2,500 A YEAR

J. W. LAMBERT, Grundy, Va., was a timber inspector, 21 years old, when he enrolled for the Surveying and Mapping Course. He attributes to this Course his rise to the position of engineer for the Russells Fork Transportation Company and the Honoker Lumber Company, and other prominent concerns. His yearly income derived from civil engineering now reaches $2,500 a year.

GRADUATE MULTIPLIES HIS SALARY

When A. B. TALMADGE, G. A. R. Building, Leavenworth, Kans., enrolled for the Surveying and Mapping Course, he was earning $45 a month. At the same time he started to work as a rodman. Spare-time study has increased his earnings and brightened his future prospects. Since obtaining his diploma he has passed a United States Government Civil Service examination and now receives $1,400 a year, having also his expenses paid when absent from headquarters.

Now Proprietor

J. L. Corbin E. K. Ramsey

CORBIN & RAMSEY
Civil and Irrigation Engineers
County Surveyor's Office
Sterling, Colo.

I had passed the ninth grade in a country school when I enrolled with you for the Civil Engineering Course, and I have not gone to school since. I can truly say that your instructions have made it possible for me to carry on the large amount of engineering connected with my position. At the time of enrolment I was foreman of a cattle ranch at the usual salary for that position. My present income is from $150 to $350 a month. I have not completed my Course, but I have studied through all of it, and use the Bound Volumes for reference. I find them to be as practical as they are complete.

J. Lloyd Corbin

www.ingramcontent.com/pod-product-compliance
Ingram Content Group UK Ltd.
Pitfield, Milton Keynes, MK11 3LW, UK
UKHW022231120925
7882UKWH00003B/270

9 781016 069083